D1538719

Silicones Under
the Monogram

By the Same Author

H. A. Liebhafsky, *William David Coolidge: A Centenarian and His Work*, John Wiley & Sons, New York, 1974.

H. A. Liebhafsky, H. G. Pfeiffer, E. H. Winslow, and P. D. Zemany, *X-Rays, Electrons, and Analytical Chemistry: Spectrochemical Analysis with X-Rays*, John Wiley & Sons, New York, 1972.

H. A. Liebhafsky and E. J. Cairns, *Fuel Cells and Fuel Batteries: A Guide to Their Research and Development*, John Wiley & Sons, New York, 1968.

H. A. Liebhafsky, H. G. Pfeiffer, E. H. Winslow, and P. D. Zemany, *X-Ray Absorption and Emission in Analytical Chemistry*, John Wiley & Sons, New York, 1960.

Silicones Under
the Monogram

A STORY OF INDUSTRIAL RESEARCH

HERMAN A. LIEBHAFSKY
General Electric Company   1934–1967
Texas A&M University   1967–

COLLABORATING EDITORS

Sybil Small Liebhafsky
George Wise

A WILEY-INTERSCIENCE PUBLICATION
JOHN WILEY & SONS, New York • Chichester • Brisbane • Toronto

WITHDRAWN

MOUNT ST. MARY'S
COLLEGE
EMMITSBURG, MARYLAND

Copyright © 1978 by John Wiley & Sons, Inc.

All rights reserved. Published simultaneously in Canada.

Reproduction or translation of any part of this work
beyond that permitted by Sections 107 or 108 of the
1976 United States Copyright Act without the permission
of the copyright owner is unlawful. Requests for
permission or further information should be addressed to
the Permissions Department, John Wiley & Sons, Inc.

Library of Congress Cataloging in Publication Data:

Liebhafsky, H.A.
    Silicones under the monogram.

    "A Wiley-Interscience publication."
    Includes Index
    1. Silicones. 2. General Electric Company
    I. Liebhafsky, Sybil Small. II. Wise, George,
    1944-   III. Title.

    TP248.S5L53   668.4'227   78-19047
    ISBN 0-471-04610-8

Printed in the United States of America

10 9 8 7 6 5 4 3 2

Here are set the names of
A. L. Marshall (1896–1974)
and
Charles E. Reed
who worked in complementary ways
to make
The General Electric Company
a better servant of mankind

Preface

Of a good beginning cometh a good end.

JOHN HEYWOOD, *Proverbs*

Silicones are unusually interesting and highly diversified materials made and sold by the General Electric Company under the GE monogram, a symbol long familiar to Americans and becoming better known the world over as the multinational character of the corporation grows.

In 1972 the General Electric silicone plant, which is in Waterford, New York, near Schenectady, celebrated its silver anniversary. On that occasion Dr. Charles E. Reed, first General Manager of the Silicone Products Department, and at present a Senior Vice President of the Company, suggested that the story of General Electric silicones be written.

The suggestion reached me that summer through Mr. R. Ned Landon, of Corporate Research and Development, successor to the General Electric Research Laboratory, where work on silicones had an unusual beginning in 1938. Dr. Reed's suggestion, attractive in itself, became more appealing for an unrelated reason.

I had just finished reading Lord Snow's *The Two Cultures: and a Second Look** for the first time. The book begins with the Rede Lecture,

* C. P. Snow, *The Two Cultures: and a Second Look*, Cambridge University Press, London, 1964. Please read this short book.

given by Lord Snow in 1959, of which the thesis, baldly stated, is as follows. The modern scientific revolution, following upon the industrial revolution, has created a gap between rich and poor nations that can and must be narrowed. Unfortunately "a gulf of mutual incomprehension" between the scientific and the literary intellectual cultures bars the way.

The violent controversy started by the lecture left no doubt that the gulf is real, and difficult to bridge. Pure science is not the main controversial issue—that distinction belongs to applied science and technology acting through products and processes, most of which are tangible consequences of industrial research, an activity brought to life by the modern scientific revolution. A better understanding of industrial research and its consequences, the need for which was implicit in *The Two Cultures* and was made explicit by the controversy, is prerequisite to a bridging of the gulf. To help meet this need, the account of early General Electric silicone research was broadened into a silicone story, and this story was placed in perspective by adding information about other industrial research. The table of contents shows what was done. The Epilogue returns to Lord Snow and concludes with a commentary upon science and technology today.

Readers who find technical information burdensome are free to pick and choose. A military history can be read with pleasure and profit even if battle maps are not dissected or plans of campaign memorized. Likewise here: those concerned with the broader implications of the inquiry might, the first time through, read the Introduction, the Epilogue, Chapter 1, Chapter 3, and the beginnings and endings of the other chapters, the bodies of which contain our military maps in the forms of equations, extracts from patents, figures with scientific or technical captions, and similar scientific abracadabra to which the reader may return if he or she wishes. First the icing, then the cake. Scientists or engineers, especially graduate students concerned about industrial careers, might well read the book straight through.

To accommodate readers with different interests, the book has been thoroughly and broadly documented. One feature deserves special mention. Original laboratory notebooks, references to which are made by name and date, have been used extensively to document the silicone project: there is no better way of seeing how discoveries were made or of following research in progress. Though no hard and fast rule could be followed, footnotes to the text are usually intended to interest the general reader; the final section, References and Notes, which deals largely with publications (references to laboratory notebooks excepted), is mainly for the experts. I have favored plain words and minimized jargon, scientific and other. I hope that this book is easier for the literary intellectual to understand than *Finnegan's Wake* would be for me.

Personalia abound in the book for good reason. General readers may not know the great extent to which research, engineering, and management are *personal* activities: different conductors interpret differently what the composer has set down. Success in these activities often seems to hinge upon Napoleon's criterion: "Is he lucky?" for selecting his generals.

I knew well most of those in the silicone project that became the General Electric silicone business. Paraphrasing Dylan Thomas, I see them as the boys of summer—boys of summer at their best. That best, now long ago, seems to have happened yesterday. I knew, or have come to know, many of those at Waterford who saw this silicone business through its Valley Forge and made it succeed in the face of respected, experienced, and formidable competition from the chemical industry.

What the book would have been like without the willing counsel and help of many, I prefer not to think. I cannot name them all. I gladly acknowledge help of various kinds from D. W. Bellamy, B. A. Bluestein, E. M. Boldebuck, D. V. Brown, A. M. Bueche, V. O. Chase, J. T. Cohen, T. H. Fitzgerald, E. J. Flynn, D. W. Fox, R. S. Friedman, G. L. Gaines, A. R. Gilbert, W. F. Gilliam, E. M. Hadsell, A. S. Hay, F. F. Holub, M. F. King, R. N. Landon, H. K. Lichtenwalner, J. Marsden, L. Navias, A. E. Newkirk, W. I. Patnode, C. E. Reed, W. L. Robb, E. G. Rochow, W. L. Roth, W. J. Scheiber, S. Seltzer, A. R. Shultz, M. M. Sprung, L. W. Steele, and J. L. Young—all employees, past or present, of the Company; and from E. A. Meyers and D. H. O'Brien, colleagues of mine at Texas A&M University, where part of the material in this book was used some years ago in a graduate seminar. Additional acknowledgments appear later. All those who helped, named or unnamed, know what they did. I thank them all.

My editors, Sybil Small Liebhafsky and George Wise, collaborated and often prevailed. Without them there would be no book.

Although the General Electric Company gave me encouragement and support, it must not be held responsible for the opinions or conclusions in this book. Infelicities and errors are mine alone.

HERMAN A. LIEBHAFSKY

College Station, Texas
May 1978

Contents

Silicones Under
the Monogram

Introduction

To see a World [of silicones] in a grain of Sand.

WILLIAM BLAKE, *Auguries of Innocence*

This quotation from a great romantic poet and its modification illustrate the difference between the two cultures, literary intellectual and scientific. Blake's short sequence of simple words suggests that nature has hidden much in a grain of sand; the modified version suggests, as has happened, that man by experiment has found a world of silicones, of which sand—more properly, silica—is the parent.

Silica and water can claim to be the most remarkable and best investigated among the multitudes of inorganic compounds that account for a large part of descriptive chemistry. Silica, SiO_2, free or combined, constitutes much of the earth's crust, and we meet it in everything from jewelry to concrete. Over the large range of pressures and temperatures studied, SiO_2 has been found in 22 forms ("phase modifications") that differ in structure but are identical in composition;[*] not many of them are found in nature. We are concerned mainly with three of these many

---

[*] Robert B. Sosman, *The Phases of Silica,* Rutgers University Press, New Brunswick, New Jersey, 1965. This and a later volume will complete the revision of an earlier classic. In the Preface Dr. Sosman sternly rejects melanophlogite as a twenty-third phase of SiO_2. For guiding me to this book, and for much other help, I thank Dr. Louis Navias.

1

forms: with quartz, a crystal of which Blake's "grain of Sand" is most likely to have been; with vitreous ("glassy") silica; and with hydrated SiO_2, in which the two most common materials on earth appear as a complex family of compounds called silicic acids. In spite of the enormous attention SiO_2 has received, its usefulness is still expanding. Silica glass of extreme purity is being "doped" with a trace of material (e.g., germanium) to prevent the escape of light, and pulled into *optical fibers,* along which pulsed lasers are expected to send our telephone messages in future lightwave communication systems [1].

Among silica's many virtues in one or another of its common forms are high electric strength; practical immunity to temperature change up to the melting point of quartz (573°C), even in the presence of air; and outstanding, though not complete, resistance to chemical attack of many kinds.

Yet there is a limit to nature's bounty. In none of these common forms does SiO_2 have certain properties highly prized because they make materials more versatile and extend their usefulness. For example, SiO_2 is not rubbery; its surface, if clean and untreated, is usually wet readily by water; SiO_2 itself cannot be brushed on to cover and protect other surfaces; and SiO_2 cannot be molded and shaped at ordinary temperatures. Many organic materials, notably the hydrocarbons, have these and other desirable properties to a much greater degree than does SiO_2. Why not find a way to improve upon nature, and have the best of both the inorganic and the organic worlds by making materials that possess in some degree the useful properties of both silica and hydrocarbons? In short, why not—as nature omitted to do (and man has done)—make silicones?

To understand what has been done, we must make simple models* of molecules. In these models we regard molecules as built of atoms (represented by their symbols) joined by chemical bonds (represented by lines). Silica has the simple formula SiO_2, but the simplicity masks a bonding surprise. Each silicon atom—each Si—is bonded to *four,* not *two*, oxygen atoms, and this bonding is possible only because each oxygen is united with *two* silicons. A crystal of silica is a single *giant*

Note. Brackets are used to enclose inserted material and references; for bracketed references see References and Notes at the end of the book.

* Models are indispensable in chemistry. The nonscientific reader should be warned that no model gives the whole truth: for example, the line representing the chemical bond comes nearer the truth if it is changed to Gilbert N. Lewis's electron-pair bond, one of the most important models in chemistry, often written as two dots. Unfortunately, however, modern physics has shown that there is no acceptable way to represent an electron. In this situation simple models might as well be used in books such as this one.

molecule extending in three dimensions and built entirely of *siloxane bridges*, each written as Si—O—Si, where each line represents one of four identical bonds to the silicon, the other three being omitted. Nothing in silicone chemistry is more important than the siloxane bridge.

In silicone molecules, organic groups (e.g., CH_3) have been joined by Si-C (silicon-carbon) bonds to silicon atoms in siloxane bridges. In two dimensions, we have

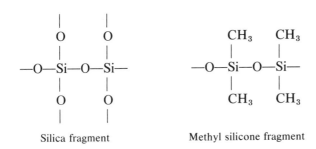

Silica fragment Methyl silicone fragment

Here the model of the silica fragment is further from the truth than the other model. Were the two silicon atoms in the body of a silica crystal, they (and all other such atoms) would be bonded in *three dimensions* to oxygen atoms. The silicone fragment could be envisioned as continuing *in two dimensions*. A three-dimensional structure of indefinite extent will be called a *skeleton*; a two-dimensional structure of this kind, a *backbone*.

Now, no good carpenter asked to put windows into a house will begin by demolishing the house. Yet that is what we must do to make silicones from silica. A solar factory in which silicones form as natural gas seeps through hot desert sands is simply not possible, even in the Mideast. We cannot leave intact the fragment of *siloxane backbone* shown above and replace four oxygens, as indicated, with CH_3 groups. If we could, the silicone story would be less interesting. To build silicone molecules industrially, we must begin either with silicon or with $SiCl_4$, attach organic groups to silicon atoms, and then carefully build the siloxane bridges needed. But the labor is worthwhile. In the end we are rewarded by silicone molecules that have the admirable properties of the silica parent to an acceptable degree, and other unexpected, interesting properties as well.

Now, a brief tribute to a great man, who will introduce us to research, is in order. In 1823 Jöns Jakob Berzelius first made silicon as an element useful in further research. He probably sought for it because he felt sure it was *there*. He did *pure research*. But silicon made in his way would be too costly for use as a raw material in the present silicone industry. About the turn of the century, F. T. Tone and others at the Carborundum Company in Niagara Falls, building on earlier pure research, found that they could make silicon on a large scale by modifying the way in which silicon carbide (SiC) was being manufactured from carbon and silica in an electric furnace. Theirs was successful *applied research*, done with the objective of making commercial silicon at low cost.

Berzelius did more to help our silicone industry. Shortly after he discovered silicon, he ignited it in a stream of chlorine and made its chloride, $SiCl_4$. Still more! In 1830 he introduced *polymer* as a name for a big molecule made up of identical small parts; thus the giant silica molecule mentioned above may properly be called *crystalline polysiloxane*; likewise, vitreous silica is *glassy, amorphous polysiloxane*: silica is an aggregate of siloxane bridges. His threefold contribution to the silicone industry is still no measure of Berzelius. He introduced "catalyst," "isomer," and "protein" into the language of chemistry; he even made a polymer in the laboratory as early as 1847! In his time he dominated chemistry to an extent probably unique in its history and out of the question today. The silicones of greatest commercial importance are *polymers* with *inorganic* (siloxane) backbones or skeletons to which *organic* groups are attached. In light of the overriding importance of the siloxane bridges, it is fitting that the most important discoveries in the General Electric silicone project should have been made by inorganic chemists.

The silicone project began in the Chemistry Section of the Research Laboratory in 1938, and its climax is conveniently placed at the silicone coming-out party on November 14, 1944, soon to be described. After the climax the project faded away as the silicone business began. Of course, silicone research continued and is continuing in connection with the business.

In the silicone project, inorganic chemists were joined by organic, physical, and analytical chemists, and by chemical engineers, with other engineers playing peripheral roles. The project began at about the time when the old guard of the Chemistry Section, most without graduate training in chemistry, was being enlarged by a vigorous PhD recruiting program. The old guard (and the new) learned polymer chemistry as this science developed.

In 1938, when the project began, two successive depressions had taught all in the Research Laboratory to value their jobs more highly, and had made them more eager for useful research results to ensure job security. As silicone discoveries were made—and they came quickly during the first 5 years—added incentive was provided by the prospect of founding a new industry and the certainty of making important scientific contributions. Nor was this all. World War II brought a 6-day week, a need for new materials, and an opportunity for further staff expansion that added new men determined to make their marks. Finally, much of the silicone work was irreconcilable with either an 8-hour day or a 6-day week. Some reactions had to go on for days on end, as did many of the distillations needed to separate the resulting products.* The resulting atmosphere of urgency, tension, and excitement was conducive to getting work done. Individual response was voluntary and necessarily uneven. In general, people did more than they had to do.

For our look at industrial research, the special circumstances just described are less important than the fact that the silicone project ran its course in an exceptional industrial research laboratory with liberal research policies that had proved their worth over 30-odd years. Chapter 1 will make this clear.

The silicone project reached its climax on November 14, 1944, when the General Electric Company held a press conference to announce silicones, from which government secrecy had just been lifted. This coming-out party inspired the press. The next morning, headlines in the *New York Herald Tribune* read: "Rubber to Last Lifetime Made of Sand and Gas: General Electric Unveils New Product; Foresees Tires to Outlast Trucks." On November 25, *The New Yorker* began its story by saying that the event was "more damn fun" and concluded with, "We fell into conversation with a chemist who admitted that this was the first press party he had ever attended. 'They said it would do me good to get out of the laboratory for a while,' he sighed, 'but I sort of wish I were back in Schenectady.' " He may have realized that much remained to be done before a successful silicone industry could be established. The difficulties that had to be overcome conflicted sharply with the statement "It is simply that technology is rather easy,"† and it even raised some doubts about "Or, more exactly, technology is the branch of human experience that people can learn with predictable results,"—the sentence that follows.

What, then, were the main consequences of the silicone project? Increased knowledge, of course. A significantly intensified commitment

---

* There was no paid overtime for the professional staff. Dr. Coolidge, when he directed the Research Laboratory, once told me; "When we hire you, we contract for your entire scientific output, which is covered by your salary." For most, salary was not the major consideration. Winton I. Patnode recently wrote me that he regarded salary primarily as a means of subsistence by the grace of which he could work on silicones.

† C. P. Snow, *The Two Cultures: and a Second Look,* Cambridge University Press, London, 1964, p. 44. Of course, in technology, as elsewhere, practice makes perfect.

Henry Boltinoff sketches the silicone coming-out party. *Collier's Magazine*, December 2, 1944, p. 61.

of the General Electric Company to polymer research and polymer manufacture. And silicones as valuable commercial products, virtually all of which stem from materials made by "direct processes" discovered in the Company.

Over $0.5 billion worth of silicones is sold worldwide each year. They are *specialty products* worth their high cost because they can outperform alternative materials in important respects. Often a little silicone makes possible a large benefit: without the silicone, we would have cases in which "for want of a nail, the shoe was lost." Silicones have measurably enriched our civilization and done it no significant demonstrable harm.

Now, a further introduction to the Company is necessary. (Henceforth, "General Electric" as part of a name will usually be omitted; other corporations will be identified.) In 1938 the Company had some 35 manufacturing locations, almost all in the United States. A location was usually called a "Works" and identified by the nearest city, the Schenectady Works being the most important. The Research Laboratory, adjacent to that Works, was housed in Buildings 5 and 37, still connected by an enclosed bridge over the road between them. The Chemistry Section occupied the fifth and sixth floors of Building 37. Managed by A. L. Marshall, the Section was responsible for *corporate** research in chemistry.

Let us conclude with some general matters. In this book, "research" means *scientific inquiry*. No inquiry, no research—no matter how important or how difficult the scientific task in hand may be. Research for which the investigator or his or her manager can formulate a *practical objective* is *applied*; all other research is *pure*. Research of either kind may be done in industry, in a university, or in a government laboratory: the ratio of applied to pure will normally be highest in industry. Berzelius did pure research to isolate silicon; Tone did applied research to make it a commercial product.

Everywhere about us, multiplicity and complexity increase at accelerated rates; meanings become foggier; and rash statements proliferate. These difficulties bedevil our inquiry. To help the reader, I have selected for him and for me a guardian triumvirate with impressive credentials.

For General Guidance: Henry Adams. In the autobiographical *Education* written in 1904, Adams formulated a "law of acceleration," according to

---

* The adjective is needed because much research was and is done also in the Company's other laboratories; we shall see the Schenectady Works Laboratory become vital to the silicone project. Corporate research is expected to benefit the entire Company.

8

When the silicone project began, Buildings 5 (left) and 37 (right) housed the Research Laboratory, the General Engineering Laboratory, and other components of the Company. All space occupied by the then Schenectady Works Laboratory (in Buildings 4, 7, and 23) is only a few steps away. Building 77 is some 1000 yards down Central Avenue, which runs toward the horizon on the left of Building 5.

Much of the Schenectady Works first learned of the silicone project when methylchlorosilanes accidentally escaped from the corner room nearest the left of the sign crowning Building 37. On the original photograph the white siliceous residues left on the red brick are clearly visible.

which energy consumption and concomitant complexities double every 10 years [2, p. 490]. Furthermore:

The movement from unity into *multiplicity*, between 1200 and 1900, was unbroken in sequence and rapid in *acceleration*. Prolonged one generation longer, it would require a new social mind . . . . Thus far . . . the mind had successfully reacted, and nothing yet proved that it would fail to react . . . but it would need to *jump* [2, p. 498; italics mine].

For Meanings: Humpty Dumpty. Our times have made Adams a remarkable prophet. Communication has become uncertain because meanings are uncertain. Humpty Dumpty can help [3]. He said, "When *I* use a word, it means just what I choose it to mean—neither more nor less." Cases in which it is advisable to be Humpty Dumptyish will be brought to the reader's attention.

Formulas of compounds will often be preferred to their names as being shorter and more precise, for example, "SiO_2" instead of "silica" or "silicon dioxide." Also, the text will be sprinkled liberally with italics in the hope of helping the nonscientific reader. The beauty of the printed page will suffer, but in a good cause.

For Judging Generalizations: Mr. Justice Holmes [4]. In 1919 the great judge wrote his friend Sir Frederick Pollock, "I always say that the chief end of man is to form general propositions—adding that no general proposition is worth a damn," to which he added in another version, "especially this one." We have had skeptics from time immemorial, but none with a greater gift for concentrated, clear, and salty expression.

Under these auspices the reader may proceed confidently to discover how the world of silicones was found "in a grain of Sand," and to take his or her look at industrial research.

Chapter One The Corporate Stage and the Laboratory Setting

I believe the industrial society of electronics, atomic energy, automation is in cardinal respects different in kind from any that has gone before, and will change the world much more. It is this transformation that, in my view, is entitled to the name of "scientific revolution."

C. P. SNOW [1, p. 30]

The electrical industry became important toward the end of the last century, when electricity began to be generated and distributed on a large scale. This event was a landmark in the industrial-scientific revolution, which Lord Snow ranks with the earlier revolution in agriculture as "the only qualitative changes in social living that men have ever known" [1, p. 23]. We shall regard the scientific revolution as the continuation of the industrial revolution through research, and we shall include man-made polymers in the "industrial society." Adopting Lord Snow's premise that dating the scientific revolution is largely a matter of taste [1, p. 29], we shall place its beginning near 1900, the year in which the General Electric Company set up its Research Laboratory for "the application of real science to industry, no longer hit or miss, no longer the ideas of odd 'inventors,' but the real stuff" [1, p. 29]. The founding of the Laboratory was announced by E. W. Rice* in language less picturesque than Lord Snow's, but little different in meaning:

Although our engineers have always been liberally supplied with every facility for the development of new and original designs and improvement of existing standards, it has been deemed wise during the past year to establish a laboratory to be devoted *exclusively* to *original research*. It is hoped by this means that *many profitable fields* may be discovered [italics mine].

By 1892 the time had come for the creation of the large electrical utility industry that generates and distributes our electricity today. The great advantages of electricity for social living were even then so obvious as to produce a compelling need for improvements in its generation, distribution, and utilization. The General Electric Company, founded in that year by the merger of the Edison General Electric Company and the Thomson-Houston Company [2], helped to fill that need more effectively than smaller manufacturing companies could have done.

A small company is unlikely "to establish . . . a [central] laboratory devoted exclusively to original research." Even for the General Electric Company of 1900, which had barely survived the financial panic of 1893, this was a bold "Adventure into the Unknown" [2]. We cannot prove who first thought of the adventure, but four men deserve major credit for the actual undertaking: E. W. Rice,[†] who became President of the Company in 1913; Elihu Thomson, head of the Lynn (Massachusetts) Works Laboratory, an eminent applied scientist and prolific inventor, whose assistant Rice had been; C. P. Steinmetz, who made invaluable contributions to the large-scale generation and distribution of alternating currents; and A. G. Davis, head of the Company's Patent Department.

Unfortunately, we cannot know what these "four wise men" thought. They probably based their conclusions on different observations and different contacts within the Company. All of them—especially Thomson, himself an inventor—must have known and appreciated Thomas A. Edison's successful laboratory in Menlo Park, New Jersey, which dated from 1876. They must have felt the need for something

---

Note. Brackets are used to enclose reference numbers and inserted material. For brackbracketed references, see References and Notes at the end of the book. Reference to laboratory notebooks is by name and date. CRDA stands for "Chemistry Research Department Archives."

* General Electric Company, *Tenth Annual Report,* p. 13.

[†] The quotation is from the "Third Vice-President's Report," by E. W. Rice, Jr. The report is a letter, dated April 15, 1902, from Rice to C. A. Coffin, President, and covers the fiscal year ending January 31, 1902. Rice had been named Vice President and Technical Director when the Company was formed.

Here, in the C. P. Steinmetz barn, General Electric's "Adventure into the Unknown" began. The barn burned 3 weeks after the Research Laboratory started work in it—cause-and-effect relationship not established!

13

The "four wise men," who deserve major credit for founding the first American industrial laboratory intended exclusively for research. Photographs taken at various stages of their careers. Elihu Thomson; others follow.

C. P. Steinmetz

E. W. Rice

A. G. Davis

different, something new. Edison, a towering figure, had shown that carefully organized, relentless, and extensive laboratory work could more than pay its way in producing new devices and processes based upon *prior knowledge*; but the words italicized in the quoted Company announcement did not reflect Edison's approach.*

Rice spoke as follows in 1935 on the occasion of the presentation of the Edison Medal to Dr. Willis R. Whitney, first director of the Laboratory:

We had learned that there was a field for all the various applications of electricity, whether of alternating current or of direct current, such as the arc lamp, incandescent lamp and the electric street railway, and we were busy welding these together. The opinion seemed to have been generally held that no radically new developments could arise. Copper was the best conductor of electricity; iron the best for magnetism; carbon the best for electrodes for arc lamps and lamp filaments and for brushes for commutators. As far as we could see it was likely that these *materials* would always remain the best for their respective purposes. Such at least was the opinion of *most engineers*. However, *there were a few who thought differently,* and it was at this period of comparative calm that the Research Laboratory at Schenectady was started. It started most modestly in a small way, with few men and few facilities, and *increased in size only as it demonstrated its usefulness* [2c, pp. 39–40; italics mine].

The "four wise men" seem to have believed that materials ought to be a prime concern of General Electric Research, and believed this long before anyone thought of calling such work "materials science." In any event their efforts led in 1900 to the founding of the Research Laboratory, with Whitney, then at the Massachusetts Institute of Technology, as the first (and part-time) member.† We shall always be in their debt.

The Edison General Electric Company had been organized to establish systems that would make possible the widespread use of Edison's incandescent light, for which there was then no suitable filament. Quite logically, Dr. Whitney chose the search for such a filament as an early task of his new laboratory. In 1910, Dr. William D. Coolidge, also from MIT, found how tungsten could be made ductile at room temperature [3],

---

* *The General Electric Story,* Vol. 1, *The Edison Era 1876–1892,* The Algonquin Chapter, Elfun Society, Schenectady, New York, 1976. On p. 47 Edison is quoted as follows from one of his notebooks: "I will have the best equipped and largest laboratory extant, and the facilities incomparably superior to any other for rapid and cheap development of an invention and working it up into Commercial shape . . . . Inventions that formerly took months and cost large sums can now be done [in] 2 or 3 days with small expense . . . ."

† Whitney accepted appointment on a trial basis and made his first visit to Schenectady in November 1900 [2c, p. 46].

W. R. Whitney as Director, Research Laboratory, shortly after its founding.

Perhaps half a century later, W. R. Whitney as lecturer on turtles to a neighborhood audience.

in a project that for a time occupied much of the laboratory. Time and other workers have not essentially changed his process, and ductile tungsten is still the preferred material for the filaments of incandescent lamps.

The ductile-tungsten triumph is a classic and significant example of successful applied research. It gave the General Electric Company a dominant position in the lamp industry. It put the fledgling Research Laboratory on a firm foundation and brought it to successful maturity. In confirming the judgment that research leading to new materials was needed, it prepared the way for research on man-made polymers. In opening the door to the Coolidge tube for the generation of X-rays, to thoriated tungsten as electron emitter, and to tungsten and molybdenum as refractory metals well suited to vacuum applications (electron tubes included), it helped launch the Laboratory into an electronics era that is still continuing as an important part of the scientific revolution. Concomitantly, it drove home the lesson—especially important for industrial research—that good research of any kind on new materials can have unsuspected and unforeseeable consequences of profound significance.

"At its beginning, the Research Laboratory was Whitney, and Whitney was the Research Laboratory" [2a, p. 5]—this statement was literally valid for longer than one might think. Initially, Whitney had expected*

---

* Whitney's expectation appears in an article on the Research Laboratory prepared by Dr. George Wise (available as Report #77-CRD273 from General Electric Research and Development Center, Technical Information Exchange, P.O. Box 43, Building 5, Schenectady, New York 12301). Dr. Wise drew upon letters in the Willis R. Whitney Collection, now on loan to the Schenectady Archives of Science and Technology, Union College, Schenectady, New York, among which are two others that deserve a place in the history of American chemistry. From them, I quote:

1. W. R. W. to Professor Jos. W. Richards, Secretary, American Electrochemical Society [A.E.S.], February 24, 1910:

". . . I have just telegraphed several friends as follows:

Could not accept presidency A.E.S. Please actively assist in defeating me. Use the 'phone.'

It is very clear to me that I should not stand as nominee for this office at the present time."

Dr. Whitney explains the telegram by saying that he refused to stand for *re-election* as President of the American Chemical Society because he had found it burdensome to fill this office while giving the Research Laboratory the attention it needed.

2. Leo H. Baekeland, President, A.E.S., to W. R. W., March 21, 1910:

"Your letter of March the 14th has been transmitted to Professor Richards. I believe that what you suggest is entirely favorable, namely that when the ballots are counted, the votes in your favor should be counted out . . . .

". . . I only regret that your many activities, just now, should prevent you from being at to set up an "experimental electro chemical laboratory," an expectation at variance with Rice's quoted announcement. The difference is not surprising. Whitney accepted Rice's offer of employment upon condition that Whitney could continue at GE the electrochemical work he had been doing at MIT, and he did not move his family permanently to Schenectady until May, 1904 [2c, p. 42]. For the first 6 months in the Laboratory, he had only one assistant (J. T. H. Dempster); the number of staff members rose to eight (some from MIT) during 1901. The important point is this: Whitney, as he became familiar with the Company, made up his mind as to what kind of Research Laboratory was needed, and then acted decisively upon what he thought, even though this meant giving priority at times to other things (such as calls for help from operating components) that were not "original research." The "four wise men" had given him a broad charter and allowed him freedom; he interpreted the first, and used the second, to the Company's advantage.

Being the first man to build, and the first to manage, a laboratory of this kind, Whitney had no formulated policies, position guides, or organization charts to rely upon; hc had to invent as he went. He admired Francis Bacon; like Bacon, he had an abiding faith in experiment as the main route to scientific progress. He had little (probably too little) respect for organization in a research laboratory; he strove to make his laboratory a happy family with him at its head. He believed in open doors through which research workers could easily seek help from one another, and through which other employees of the Company could have access to scientific advice and to new materials for which the Laboratory could help them find applications. He wisely encouraged travel to Europe on Company business (Whitney, Coolidge, and Langmuir all did PhD work there). Whitney's thoughts, ideas, and policies influenced the Laboratory until long after 1932, when he retired as its director. He established the "Whitney atmosphere" for industrial research.

A research worker in the Laboratory after the end of the 1937 depression could expect the following: to punch a time clock during the regular work week, which had risen to 40 hours after having been cut (along with pay) from 45 to 32 hours during the depressions; to work extra hours (without added pay) if he wished; to have available excellent

---

the head of our Society, and I also expect that the time will soon arrive when you will have no further objection of letting your friends honor you in that way."

These quotations are enough to indicate the standing of Whitney and the Research Laboratory a bare decade after its founding, and they introduce Baekeland, whom we shall soon meet again.

I particularly thank Dr. Wise for help with this footnote.

glass-shop and machine-shop services; to do his own experiments (assistants were scarce); to wash his own laboratory glassware; to have difficulty in getting work done for him by the Laboratory's analytical chemists; to be free of budget pressures; to have some say in choosing his research field; to have all the freedom he could advantageously use in planning and doing his research; to find competent colleagues in a wide range of disciplines glad to discuss his work with him; to make business trips when they were useful, and to attend scientific meetings; to have close contacts within the Company; and, if he so wished, to do the best scientific work of which he was capable. Pay and fringe benefits, though small by present standards, were among the best available during those postdepression times.

Chapter 1 to this point invites comparison with the early history of research in the Badische Anilin- & Soda-Fabrik, BASF,* a great chemical manufacturing company founded in 1865 to continue the manufacture of fuchsine (starting material, *aniline*), a dye made by a predecessor firm, and to begin making *soda ash* (Na_2CO_3), an important industrial chemical. The italicized words explain the firm's name and mark the beginning of BASF diversity, which has continued as a conscious policy that antedates General Electric's. In 1870, BASF employed 5 chemists; in 1884, 61. In 1889 the "big main [research] laboratory" was built in Ludwigshafen, West Germany. The first Director of Research for BASF was Heinrich Caro, a leading dye chemist even when he joined the firm in 1868. We cannot record here BASF's many research triumphs—a matter of regret because many of them have come in the field of polymers, a principal concern of this book. We shall, however, say a few words here about synthetic indigo and discuss synthetic ammonia in Chapter 9.

Adolf von Baeyer (Nobel Prize, 1905) had become fascinated by indigo at age 13 and had devoted years to pure research on this dye. By April 1880, Caro had negotiated a contract with von Baeyer. Seventeen years later, "Indigo pure BASF" went on the market after it had been found necessary to develop four methods by which the dye could be made. Ductile tungsten was General Electric's "indigo."

One difference between Caro and Whitney must be noted. By and large, Whitney could rely upon universities only for recruiting, not for research. By contrast, the relationship between BASF and von Baeyer was classic of its kind. It began before the indigo contract and continued for decades. Of the first meeting, von Baeyer wrote, "Never before in

---

* The information about BASF is taken from *In the Realm of Chemistry: Pictures from Past and Present*, Econ-Verlag, Vienna, 1965, a splendid volume that commemorates BASF's centenary.

The first experimental entry in a historic laboratory notebook: "April 1901. Tried following dyes dissolved in gelatine and coated upon glass . . . as to the effect on the light of the mercury arc. . . . Rhodamine 6B (plate E) . . . gave a feeble band in the red . . . ."

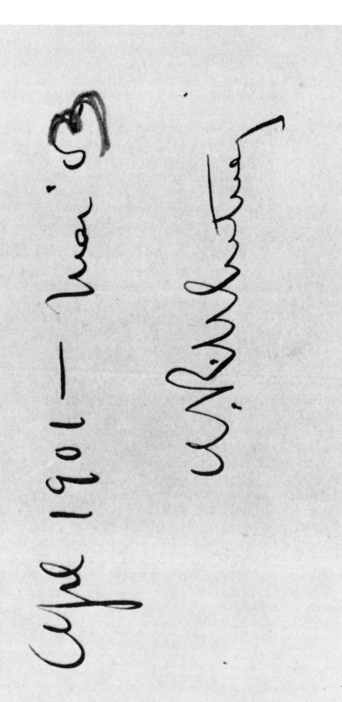

The first entry in a historic laboratory notebook.

LIBRARY
OF
MOUNT ST. MARY'S
COLLEGE
EMMITSBURG MARYLAND

25

my life had I seen a technologist and I gazed at them [Caro and two colleagues, who had come to von Baeyer's laboratory] as if they were beings from another world." Of the hundreds of letters addressed to Caro by leading German chemists and kept in the German Museum at Munich, von Baeyer wrote more than 200. There seem to have been no von Baeyers in American universities in the early part of this century; hence the Research Laboratory was forced to rely upon itself to a greater degree than would otherwise have been necessary. A comparison, based upon original sources, of how research developed in these two companies would improve our understanding of industrial research.

ELECTRICAL INSULATION AND POLYMERS

In the address quoted on p. 18, E. W. Rice failed to mention *insulating materials,* a class as vital to the electrical industry as conductors or magnetic materials. He may have omitted them because they were then diverse and unsatisfactory. Man-made polymers, a class of materials unknown when the Laboratory was founded, have since revolutionized the electrical industry.

Flowing electrons produce an electric current. For the satisfactory operation of most electrical equipment, such currents must be confined to electrical conductors, usually metals—most often, copper. Insulation accomplishes the confinement. Electrical insulators can be solid, liquid, or gaseous.

Consider magnet wire, a product vital to the industry. It forms the electromagnetic coils placed in rotating machinery so that mechanical motion of an electrical conductor in a magnetic field can produce an electric current in generators, and so that the process can be reversed in motors. Magnet wire plays a crucial role in making useful to mankind the fundamental discoveries about electricity, magnetism, and motion made by Hans Christian Oersted and Michael Faraday, and put into mathematical form by James Clerk Maxwell a few decades before the electrical industry began.

Magnet wire, almost always copper, must of course be insulated. If there were no confinement of electrons by insulation, they would jump the narrow gaps between adjacent turns of wire in the electromagnetic coils; a short circuit, always undesirable and sometimes fatal, would result. The gaps are narrow because the coils must be highly compact if the equipment is to be compact also. The insulation on magnet wire is called "wire enamel," and it can claim to be the most important among the many insulating materials in common use. An improved wire enamel

Copper wire, the electrical industry's most important conductor, being coated with wire enamel, perhaps its most important polymeric insulating material. The copper wire from the large spools on the left passes through the large, cylindrical, horizontal furnace on the right, and the finished wire is wound on the smaller spools about 4 feet above the floor on the right after it has been coated with the enamel, which is "cured" by heating. Larger wires are enameled in vertical furnaces, called "towers." The enamel, tightly adherent and sometimes exasperatingly difficult to remove, is often so thin as to escape notice by a casual, unskilled observer.

means improved equipment that is easier to sell. Few scientific accomplishments give research a better name within the industry than a good new wire enamel.

Wire enamel must have at least the *electric strength* needed to prevent the objectionable short circuits. Now, high electric strength in bulk is an intrinsic property of many materials; it is not hard to achieve. But adequate electric strength in a wire enamel is another matter. The ideal wire enamel must be of adequate electric strength *when in place,* where it must be thin, uniform, and "permanent." A useful enamel must retain its electric strength for an adequate time—preferably forever—no matter how bad the conditions. It must be easily applied. It must survive the torture attendant upon mechanical winding. Finally, it must not unduly increase the cost of equipment.

As an example of the requirements hidden in this brief description, consider the matter of operating temperature. If magnet wire can carry higher currents satisfactorily, motors and generators can be made smaller and lighter. Higher currents mean that heat is generated more rapidly. This heat must be eliminated, or the equipment will fail. The higher the allowable temperature inside a coil, the easier it is to remove the heat. Hence a wire enamel that has an adequate life at a higher temperature is essential to the building of satisfactory motors and generators of reduced size and weight. There is a further consideration. Equipment is sometimes overloaded, and the temperature during periods of overload will rise above normal. An ideal wire enamel not only permits a higher normal temperature but also satisfactorily resists deterioration during overload.

The importance of wire enamel among insulations and the drastic requirements that a satisfactory wire enamel must meet make wire enamels of primary concern in the electrical industry. When a new insulating material is mentioned, it is an almost reflex reaction to think, "But is it any good on magnet wire?" Of course, as common sense indicates and Table 1-1 shows, the insulation problem is much broader.

Table 1-1 conveys the following message. Almost every property of a material must be known before its usefulness as insulation can be established. Not every application requires major emphasis on every property; and compromises (trade-offs) are usually made in selecting the material that gives the best dollar value for the application in hand. In the absence of previous experience, the selection is best approached empirically, that is, by "life-testing" proposed materials under "appropriate" conditions, not merely under operating conditions alone. (Here the quotation marks hint at Adamsian multiplicity and complexity.)

Man-made polymers, because of their diversity and the controllability of their properties, are by all odds the best class of materials for

Table 1-1 [4]. Significant Properties of Insulating Materials[a]

Mechanical Properties	Electrical Properties	Thermal Properties	Chemical Properties	Miscellaneous Properties
Strengths	Electric strength (volume)	Thermal conductivity	Sensitivity to:	Specific gravity
Tensile	Breakdown strength (surface)	Thermal expansion	Chemical attack	Refractive index
Compressive	Liability to tracking	Primary creep	Electrochemical attack	Transparency
Shearing	Resistivities	Flow	Aging	Color
Bending	Volume	Plastic	Oxidation	Porosity
Impact	Surface	Thermal	Solvent- crazing	Permeability to:
Tearing	Permittivity	Flame resistance	Effects on adjacent materials	Gases
Elastic moduli	Loss angle	Melting point	Composition	Vapors
Hardness	Resistances	Pour point	Crystallinity	Moisture sorption
Viscosity	Insulation	Vapor pressure	Solubility	Surface adsorption of water
Extensibility	Spark		For polymers only	Resistance to fungus
Flexibility	Arc		Molecular weight	Resistance to change by light
Machinability			Molecular weight distribution	
Sensitivity to:				
Fatigue				
Stress- crazing				
Abrasion				

[a] Significant changes in properties with changing electrical frequency and with changing temperature must also be known.
Note. The list of properties in the table is comprehensive enough for most applications of polymers. A "property profile" can be constructed from data on the properties of each polymer. Polymers that have acceptable and reproducible property profiles significant for mechanical and structural applications are coming to be called *engineering polymers*. Such polymers generally serve markets much larger than those for the same polymers as insulating materials.
References to this table will occur throughout the book.

meeting insulation needs. To see whether these needs can be met, so many of the properties of a polymer must be measured that other possible uses of the material, as in structural or mechanical applications, become evident as its value for electrical applications is being assessed. Research on polymers for insulation did in fact provide the Company with an entry into the chemical business and led to its making *engineering polymers* (see Table 1-1), which today far exceed Company wire enamels in dollar value.

Today the Company is at once the world's largest maker of electrical equipment, the world's most diversified manufacturer, and an important producer of chemicals, many of which are—or are related to—unusual man-made polymers. Most of these chemicals find nonelectrical applications, whence the chemical activity of the Company has become a major contributing factor to its diversity.

The concern about materials evident from the first in the young Research Laboratory presaged an important role for chemistry in the Company's future. But it did not ensure, for example, that the Company's external chemical business would eventually dwarf its metallurgical counterpart. To see how this started to come about, let us return to the early concern about electrical conductors, magnetic materials, and insulation.

We will first consider conductors. The high cost of silver and the high resistivity of most other metals have long made certain that copper will always be the prime electrical conductor. Once this metal is adequately purified, the applied research remaining to be done deals with problems such as devising improved methods of making wire and of joining the copper to other metals.

Next, iron was the logical starting point for the magnetic material needed to make the cores of transformers, motors, and generators. Here long, difficult, and costly applied research in the Company showed that iron alloyed with silicon and formed into thin sheets would give magnetic cores in which energy losses were permanently low. In distribution transformers, which are near almost every house, such losses continue night and day. In these days, when waste of energy is of national concern, the metallurgical research done in the Company to develop low-loss silicon steel deserves renewed recognition.

And yet, this metallurgical research did not change the Company as much as did research on organic insulating materials. For this, the emergence of man-made polymers was largely responsible: man outdid nature or improved upon it. The early materials of natural origin (e.g., paper, cotton, linen, silk, rubber, wood, asphalt, and shellac) were defective in respect to many of the properties listed in Table 1-1. Man-made polymers, which came into existence shortly after the Research Laboratory was founded, soon gave promise of almost endless variety, and of properties controllable to a much greater degree than those of substances from natural sources. Accordingly, research on man-made polymers was almost certain to result in products that would find uses—sometimes their main uses—outside the electrical industry. No wonder the Research Laboratory came to have an Insulation Section! Of course, that alone could not ensure the Company's becoming a chemical manufacturer. The research had to be successful. Moreover, the management of the Company had to be willing to take the risks attendant upon entry into the chemical industry, where it would not necessarily benefit from its success as an electrical manufacturer. These risks probably seemed more tolerable because the Company often had important internal uses, not all electrical, for the products involved. When it came to a "make-or-buy" decision, "make" often won. But, as we shall see, the Company did not become an important chemical manufacturer overnight.

BAKELITE AND THE COMPANY

It is thus no wonder that the invention of Bakelite, the first commercial man-made polymer,* must be regarded as a landmark in General Electric history. The invention was made (outside the Company) by Leo H. Baekeland, who on July 13, 1907, applied for one of the most famous of all U.S. patents, "Method of Making Insoluble Products of Phenol and Formaldehyde," which was granted as No. 942,699 on December 7, 1909.

Baekeland will always rank among the world's most successful practitioners of applied research. A scholarship student at the University of Gent (Ghent), Belgium, he was awarded the ScD with highest honors in 1884 at age 21; and by his twenty-eighth year was an independent consultant in the United States. For Mr. Elon H. Hooker he directed work on the development of the Townsend cell for the production of NaOH and Cl_2, the eventual success of which led to the founding of the Hooker Electrochemical Company [5d].

As early as 1883, Baekeland had worked to perfect a photographic paper, later called "Velox," developable in ordinary light. In 1893 he became a partner in the Nepera Chemical Company, formed to make and sell this paper. In 1899 Mr. George Eastman, acting for his company, bought the entire business for $1 million, his opening offer. Baekeland

---

* We shall consider a polymer man-made only if the polymerization has been done by man. Celluloid, which dates from 1870, does not qualify because cellulose, its principal component, occurs in nature.

had expected to ask $50,000 and was prepared to retreat to half that amount. Eastman spoke first. Baekeland accepted [5b,d].

With part of the proceeds from that sale, Baekeland built a research laboratory in which he began to look for a man-made equivalent of shellac, a natural resinous polymer secreted by certain insects. After importation from India, shellac was at that time dissolved in volatile solvents to make varnishes. Adolf von Baeyer, whom we have already met, had found in 1872 that phenol (hydroxybenzene; carbolic acid) would react with aldehydes to make polymeric materials resinous in character. Baekeland hoped to find some such material that could be suitably dissolved to make a varnish. What he got were hard, transparent, complex polymeric substances, difficult to melt, impossible to dissolve, and hopeless as replacements for shellac. Making a virtue of necessity, Baekeland *changed his objective to fit his product.* He found *needs* the product might fill. He learned to make his polymer take useful shapes (fountain-pen barrels, tubing—what have you) as it formed from phenol and formaldehyde in the presence of alkaline substances (catalysts) added to speed the reaction. Baekeland's announcement of Bakelite to the Chemist's Club, New York City, in 1906 may be taken as the beginning of the man-made polymer era [5a].* This oversimplified account will be supplemented in Chapter 2.

The Company moved quickly to assess the usefulness of Bakelite in the electrical industry. During 1909, the year in which the first Bakelite patent was issued, Dr. Whitney and Mr. L. E. Barringer,[†] an insulation engineer then in the Research Laboratory, visited Dr. Baekeland at his estate in Yonkers to learn more about this new material, which soon proved its worth in the manufacture of molded parts to serve insulating, mechanical, and structural needs in electrical equipment. Production of Bakelite, presumably under license, began in Schenectady within a year of the visit. Even at the high cost of $1 per pound for material up to half of which was a cheap filler (usually wood), Bakelite could economically

---

* Bakelite and ductile tungsten, which may be regarded as two of the earliest triumphs of applied research in the scientific revolution, have been combined in a surgical device that continues to benefit aging men. The device is the *resectoscope,* used in the prostate-gland operation called *transurethral resection* (TUR), and it consists of a "Bakelite sheath and a working element. The working element carries a movable loop of fine tungsten wire which cuts easily through the soft tissue of the [prostate] gland when the high frequency current is turned on" [6].

† Barringer later became well known in the Company as the father of "Barringer Red," a varnish pigmented with iron oxide, which was widely used to paint electrical and other equipment, and was made from Glyptal®. Whitney and Barringer, in another notable collaboration, brought Louis Navias into the Laboratory; see Chapter 3.

replace traditional materials that had constituents such as rubber cement, long-fiber asbestos, paper, cloth, sulfur compounds, and shellac [7]. No wonder Mr. Rice did not mention insulating materials in his address!

Polymers based on phenol, as is Bakelite, were being investigated in the Pittsfield Works Laboratory in Massachusetts by De Bell in 1925, and manufacture of such polymers was begun there immediately upon expiration of the first Bakelite patents [7]. Pittsfield is today the headquarters of the Plastics Business Division, largely because of the successful work done there on man-made polymers.

Even earlier, work had been undertaken to discover whether useful man-made polymers could be formed by reacting polyhydric alcohols* with polybasic* organic acids. Such polymers could be regarded as distant relatives of Bakelite; they are *polyesters.* On May 21, 1912, Steinmetz wrote the letter excerpted below:[†]

Mr. C. C. Chesney,
Pittsfield Works

My dear Mr. Chesney:

As you know, in the last weeks a very material and radical advance has been made by Mr. Gifford in the field of artificial rosins [resins, man-made polymers] of high resistivity, in the production of the first thoroughly flexible material of this kind, which has the further important property, that it can be combined with other artificial and natural rosins, and gives them a very great toughness and strength.

. . .

I believe we could well afford the employment of *several chemists,* in view of the importance of the subject, and the various directions, in which the investigations should be carried simultaneously, as for instance:

(a) The study of *all* the compounds of polybasic [polyhydric] aliphatic alcohols (and aldehydes) and di- or poly carboxylic [polybasic] acids.

* A Humpty-Dumpty footnote. Here "poly" means from 2 to 4; "hydric" identifies the OH group; "basic" is linked to "acid hydrogen." Please do not confuse this "poly" with that in "polymers," where the same prefix usually signifies thousands, hundreds of thousands, or even more. "High" polymers have big molecules. See Chapter 2.

[†] Mr. Chesney probably was Pittsfield Works Manager. Mr. [A. McKay] Gifford had a distinguished career as head of the Pittsfield Works Laboratory. Copies of the letter went to E. W. R[ice] and A. Mc. G., not to the Research Laboratory. On May 3, 1960, however, A. L. Marshall, then Manager, Chemistry Research, sent a copy to Dr. C. G. Suits, then Vice President, Research, and in charge of the Laboratory. How and when this copy reached Marshall, I do not know.

(b) The investigation of the *chemical structure* of these new artificial rosins, and of the *structural change* which occurs during the final condensation. . . .

(e) Development of a cheap method of production of succinic acid. . . .

(d) Considerable work has been done during the last months here and in your factory on linseed oil, as it enters most of our important insulations, and the electrical characteristics of these insulations are rather unsatisfactory, the insulation being greatly impaired by temperature rise . . . .

I believe that this work should be carried out in close co-operation with our Research Laboratory, but should be done in Pittsfield . . . to be successful, the work has to be done in very close and continuous co-operation with the engineering departments [italics mine].

On May 25 Mr. Chesney replied, promising additional facilities about June 15, and the addition of "probably three good men" [*sic*] even though "The difficulty will be to get satisfactory chemists." To carry out the ambitious Steinmetz program, *eminently* satisfactory chemists would have been needed in considerable number, especially in 1912, when—as we shall see in Chapter 2—polymers were poorly understood.

Since these early days, Company interest in man-made polymers has never flagged, and Company research on them has continued.* The kind of work described in the Steinmetz letter eventually resulted in Glyptal® resins, which we have already met as a constituent of Barringer Red, and which form when *gly*cerine (a trihydric alcohol) reacts with *ph*t*ha*l*ic* anhydride (the anhydride of a dibasic acid); the italicized letters form the acronym. Roy H. Kienle, then of the Research Laboratory, became an authority on polymers of this kind, which belong in the class of *polyesters*. He coined for them the general name "alkyd," and greatly increased their usefulness by incorporating in them the unsaturated fatty acids (acids containing a carbon-to-carbon double bond and hence reactive toward the oxygen in air) best suited for "air-drying," that is, for forming durable films on exposure to air. In the electrical industry such films suggest wire enamel. Alas, no satisfactory alkyd wire enamel based on Glyptal ever came into Company-wide use, though not for lack of due diligence. But all was not lost! The alkyd polymers turned out to make very good paints—so good, in fact, that they had their day of glory in the automobile industry, in which the finishing of automobiles is as important as the insulating of magnet wire in the electrical industry. The Company,

---

* Today, corporate polymer research is managed by Dr. Allan S. Hay, Corporate Research and Development Center, to whom I am indebted for help and advice. We shall meet him again in Chapter 9.

then, began to manufacture paints. The air-drying alkyds were in the factory by 1933, over *20 years* after the research on polyesters had been begun. By 1950 the General Electric production was 17.5 million lb—only about *one-twentieth* of that of all industry. The electrical manufacturer had begun to be a chemical manufacturer as well, but in a way that might be described as too little and too late. This is not surprising. After all, the alkyd resins resulted from research done by this manufacturer primarily to obtain materials useful in his industry. Other uses for these products became dominant, but the Company at this stage in its history was not prepared to capture a strong position in the alkyd resin market. It could not yet see large chemical plants in its future.

A. L. MARSHALL (1896–1974)

At this point we must meet the man responsible for the conduct of corporate research in chemistry from the founding of the Chemistry Section in 1933 until his retirement in 1961.

A. Lincoln ("Abe") Marshall was born in Victoria, British Columbia; completed undergraduate work in chemistry at the University of British Columbia in 1918; and received the MA from the University of Toronto in 1920, and the PhD in physical chemistry in 1922 from the University of London, where he was an 1851 Exhibition scholar. He first did chemical work in 1917, when he was employed by the Consolidated Mining and Smelting Company at Trail, a small town on the Columbia River. At Toronto he came under the influence of Professor W. Lash Miller, a formidable teacher who gave his students the alternatives of learning chemical thermodynamics as he taught it, or going under. Marshall began his research career in electrochemistry [8]. After completing graduate work in that field, he went to Princeton University as instructor and soon turned to photochemistry, in which he collaborated with Professor Hugh S. Taylor, internationally known as a physical chemist. The scientific literature of that time indicates that a promising and productive career in pure research was within his grasp. The following quotation [9] shows how he made the transition from Princeton to the Research Laboratory, a transition begun during two summers and completed in 1926:

Following upon the work of Taylor and Marshall, I [A. L. M.] have designed an all-quartz apparatus which will enable us to study the quantitative aspects of these reactions [reactions such as that between hydrogen and oxygen under the

A. L. Marshall at about the time he became head of the Chemistry Section, Research Laboratory. He is one of 17 posthumous members of The Plastics Hall of Fame, which was established in 1973 by *Modern Plastics Magazine* in cooperation with the Society of Plastics Engineers.

influence of light absorbed by mercury vapor] more conveniently. The reaction chamber . . . was made by Mr. Henry Wayringer of the Research Laboratory.*

Marshall turned very quickly from pure to applied research in physical chemistry that was useful to the Company. Soon he was supervising a group of five, among whom were Francis J. Norton and Maynard C. Agens; both will reappear later in this narrative. The group worked on problems such as the generation of ozone at high concentrations in air, atmospheres for the tarnish-free annealing of metals, and the vapor pressures of various important substances of low volatility—graphite, in particular.

Soon after 1933, Marshall's scientific interests shifted again—this time to man-made polymers—and this shift led to the most important decision of his industrial career, namely, to make applied research on such polymers, and their pilot-plant production, the principal concern of the Chemistry Section. About 1945 he saw that Section become a Division. Approximately 5 years later the Chemistry Division was merged with the Mechanical Investigations Division to form the Chemistry Research Department, of which Marshall was Manager when he retired.

Marshall was an exceptionally brilliant man who had many other important qualities to a superlative degree. He was energetic, strong-willed, reticent, complex, and gifted with intuition and insight. He had a memory so good that he seldom needed to turn to books and could operate largely on the basis of what he had been told: he was a superb listener, with a quick mind that seldom rested. Though he seemed at times impatient, impulsive, and enigmatic, he was fundamentally conservative and his judgment was almost always good. He believed in extensive business travel, preferably in his automobile, as the best way to establish liaison with Company components and to obtain information not otherwise readily available; he was a skillful driver who often thought about his work as he drove. Especially at the height of the silicone project during World War II, he carried a load beyond the capacity of most men; he was away from Schenectady much of the time. He normally expected others to meet the high standards he set for himself. In the pursuit of chosen objectives, scientific or business, he could be unsparing of himself and others; he always kept first things first. He was never eager to put all his cards on the table, and he was reluctant to write or to sign anything. Often he seemed harder to convince than he really was. He

* Henry Wayringer deserves to be remembered for his great skill as a quartz-glass blower at a time when such blowing was an art with practitioners so rare they could afford to be prima donnas. Henry was.

did not suffer fools gladly, but he bore them no grudge. Among those who found him difficult at one time or another were many who would have been surprised and gratified by his strong support of them when it counted and when they were absent.

Marshall served under two vice presidents, Dr. Coolidge and Dr. C. Guy Suits, and had the strong backing of both. I regard Marshall's value to the Company as comparable with that of anyone else active in the Research Laboratory during its history. The accomplishments of corporate chemical research under his management were outstanding by any standard; he directed applied research that resulted in* Formex® wire enamel, Flamenol®, silicones, the Permafils, Alkanex® wire enamel, Irrathene®, Vulkene®, Lexan®, Man-made ™ diamonds, Borazon®, PPO®, and Noryl®. (One of the Permafils, the "anaerobic" variety, led to the founding of the Loctite Corporation, an independent and highly regarded company.)

THE CHEMISTRY SECTION

In 1925 the importance of insulating materials to the Company had influenced Dr. Whitney to form a group that by 1933 had become the Insulation Section. On June 9 of that year, Dr. Coolidge, who had succeeded Dr. Whitney as Director in 1932, wrote an important letter, which read in part as follows:

Dear Marshall:

Mackay is leaving us, to take charge of the research work of the Cyanamid Company. His work with them begins July tenth.

I want you to take his place as head of the Insulation Section. (We may later want to broaden the title, as the field of activity which I have in mind includes such other chemical activities as those which you have been carrying on in the

---

* Adamsian multiplicity and complexity sooner or later frustrate all who must write about commercial products. In the list given above, "silicones" is the *scientific name* of a class of materials; Permafil is a *trade name*; Man-made™ is a *trademark*; and Formex® is a trademark *registered* in the U.S. Patent Office, from whom the reader can obtain further information.

Courtesy and conscientiousness require an author to name commercial products properly, but the effort entailed can be prohibitive. In this book ® and ™ as appropriate will be repeated at intervals. I hope to avoid blunders comparable to the immortal "General Electric Frigidaire." For industrial research trademarks and the like have an importance approaching that of patents.

past.) . . . May I congratulate you on your past achievement, which has made it so easy for me to satisfactorily fill this important position.

Probably to no one's surprise, the letter had to be sent to Marshall in Philadelphia, where he was using the high-speed presses of the Curtis Publishing Company to learn whether his ozone generator could usefully accelerate the drying of printing inks. Alas, the ozone process could not compete with the improved inks that soon came into use.

Marshall must have begun to think about the future Chemistry Section even before he returned to Schenectady. His group of five, mentioned above, was merged de facto with the analytical chemists, headed by Charles G. Van Brunt, and of course with the Insulation Section, which included J. G. E. (Gilbert) Wright, a most unusual scientist who was to leave his mark on the silicone project. Though the Great Depression was scarcely over, a vigorous PhD recruiting program was begun in 1933 with Dr. Coolidge's approval; Winton I. Patnode (1933), Murray M. Sprung (1933), and Eugene G. Rochow (1935)—all of whom later did vital silicone work—were among those who joined the Chemistry Section, which appears as a unit in the 1935 Organization Directory of the Company. Figure 1-1 shows the increase in professional personnel ("research associates" and "research assistants," as they were known in the early days) of the Chemistry Section and its successors over the years from 1933 to 1952.

We have seen that Marshall's major interest in corporate research shifted from applied physical chemistry to insulating materials, and finally to man-made polymers that could launch a chemical business. Patnode, through influence and advice, played a major role in bringing about the final shift.

Hope for improved wire enamels was an ever-present inducement for the change in emphasis. As the chemical business grew, cost, which is not primarily important in a promising young wire enamel,* became more important overall. Marshall's last shift had to survive testing in the marketplace. Because the time of early trial came in the first years of the Chemistry Section, a brief record of the Section's polymer activity in those years is of general interest in the history of man-made polymers, and of special interest here: the success of the Section's earlier polymers made the silicone project possible.

Work on *flexible* alkyd polymers, which were intended for use as insulation in electric cable and for other applications unrelated to those

---

* Formex® wire enamel, when introduced at a price of $2 per gallon, easily displaced much cheaper materials. Its superior performance led to a successful trade-off.

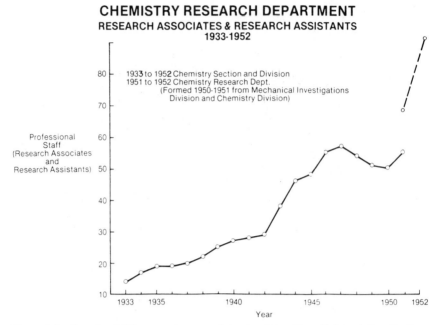

Figure 1-1   Growth of GE corporate research in chemistry. Compiled by Arthur E. New-kirk.

served by *air-drying* alkyds such as Glyptal®, was discontinued in 1934. The flexible alkyds became victims of two hazards that often bedevil applied research on new materials: inherent weaknesses and new competitors; for example, a Glyptal oilproof rubber was displaced by Du Pont's Chloroprene®. Work on condensation polymers related to Glyptal eventually produced the Permafils.

In the early 1930s, *polyvinyl chloride* was an almost useless laboratory curiosity. By 1934 Wright had learned how to make it flexible—how to "plasticize" it—by admixing and incorporating suitable organic liquids. In 1936, thanks largely to Moyer M. Safford, also in the Chemistry Section, the plasticized polymer was being commercially extruded at the Bridgeport (Connecticut) Works, under the product name "Flamenol®,"* to make electrical cable. The U.S. Navy used a great deal of such cable in World War II.

---

* The name was coined to convey the message that Flamenol is flame resistant. The confusion about "flammable" and "inflammable" is confounded here because the suffix "ol" denotes an alcohol in organic chemistry, and alcohols are often flammable. Page Humpty Dumpty!

Marshall's faith in physical chemistry strongly influenced the applied research done on Flamenol®. In addition to the usual life-test experiments, which included work on stabilizers, and other measurements related to Table 1-1, the losses of plasticizers (e.g., tricresyl phosphate) from the product and the vapor pressures of the plasticizers were measured and treated in fundamental fashion [10]. At the time such an approach was rare in applied research on polymers. We shall regard it as *augmentation of discovery*, a kind of activity that was to become prominent in the silicone project.

Marshall and Patnode regarded the polyvinyl chloride experience as a missed opportunity to establish the Company in a lucrative chemical business. Here are the facts to support their view. Plasticized polyvinyl chloride is among the ranking low-cost, common, man-made polymeric materials: it is unusual for anyone in this country to go through a day without encountering it in some form. The Company has never made it.* Marshall and Patnode saw the Company use it at annual rates measured in millions of pounds, and the nation consume it at rates many times greater. The Chemistry Section had been among the first in the field and had done proportionately much more research˙than was warranted by the consumption of the material in the Company. The profit was to be found mainly in the nonelectrical, as opposed to the electrical, market. Ergo, the Company should have entered further into the chemical business by manufacturing polyvinyl chloride. This case was even more poignant than that of the alkyd resins, for the Company was at least making Glyptal®.

Next, the Marshall-Patnode position was bolstered by Formex®, a brilliantly successful wire enamel. We have seen what such a success means to the electrical industry. Marshall and Patnode realized that Formvar, a Canadian polymer then almost useless, might be the basis of an advanced wire enamel if suitable solvents for it could be found. In this connection Marshall's Canadian origin and his Canadian contacts proved helpful. Work began in the Chemistry Section in 1933; by 1938 Formex, the enamel based on Formvar, was in production. Initially a costly product, it had by 1953 virtually driven other wire enamels off the market and given the Company a dominant position. The patent royalties were handsome and long continued. In its impact on the industry, Formex ranks with ductile tungsten [2]. Both annihilated the competition. But

---

* The research done by the Chemistry Section did not give the Company a significant patent position in the manufacture of polyvinyl chloride, although Moyer M. ("Mac") Safford did patent its stabilization by lead salts. Members of the Section appreciated its usefulness. One tested it as soles for his shoes; another established that it was far less subject to cracking on exposure to the sun than was rubber hose.

one great difference did develop: ductile tungsten is still king, whereas Formex® has been partially displaced by enamels that can withstand temperatures above 105°C. Formex was a great success; but, from the Marshall-Patnode point of view, the success would have been greater had Formvar been manufactured in the Company. They continued as strong and persistent advocates of the policy that the Company should become an important chemical manufacturer capable of quickly exploiting the results of extensive applied research on new materials. They felt sure that all the important man-made polymers had not yet been discovered.

COMMENTS IN CONCLUSION

• The importance speedily assumed by man-made polymers after Bakelite was announced in 1906 warrants their inclusion in the modern scientific revolution defined by Lord Snow [1, p. 30]. Even so, they are less glamorous than the polymers (nucleic acids, proteins, polysaccharides, i.e., sugar polymers) needed for life: Lord Snow speaks [1, p. 73] of "the leap of genius by which Crick and Watson [11] snatched at the structure of DNA [deoxyribonucleic acid] and so taught us the essential lesson about our genetic inheritance." I urge two books upon the general reader: *Organic Molecules in Action* (especially Chapter IV) by M. Goodman and F. Morehouse [12], and the more technical, but admirably clear and simple, *Big Molecules* by Sir Harry Melville [13].

• The General Electric Research Laboratory was established "exclusively [for] original research in many . . . fields." It was the first (perhaps the only) industrial laboratory to have so broad a charter—a charter that stemmed from the firm conviction held by influential men that the future of the electrical industry could be improved by closer coupling to science and to research, particularly research related to materials. The charter gave Willis R. Whitney the freedom to build the kind of laboratory he considered best for the Company, and to undertake any research that he though might prove profitable. The discovery of tungsten ductile at room temperature and the successes stemming therefrom gave powerful support to the charter and to Whitney's policies. In this situation man-made polymers were bound to become a principal concern of the Laboratory.

• Whitney was fortunate and successful as a recruiter of exceptional men. He appreciated that European scientific training was then the world's best. Before 1910 the Laboratory record reads: Whitney, PhD Leipzig (Ostwald), 1896, joined Laboratory, 1900; Coolidge, PhD Leipzig (Wiedemann and Drude), 1899, joined Laboratory, 1905; Irving Langmuir, PhD Göttingen (Nernst), 1906, joined Laboratory, 1909. All three men became internationally recognized while the Laboratory was still young. Though their training proved invaluable, there was no detectable foreign influence on Laboratory policies; these were set by Whitney. The high competence of the staff made it unnecessary to contract for outside research and promoted the well-rounded growth of the Laboratory into an American institution.

• In the beginning, corporate research was unorganized. Growth made increasing organization necessary. The 1939 Organization Chart* of the Research Laboratory identifies the 98 staff members on its monthly payroll. (Those on the hourly payroll are not given.) In the "front office" were Dr. Coolidge, Director; Laurence A. Hawkins, Executive Engineer; Leland M. Willey—General (i.e., factotum); and their secretaries. That was all. These three managed the organization. They used centralized Works services (e.g., purchasing, treasury). Section heads were more nearly project leaders than managers. A. L. Marshall's responsibility was described as "Chemistry-Insulations, etc."; this left him ample elbow room, of which he took full advantage. I do not remember that Marshall during the early years ever held meetings of the Chemistry Section, or even of key people representing the Section. The first meeting of the Section came, I think, in 1943 and was primarily an information session, question-and-answer period included, for the benefit of newcomers. Similar meetings followed at irregular intervals of a few years. After the Chemistry Department was organized, Marshall began to hold weekly meetings with the managers reporting to him.

A quotation from Lord Snow applies with even greater force in the early days than now, when organization has become more complete:

Yet the personal relations in a productive organization are of the greatest subtlety and interest. They are very deceptive. They look as though they ought to be the personal relations that one gets in any hierarchical structure with a chain of command, like a division in the army or a department in the Civil Service. In practice they are much more complex than that, and anyone used to the straight chain of command gets lost the instant he sets foot in an industrial organization [1, p. 31].

* Organization charts of this period are difficult to interpret.

• When the silicone project began, organization in the Research Laboratory was still minimal, responsibilities were not spelled out, and management was not emphasized; yet nobody seemed to mind. The Laboratory had been successful. Interactions among its staff were generally easy, frequent, informal, and effective. There was pressure, of course, but it was largely self-generated now that the recent depression was over. One thinks of a Camelot with Dr. Whitney, then retired, singing his favorite song, "When I grow too old to dream . . . ," slightly off-key in the hallways as he made the rounds of laboratories in which the work interested him. But Camelots cannot survive unchanged in the face of growing and unavoidable Adamsian multiplicity and complexity.

• A different company would have set up a different Research Laboratory in which there would have been a different Chemistry Section. "Of a good beginning cometh a good end." By 1938 this Chemistry Section was a going concern that had recruited enough PhDs* so that it could successfully undertake chemical research of any kind. The stage was set for a new polymer project.

* Dr. Coolidge once told me, "If you recruit only PhDs, you'll soon have assistants enough."

Chapter Two A Guide to Chemistry, Chemists, and Pure Research

There is no royal road to geometry [or to silicones!].

EUCLID (said to Ptolemy I)

Chapter 2 was written to prepare the reader for what follows and to give him or her a feeling for how pure research is done and by whom, with emphasis on the history of certain important ideas.

Each chapter in this book aspires to resemble a good, orderly meal, acceptably served to the reader, and topped by a dessert of comments in conclusion. Among them all, Chapter 2 is furthest from this ideal. It resembles rather a jumbled buffet from which the reader must choose according to his or her appetite. Unfortunately but unavoidably, some of the dishes are difficult to digest. They may be skipped the first time through the food line. The reader who does not take every dish will be more fortunate than Ptolemy: skipping is easier in chemistry than in geometry, chemistry being the less orderly science.

THE READER'S GUIDE TO CHEMISTRY

General. The body of chemistry consists of experimental observations that men and women continue to enlarge, and try to refine by using models and making theories (generalizations) about which Holmesian skepticism is warranted in varying degrees. Most simply, chemistry is divided into inorganic and organic (or carbon compound) chemistry.* Each has its enormous literature. As the siloxane (Si—O—Si) bridge, so vital to silicones, is inorganic, we shall let *Gmelin* [1] content us here.

Gmelin is the bible of inorganic chemistry, presented in fine print and in the tersest way conceivable. It probably consists of over 100 volumes by now and occupies some 200,000 cc; it grows so fast that numbers are soon out of date. Keeping *Gmelin* up to date is a Sisyphean labor. It is the best possible evidence that "the chemical elements are all chemical individuals, and their behavior is governed by many factors" [2].

Let us look at two volumes of *Gmelin* that contain information relating to silicon [3a] and silicones [3b]. The first summarizes the chemistry of the element and its inorganic compounds; of its 922 pages, pp. 269–582 deal with SiO_2. The second contains information about the organic compounds of silicon, silicones included. The first volume, published in 1959, covers the literature through 1949. The second, published in 1958, covers the literature until 1953, when the silicone industry was young. In Chapter 8 we shall see what faces *Gmelin* now that the industry has reached early maturity.

Gmelin is truly an example of Adamsian multiplicity and complexity. The multiplicity is evident from the bulk. The complexity is inherent in the science itself, and is exacerbated by contradictions and unreliabilities in the reported observations. Nevertheless, *Gmelin* is indispensable.

Fortunately, the periodic table, which we may regard as a model of descriptive chemistry, helps one to cope with this multiplicity and complexity. Lothar Meyer (1868) and Dmitri Mendelejeff (1869) deserve the great credit that attaches to its discovery, though the table in its modern version (Figure 2-1) differs greatly from theirs. Mendelejeff in 1905 summarized his thinking as follows: "there must be some bond of union between mass and the chemical elements; and as the mass of substance is ultimately expressed in the atom, a functional dependence should exist and be discoverable between the individual properties of the elements and their atomic weights" [4].

Today, atomic numbers have replaced atomic weights, and the no-

---

Note. Brackets are used to enclose references and inserted material. For bracketed references, see References and Notes at the end of the book. Numbers in parentheses on the right of pages are often used in the text to identify equations and the like: thus (2-1) means Equation 2-1. Reference to laboratory notebooks is by name and date. CRDA stands for "Chemistry Research Department Archives."

* Henry Adams would point out that many kinds of chemistry are recognized today, that each has at least one journal, and that most have more.

Group	Ia	IIa	IIIb	IVb	Vb	VIb	VIIb	VIII			Ib	IIb	IIIa	IVa	Va	VIa	VIIa	O
1st period	1 H													NON-METALS HERE				2 He
2nd period	3 Li	4 Be		METALS									5 B	6 C	7 N	8 O	9 F	10 Ne
3rd period	11 Na	12 Mg		ON	THIS	SIDE							13 Al	14 Si	15 P	16 S	17 Cl	18 Ar
4th period	19 K	20 Ca	21 Sc	22 Ti	23 V	24 Cr	25 Mn	26 Fe	27 Co	28 Ni	29 Cu	30 Zn	31 Ga	32 Ge	33 As	34 Se	35 Br	36 Kr
5th period	37 Rb	38 Sr	39 Y	40 Zr	41 Nb	42 Mo	43 Tc	44 Ru	45 Rh	46 Pd	47 Ag	48 Cd	49 In	50 Sn	51 Sb	52 Te	53 I	54 Xe
6th period	55 Cs	56 Ba	57* La	72 Hf	73 Ta	74 W	75 Re	76 Os	77 Ir	78 Pt	79 Au	80 Hg	81 Tl	82 Pb	83 Bi	84 Po	85 At	86 Rn
7th period	87 Fr	88 Ra	89 Ac															

METALLOID

DIAGONAL

*Lanthanide series begins here; not shown.

Figure 2-1 Modern version of the periodic table with metals, metalloids (shaded according to Rochow [5a]), and nonmetals. Atomic numbers, shown above the symbols of the elements, have replaced atomic weights. We may disregard the inert gases (Group O) and the entire 7th period. Fluorine (atomic number, 9) is the most electronegative element, and cesium (atomic number, 55) the most electropositive, among those considered; they form the purest ionic bond. The metalloids lie adjacently along a diagonal that separates metals and nonmetals. As the group number shows, carbon, silicon, and germanium are much nearer the center of the table than a quick glance might indicate. Because of this central position and their location along the upper part of the diagonal, carbon and silicon tend to form covalent bonds and to follow the tetrahedral habit in their chemistries. The position of silicon in the table is in accord with its behavior as the leading semiconductor; it is used in many of the solid-state devices important in the scientific revolution.

47

tion of "functional dependence" is still undergoing refinement. The great truth revealed by the table continues to be that the groups in Figure 2-1 contain elements that belong together because of their physical properties and chemical behavior, with the physical relationships more reliably established. This truth is great but not simple; it invites Holmesian skepticism. Were the truth simpler, *Gmelin* would be a tidier package.

Rochow* [5a] has emphasized that the metalloids may be regarded as a diagonal around which the entire table turns—metals on one side, nonmetals on the other. One ought to look at all the neighbors of an element in using the table; we must look mainly at boron, carbon, and silicon. The first two establish the upper end of the diagonal axis, and carbon is the upstairs neighbor of silicon. The importance of organic (i.e., carbon) chemistry and the significance of the metalloids (e.g., silicone chemistry) attest to the need for understanding this region of Figure 2-1.

The periodic table cannot be expected to cope with all the bewildering wealth of *Gmelin*. Least reliable in its predictions of chemical behavior, it yet fulfills surprisingly well the prime function of a scientific model, which is to help the mind reach conclusions for test by experiment if necessary.

Chemical Bonding.   It bears repeating that molecules exist because each atom in them can be *bonded* (linked) either to another atom or to a group of atoms. A knowledge of chemical bonding supplements the periodic table and adds meaning to it. Chemical bonds result finally from the interactions of electrically charged particles. We are concerned mainly with *covalent* and with *ionic* bonds. As in the Introduction, we represent the former by a line joining the bonded atoms; thus H—H is the hydrogen molecule, which may be taken as the prototype of a molecule containing a covalent bond. Each such line represents an *electron pair* shared by the two atoms, that is, H:H in H—H. In the *ionic bond* (prototype Cs^+F^-), the binding is electrostatic and governed by Coulomb's law; accordingly, such bonds are often represented by placing signs on the bonded ions to indicate their charges. Most actual bonds fall between these two prototypes. Quite often, with Humpty Dumpty's consent, bonds even considerably removed from pure covalence are called "covalent": to borrow from Orwell's *Animal Farm*, some "covalent" bonds are more covalent than others.

* Professor Rochow has written a second book, *The Chemistry of Silicon* [5b], that is required reading here. Had I seen it in time, this would have been easier to write. I thank Professor Rochow for valuable suggestions concerning the first part of *Silicones under the Monogram*.

The complexity of chemical bonding makes this Orwellian position desirable for the general reader. For the purposes of this book, the treatment of bonding [6, 7, 8] by Linus Pauling (Nobel Prize, 1954) is more suitable than the more theoretical approach that relies on molecular orbitals.

If chemical bonds are real, they must determine the structure of molecules in three-dimensional space. Modern physical methods, notably X-ray diffraction, in which X-rays scattered by crystals establish structure, have made it possible to fix the directions of bonds and the locations of combined atoms in space with high precision [7]. Modern X-ray diffraction equipment has been improved and computerized, and it is now possible to pinpoint the intricacies of huge polymer molecules such as hemoglobin, an accomplishment that far transcends the establishment of bond angles and bond lengths in simple crystals. These *bond lengths* and *bond angles* are invaluable experimental data superior to early information about bonding, which rested on observed chemical behavior. The X-ray evidence has made possible the construction of reliable models of molecules; one such model, that of a fragment of *polydimethylsiloxane* (the most important silicone of all), is shown in the accompanying plate.

The important siloxane (Si—O—Si) bridge is built of two (Orwellian) covalent single silicon-oxygen (Si-O) bonds (see plate). Covalent bonds need not be *single*: carbon atoms, for example, are often *doubly* (C=C) or *triply* (C≡C) bonded; here, also, the number of lines shows the number of electron pairs shared by the bonded atoms. The bonds to silicon atoms in silicones are always single, always covalent, and always four to a silicon atom.

Chemical Nomenclature.   Let us expand the Introduction. Chemical *formulas* give the composition of compounds; chemical names do so far less obviously, if at all. Chemical names can be unwieldy, and usually take more space; compare "polydimethylsiloxane fragment" with —$[(CH_3)_2SiO]_n$—, or with —$(\mathbf{D})_n$—, a shorthand version to be explained later. Also, the reader not familiar with silicones will find formulas quicker to catch his or her eye and easier to remember. Although formulas will generally be preferred to names,* we cannot avoid using "silicones" as a class name. To help the reader, we plan to write it as a separate word (e.g., methyl silicones, phenyl silicones, methylphenyl silicones) instead of writing the complete name as a single word (e.g.,

---

* Gibbon in his *Decline and Fall* speaks of German tribes "whose obscure and uncouth names could only serve to oppress the memory and perplex the attention." What would he have thought of nomenclature in chemistry?

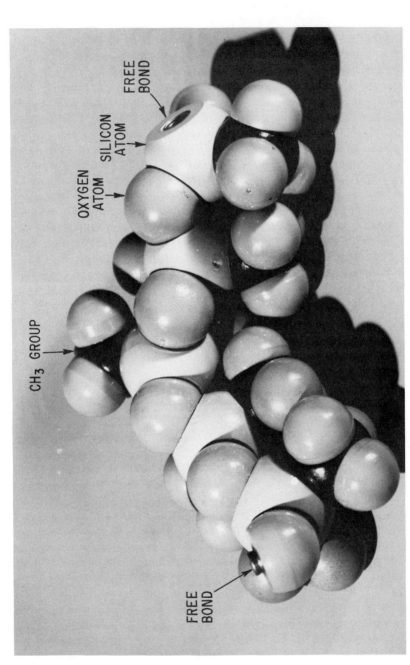

An organic chemist's model of —[(CH₃)₂SiO]₅— with a CH₃ group and free terminal bonds indicated. Such models are useful approximations that show whether molecules are likely to be stable. The backbone fragment shown could be made into a ring by joining the free bonds, but a fragment containing two or three silicon atoms could not: the various atoms and groups would interfere with each other because of lack of room. In an actual silicone molecule there would be continuing internal motions, mainly rotations about various bonds.

methylphenyldichlorosilane, which is not a silicone). I trust that gradual immersion in the cold water of nomenclature will ease the shock.

We cannot get along without models of molecules. True models would have to be three-dimensional and portray the complex motions within the molecule. For us, simple two-dimensional models will suffice. Thus

$$
\begin{array}{c c c c c}
CH_3 & CH_3 & CH_3 & CH_3 & CH_3 \\
| & | & | & | & | \\
-Si-O-Si-O-Si-O-Si-O-Si-O- \\
| & | & | & | & | \\
CH_3 & CH_3 & CH_3 & CH_3 & CH_3
\end{array}
$$

represents $-[(CH_3)_2SiO]_5-$, which is the polymer fragment in the plate.

A GUIDE TO POLYMERS

When Berzelius introduced the polymer concept into chemistry in 1830 [9], he was principally concerned with distinguishing between different classes of molecules all of which had the same composition (percent by weight). In *isomers,* one class, properties could vary, but the molecular weight remained the same. In *polymers,* properties and molecular weight could both vary: this was to be expected if polymers were indeed big molecules made up of many identical parts.

We shall regard *linear* polymers as having *backbones* to which atoms or groups of atoms are attached; a collection of such molecules brings to mind a can of spaghetti: the strands are intertwined but not tied together. We shall regard *three-dimensional* polymers as having *skeletons* with like attachments. These skeletons (or networks) resemble the structural-steel skeleton of a skyscraper more closely than the skeleton of an animal. The book will be filled with backbones and skeletons.

The most common polymers, natural or man-made, contain surprisingly few elements. Their backbones and skeletons usually contain not more than three elements from among the following five: carbon, nitrogen, oxygen, silicon, and sulfur. To finish the polymer, its backbone or skeleton must be fleshed out by attaching elements or groups such as hydrogen; carbon and hydrogen in groups, often represented by R, such as methyl (CH_3) or phenyl (C_6H_5); fluorine; or chlorine. These attachments can give the polymer desirable physical, chemical, electrical, or mechanical properties. When one is lucky, more than one kind of property (see Table 1-1) can be improved by making a single change in composition; in exceptional cases the entire property profile improves. All eight elements named above are relatively light. Simple atoms, of about the same size, that form strong covalent bonds make the best polymers.

What Berzelius thought of in 1830 is easy to visualize. Suppose that a small molecule of molecular weight M can be made to react with itself to give a big molecule of molecular weight P according to the equation

$$nM = P \qquad (2\text{-}1)$$

Here M represents both the formula of the *monomer* and its molecular weight; likewise, P for the *polymer*; and n is the number of monomer molecules it takes to make the polymer molecule. *E pluribus unum*!

Compare the simplicity of (2-1) with an authoritative definition of "polymer":

A polymer is a substance composed of molecules characterized by the regular or irregular repetition (ends, branch junctions, and other minor irregularities being neglected) of one or more types of chemical units. In general, a linear polymer of a given series may be considered a *high polymer* if its physical properties (especially its viscoelastic properties) do not vary markedly with the degree of polymerization. In macromolecular chemistry, "polymer" is usually considered to mean "high polymer" [10a].

No wonder Reference 10 consists of 16 large volumes that took 8 years (1964–1972) to appear!

In Chapter 1 "poly" meant "from 2 to 4" when we spoke of the acids and alcohols from which Glyptal® is made. What does the prefix mean when we speak of Glyptal itself or of another *polymer*? There is no firm answer. It is customary to speak of *monomers, oligomers,* and *polymers* (think of "monarchy," "oligarchy," and "polyarchy") with no precise dividing line between the last two. Invoking Humpty Dumpty, we shall avoid using "oligomers" and related words. We shall designate individual oligomers by using "dimer," "trimer," and so on; in these words the prefix gives the number of identical parts ("mers") present in the molecule. We shall use "poly" in place of "oligo" as a general prefix, as is often done, Lichtenwalner and Sprung [11] being noteworthy exceptions; their precise usage is justified by the unusual importance of oligomers* in silicone chemistry.

---

* Many silicones are properly called "oligomers" if the word is stretched to cover more "mers" than the report [12] by a distinguished Nomenclature Committee permits. The report says "An *oligomer* is a *polymer*, composed of molecules containing only two, three, or a few mers." The opening clause, seemingly drawn with Humpty Dumpty's help, supports us.

The molecular weights of "high" polymers can reach into the millions; in such polymers n may reach into the hundreds of thousands. If, as is entirely logical, a diamond or crystal of SiO_2 is regarded as a single polymeric molecule of formula C_n or $(SiO_2)_n$, then n can be truly enormous because it is limited only by the size of the crystal.

Berzelius, to whom chemical bonding must have been a mystery [13], could not have fully appreciated (2-1). The distinction between atoms and molecules remained blurred until Stanislao Cannizzaro made it clear in 1858. Friedrich August Kekulé von Stradonitz and A. S. Couper introduced chemical bonding very soon thereafter [8]. Once Berzelius had introduced the polymer concept, however, it became possible to see that important substances such as starch, cellulose, and rubber were in fact natural polymers; yet it took a long time for (2-1) to become widely accepted. By the end of the last century, the (erroneous) belief had grown that polymers were clusters of small molecules held together by forces considerably weaker than the chemical bonds within the molecules themselves; polymers and colloids were thought to be at least closely related— perhaps even identical. The cluster model has limited industrial appeal; in most solid polymeric materials, cohesive forces need to be strong, not weak. Fortunately the model was wrong. About 1920 Hermann Staudinger (Nobel Prize, 1953) began to insist that true polymers are covalently bonded throughout. In 1929 Wallace H. Carothers at the Du Pont Company started the important pure research in which he made polymer molecules of known structures, correlated these structures with the properties of polymers in bulk, studied polymerization reactions, and, happily for us, classified them as *additions* and *condensations*. These two men are outstanding among the many to whom we owe our present knowledge of man-made polymers.

Polymers thus did not begin to be well understood until some 20 years *after* Baekeland launched the man-made-polymer industry. Although this understanding still does not enable us to make reliable predictions of polymer properties, it sufficed to spur applied research on silicones in the Chemistry Section. The Section and its members deserve credit for appreciating quickly that new polymer knowledge was of first importance to the Company.

MODELS OF POLYMER COMPLEXITIES

Polymer molecules, even those made by man, vary greatly in size and shape, and exist in three dimensions. The easiest way to become familiar with them is via models. Two kinds of models are essential: models of the molecules themselves, and models of the reactions by which they are formed. The latter are called mechanisms, and they are more elusive than the others. As preparation for the silicone story, we shall look at models of polymers introduced earlier. We shall be concerned with the following:

1. Backbone formation and lengthening.
2. Branching to form skeletons.
3. Cross-linking to form skeletons.
4. Ring (= cyclamer) formation.
5. Termination and end groups.

Other processes exist, but these five suffice us.

*Condensation polymers** will be our main concern. *Addition polymers,* Carother's other class, though of great commercial importance, will be introduced here as the simplest illustration of backbone formation and lengthening (Process 1).

Linear polyethylene ("polythene") is a typical addition polymer, formed at suitable temperature and pressure in the reaction

$$(n + 1)\ C_2H_4 = -(C_2H_4)_n - \overset{\displaystyle H}{\underset{\displaystyle H}{C}} - \overset{\displaystyle H}{\underset{\displaystyle H}{C}} - \qquad (2\text{-}2)$$

in the presence of an effective catalyst. Note that the polymer molecule is *unfinished; end groups* are lacking: neither (2-1) nor (2-2) gives all the *mechanism.* Note also that the polymer backbone is —C—C—C—C— etc. (H atoms omitted). There is more about end groups below.

The linear polymer segment formed in (2-2) is representable by

$$\wedge\wedge\wedge\wedge\wedge\wedge$$

This zigzag model of a "linear" polymer segment will find frequent use. Each line in the model joins two atoms in the backbone. A backbone and the polymer it represents are "linear" by courtesy of Humpty Dumpty: they actually exist in three dimensions and have complex internal motions. But "linear" serves to distinguish them from models of polymers that have skeletons and are patently three dimensional.

---

* The first condensation polymer was made by Berzelius himself, when he reacted tartaric acid with glycerol in 1847 and obtained a brittle, resinous product [14], about 14 years after he originated the polymer concept.

In condensation polymerization, small ordinary molecules (e.g., H_2O, HCl, NH_3) are produced as the big molecules form. Glyptal®, a condensation polymer that is a polyester, is a good introductory example. Let us begin with the formation of a simple ester:

$$C_6H_5-\overset{\overset{\displaystyle O}{\|}}{C}-OH + CH_3-\overset{\overset{\displaystyle H}{|}}{\underset{\underset{\displaystyle H}{|}}{C}}-OH = C_6H_5-\overset{\overset{\displaystyle O}{\|}}{C}-O-CH_3 + H_2O$$

$$(2\text{-}3)$$

Benzoic acid   Ethyl alcohol   Ethyl benzoate (an ester)

At first sight (2-3) seems analogous to the neutralization of a strong acid (e.g., HCl) by a strong base (e.g., NaOH), which may be regarded as a reaction between two ions (H^+ and OH^-) to form H_2O. It is actually quite different. It is best described by saying that an *active site* on C_6H_5COOH and an *active site* of a different kind on C_2H_5OH make the reaction possible. The active site on C_2H_5OH exists because a lone (unshared) electron pair on the oxygen atom can form a covalent bond with the carbon in the COOH group by a mechanism we shall not try to describe.* Both active sites are annihilated when the ester in (2-3) is formed. Consequently, this ester cannot polymerize: it is dead.

Let us now see what happens when acid and alcohol have two active sites on each molecule as in

$$HO-\overset{\overset{\displaystyle O}{\|}}{C}-C_6H_4-\overset{\overset{\displaystyle O}{\|}}{C}-OH + HO-\overset{\overset{\displaystyle H}{|}}{\underset{\underset{\displaystyle H}{|}}{C}}-\overset{\overset{\displaystyle H}{|}}{\underset{\underset{\displaystyle H}{|}}{C}}-OH$$

Phthalic acid   Ethylene glycol

$$= HO-\overset{\overset{\displaystyle O}{\|}}{C}-C_6H_4-\overset{\overset{\displaystyle O}{\|}}{C}-O-\overset{\overset{\displaystyle H}{|}}{\underset{\underset{\displaystyle H}{|}}{C}}-\overset{\overset{\displaystyle H}{|}}{\underset{\underset{\displaystyle H}{|}}{C}}-OH + H_2O \quad (2\text{-}4)$$

Ester with cumbersome name

* Unshared electron pairs of this kind make possible active sites in many molecules that form condensation polymers; $C_2H_4(OH)_2$ and $C_3H_5(OH)_3$ are examples. They are important also in the formation of Si—O—Si bridges in silicones.

In (2-4) an active site remains at each end of the ester molecule that is formed. Obviously, an ethylene glycol molecule could react with the left-hand end, or a phthalic acid molecule with the right-hand end, the result being a longer molecule still blessed with active sites that permit growth to continue. Eventually, the backbone of a *linear polyester* would result.

If we substitute glycerine:

$$
\begin{array}{ccc}
\text{H} & \text{H} & \text{H} \\
| & | & | \\
\text{H---C---C---C---H} \\
| & | & | \\
\text{O} & \text{O} & \text{O} \\
\text{H} & \text{H} & \text{H}
\end{array}
$$

for $C_2H_4(OH)_2$ in (2-4), we can form a *polymer skeleton* in three dimensions, and the result will be a polyester molecule related to those in Glyptal®

Phenol-Formaldehyde Resins.   Chapter 1 gave little hint of the complexities Baekeland encountered when he studied the reactions between C_6H_5OH (phenol) and CH_2O (formaldehyde) in his successful search for the first commercial man-made polymer. Here these complexities (appropriately simplified) become a virtue because they provide a good introduction to silicones. Likenesses and contrasts to silicone will be enclosed in brackets. We shall rely on the authoritative discussion of phenol-formaldehyde resins by Keutgen [10b, p. 1].

We must first meet the benzene ring, that Mona Lisa among organic models:

| Benzene, C_6H_6 Kekulé (1865) | (Atomic symbols omitted) Benzene, C_6H_6 Thiele (1899) | (* = active sites) Phenol, C_6H_5OH (Hydroxybenzene) | (*Two kinds of* active sites: * and ⊛) Saligenin $C_6H_4(OH)(CH_2OH)$ |

(2-5)

In (2-5) the active sites are marked. Note that they are of two kinds; the kind marked ⊛ resembles those in C_2H_5OH, $C_2H_4(OH)_2$, and $C_3H_5(OH)_3$. Trying to understand benzene and its derivatives is like trying to interpret the famous smile. No one model tells the whole truth. We shall use Thiele's, which is easy to draw, and we shall speak of the benzene ring even when some of its hydrogen atoms have been replaced.

When one hydrogen atom in a benzene ring has been replaced by OH to make C_6H_5OH, three of the remaining hydrogen atoms (designated above by *) become *active sites*. Ask Mona Lisa why! (Actually, theoretical chemists do have explanations.) As C_6H_5OH is acidic, common sense prescribes an asterisk for the OH group and predicts that this group will be active in condensation reactions: we saw above that acids form polyester resins such as Glyptal®. Baekeland used common sense. He was wrong, but by the time others had shown that this OH group is *not* an active site in the reactions that form phenol-formaldehyde resins, Baekeland—wrong model and all—had managed to open the era of man-made polymers, to his considerable profit (and ours). [Attached to Si, OH groups are *highly reactive,* and silicones can form when they condense.]

Saligenin, shown in (2-5), is a crystalline *addition product* that results from the reaction of one molecule of formaldehyde (HCHO or CH_2O) with one molecule of C_6H_5OH. The reaction is not simple: Mona Lisa once more!

When C_6H_5OH is mixed with an equal volume of CH_2O dissolved in water (37 to 40% CH_2O by weight), allowed to stand, or even boiled, for days or weeks, *nothing much happens.* When a *catalyst, acid* or *alkaline,* is added, however, reaction quickly begins; it can be violent and form products sensitive to conditions and catalyst. [Acid and alkaline catalysts accelerate the formation of silicones.]

With a suitable catalyst the nature of the resulting phenol-formaldehyde resin can be controlled by changing the proportion of the two reagents. Equation 2-5 makes this seem reasonable: when CH_2O is in short supply, the tendency is to form saligenin and compounds to which saligenin is a *precursor*; with CH_2O in abundance (more precisely, with the molar ratio CH_2O/C_6H_5OH greater than unity), all three active sites on C_6H_5OH come into play. Phenol-formaldehyde resins can thus to some extent be *tailor made.* [Silicones can be tailor made in similar ways, and silicones have precursors also.]

With CH_2O in abundance, the compounds in (2-6) all appear as early products of the CH_2O—C_6H_5OH reaction:

$$(2\text{-}6)$$

Note how the reactive sites remain at three *but change in character.* [The reactive sites marked with circled asterisks have counterparts in silicone chemistry (see the **M, D, T** shorthand to be explained later).]

Now, we consider *idealized* phenol-formaldehyde polymer building according to Processes 1 to 5 (see above).

1. Backbone formation or lengthening by condensation (continued). See (2-2). With a dearth of CH_2O and an acid catalyst, Process 1 begins with (2-7):

(Bonds omitted)   (Some bonds omitted)
(* sites not shown)   (*sites not shown)

$$(2\text{-}7)$$

If enough CH_2O is provided, the backbone started in (2-7) grows until it contains 15 to 20 benzene rings, and reaction stops when the CH_2O is gone. The active sites not shown are useless in the absence of CH_2O. [Equation 2-7 has no counterpart in silicone chemistry: silicon is not carbon.]

With alkaline catalysts under special conditions [10b, pp. 5, 6], the link —CH_2—O—CH_2— may occasionally join two benzene rings, but this link is decomposed by moderate heating. [The siloxane bridge, Si—O—Si, the most important link in silicones, is superficially similar to the organic link just written. Both links can result from the condensation of two OH groups. There, however, the resemblance ends. The siloxane bridge, for practical purposes, *cannot* be decomposed by heating.]

2. Branching to form skeletons. Let us assume that (2-7) has formed a backbone ⋀⋀⋀⋀⋀. If this backbone results in part from the condensation of molecules such as the last in (2-6), added condensation is clearly possible at the other two circled-asterisk sites. This continued condensation will form *branched chains* representable by

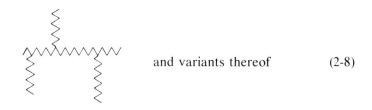

and variants thereof   (2-8)

Such branching in three dimensions will lead to *skeletons*.

3. Cross-linking to form skeletons. Sometimes desirable properties in a polymer can be obtained by linking backbones to make structures such as

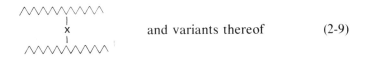

and variants thereof   (2-9)

Here the *cross-link* **X** is usually made by adding specially reactive substances in the proper amount to a (linear) polymer backbone. Radiation can often cross-link also. [Cross-linking is important in silicones, particularly in silicone rubber.]

4. Ring (= cyclamer) formation. Hultzsch [10b, p. 16] succeeded in making the *cyclamer*

or $[C_6H_2R(OH)CH_2]_4$ (2-10)

by linking four (substituted) benzene rings through CH_2 groups. He succeeded, however, only by replacing unwanted asterisked hydrogen atoms
with inert alkyl R groups, in this way "blocking" unwanted active sites
and stopping chaotic growth. In organic chemistry, as in football, blocking can lead to victory. [In silicone chemistry, the compound in (2-10)
would be called a *tetramer*; note the subscript "4."]

     5.   Termination and end groups. How do polymer molecules stop
growing? There are three ways:

     1.   By exhaustion of an essential reagent, for example, CH_2O to
make the compound needed for (2-7).

     2.   By becoming so elephantine that they cannot grow further; in
such cases, the end groups on the molecules, though still active, are so
few and so far apart that reaction stops, and polymer properties are
virtually unaffected by the presence of these groups.

     3.   By uniting with *inert* end groups E so that inert polymer mole-
cules such as

         E ⋀⋀⋀⋀⋀⋀ E   and variants thereof; cf. (2-2)   (2-11)

are produced.

[Silicone chemistry includes elegant ways of dealing with end-group problems.]

Phenol-formaldehyde compounds range from crystals of simple molecules (e.g., saligenin), through water-soluble resins of low molecular weight, through products of higher molecular weight that dissolve in organic solvents, up to solids in which the molecular weight is too high to be measured. The resins can be used commercially in pure form, in solution, as solids, or as solids containing fillers. Some can be hardened by simple heating; others harden only upon addition of a catalyst called "hexa," and these resins are useful when they must be applied as liquids and become solid in place. "Phenolics," as phenol-formaldehyde resins are usually called, are used as molding compounds, coatings, and bonding agents. In these ways, among others, they serve many industries in addition to the electrical one [10b]. [If anything, silicones are more diverse and more versatile than phenolics.]

Around 1965 some 1 *billion* lb of phenolics were being produced annually in the United States, and some 700 million lb abroad, the U.S.S.R. and eastern Europe excluded [10b]. Let us arbitrarily put the world's annual production of phenolics at over 2 billion lb, worth about $1 billion in some recent year. Likewise, let us say that the world in that year made 200 million lb of silicones, worth over $0.5 billion. That puts the two industries in rough perspective.

Baekeland of course did not build the phenolics industry alone [10b], but he did a great deal. He defined the differences between the actions of acid and of alkaline catalysts, and he learned to use the CH_2O/C_6H_5OH ratio as a way of controlling the reaction product. This was admirable applied research of which any university laboratory might have been proud. But his most important contribution was decidedly industrial. He discovered that the application of pressure during the curing (heat hardening) of Bakelite resins prevented foaming and the forming of bubbles caused by the evolution of water: that discovery founded the phenolics industry. Baekeland clearly had a rare combination of research abilities. Yet he did not understand the mechanisms of his reactions or the molecular structures of his products.

Four years before Baekeland's "heat and pressure" patent expired in 1926, his company had merged with two competitors to form the Bakelite Corporation, a holding company, which in 1924 became an operating company that was acquired by the Union Carbide Corporation in 1939 upon Baekeland's retirement.* Today more than 50 American

---

* Byron M. Vanderbilt's *Thomas Edison, Chemist,* American Chemical Society, Washington, D.C., 1971, supplements Reference 10b as regards the phenolics industry and is of further interest here because it deals with Edison. Both sources reveal that controversy and patent litigation marked the phenolics industry. This situation is normal for successful industrial research. Often the stakes are high, and the issues complex. Each industry has its own story.

companies, the General Electric Company among them, make phenolics. [The silicone industry developed in a quite different way, as Chapters 3, 7, and 8 will show.]

Two Final Items. First, the repeating units ("mers") called for in (2-1) do not exist as monomer (M) molecules in condensation polymers, phenolics and silicones included. For example, in $(CH_3)_3SiOSi(CH_3)_3$, hexamethyl*di*siloxane, the "mer" would be half the molecule—an impossible fragment. Instead, the molecule has a *precursor*, $(CH_3)_3SiCl$; of this, more later. The Greek prefixes "di," "tri," "tetra," . . . indicate the number of silicon atoms in the siloxane bridges of a silicone.

Second, for reasons not obvious from (2-5), silicones benefit from the incorporation of benzene rings (i.e., C_6H_5 groups). The greatest benefit is an increased resistance to oxidation over that of methyl silicones, which are already good in this respect. As the electrons in benzene rings move around easily, it seems strange that these rings should resist oxidation. The modern explanation is that the existence of the benzene ring in various forms makes possible a *resonance* among those forms, and that this resonance promotes inertness. The benzene ring will appear frequently from here on.

CARBON. THE TETRAHEDRAL HABIT

Even before the periodic table came into being, the belief was growing that the chemistries of carbon and silicon closely resembled each other. This belief was to influence the pure research that eventually established silicone chemistry. Here we shall concentrate on the aspect of carbon chemistry for which the belief is most nearly valid. We shall call this aspect the "tetrahedral habit," by which we mean, for an atom, that it forms four covalent bonds directed toward the apices of a regular tetrahedron.*

We shall say that it took three steps to establish the tetrahedral habit in carbon chemistry.

During Step 1, compounds were prepared and analyzed to establish their compositions. It is natural now, when we have much greater knowledge and sophisticated analytical methods at our disposal, to underestimate what it took to accomplish Step 1 [15]. Much work of this kind had

---

* Humpty Dumpty once more. "Habit" of course has many meanings, and "crystal habit" denotes the characteristic form in which a crystal occurs. There should be small risk of confusion with "habit" as used above.

to be done before Sir Edward Frankland in 1852 could have the insight (or "flash of genius") which led him to suggest that elements have a definite valence, or bonding capacity [8]. This revelation was followed in 1858 by others, which came independently to Kekulé and to Couper, namely, that valence is evidence of bonds between atoms, that a carbon atom usually has *four* such bonds, and that carbon atoms can be united to each other. (The kind of bonding represented by C=C and C≡C, immensely important in organic chemistry, concerns us chiefly in a negative way: silicon does not show it in stable compounds such as silicones.)

The next revelation crowned Step 2 and lent substance to stereochemistry [16], a subdivision of structural chemistry. It came independently to Jacobus Hendricus van't Hoff (first Nobel Prize winner in chemistry, 1901) and Jules Achille Le Bel when they were mere youngsters— under thirty. It explains the optical activity shown by molecules in which *four different* atoms or groups are connected to a central atom as in Example 4, Figure 2-2. The other examples do not show optical activity.

Optical activity was discovered to be a property of certain dissolved substances by Jean Baptiste Biot in 1815. He found that these substances, identical in all other properties, sometimes rotated a beam of plane-polarized light to the *left*, sometimes to the *right*. Louis Pasteur discovered in 1848 that a salt of tartaric acid could give "left-handed" and "right-handed" crystals (somewhat like the left and right shoes in a pair). Tartaric acid itself is optically active in solution. Pasteur's discovery indicates that optical activity on a molecular scale is a three-dimensional phenomenon.

Van't Hoff and Le Bel were familiar with the results of Step 1 for carbon compounds. They were the first to realize that optical activity was shown only by substances with molecules analogous to Example 4, Figure 2-2. From these experimental facts they made a brilliant deduction. They postulated the tetrahedral habit, which is illustrated for carbon on the right of Figure 2-2. On a molecular scale the difference between the two tetrahedra shown there is as great as the difference between the right and left shoes in a pair: each foot hurts in the wrong shoe. The tetrahedral habit gains importance because it is followed also by molecules *that are not optically active*; it is widely applicable in organic chemistry.

Step 3 consists in a physical verification of the tetrahedral habit. In the tetrahedra of Figure 2-2, there are four equivalent covalent bonds that form six plane angles (aCb, aCc, aCd, bCc, bCd, and cCd). Each angle should measure 109°28'. Exactly this value is found by X-ray diffraction in the simplest case, that is, when (as in CH_4) the electrical forces that produce chemical bonding do not distort the tetrahedron.

STEP 1   STEP 2

Example 4 from Step 1

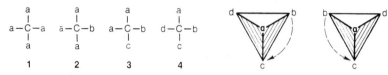

Figure 2-2   The first two steps in establishing the tetrahedral habit illustrated for carbon. Explanatory notes follow.

Step 1

1.   The letters a, b, c, and d represent elements or groups bonded to a central carbon atom, C. The lines are covalent bonds.

2.   When the tetrahedral habit exists, the number of bonds to the central atom is four.

3.   Optical activity is observed *only* for Example 4.

Step 2

1.   The existence of optical activity proves that the molecules of Example 4 exist in the two versions shown.

2.   The centers of the atoms or groups a, b, c, and d are at the apices of regular tetrahedra. The central carbon atom and the bonds *are not shown*. The lines that outline each tetrahedron connect the centers of a, b, c, and d.

3.   To see that the two tetrahedra are different, imagine yourself standing on apex a and looking down at b. To see the other two atoms in alphabetical order, your eye will have to move as indicated by the arrows. As the arrows have different directions, the molecules must be "different." Only Example 4 has "different" molecules.

4.   The molecules in Examples 1, 2, and 3 can also be represented by regular tetrahedra, but only by one tetrahedron for each example.

5.   Molecules representable by regular tetrahedra have an angle of 109°28' between each pair of chemical bonds.

6.   Neither the models on the left nor those on the right of the figure give complete pictures of a molecule. The two kinds of models give a better picture than either kind alone. Spheres could be used to represent atoms.

The story comprising Steps 1, 2, and 3 has been told here for carbon not only because it involves one of the most brilliant and productive examples of pure research in any science and shows the usefulness of models, but also because Kipping's attempt to carry out Step 2 for silicon was the beginning of silicone chemistry and led directly to the silicone industry.

SILICON. THE TETRAHEDRAL HABIT. STEP 1

The history of the human race and progress in chemical research are revealed to a surprising extent by the advances in the chemistry of silicon, compounds of which make up most of the earth's crust. The use of sand, granite, and clay in remote antiquity, and the later discovery and use of pottery, vitreous enamel, glass, and cement in that order—all preceded modern silicate chemistry, which began with the introduction of water glass (sodium silicate in solution) by Johann Rudolf Glauber in 1648 [17]. The simplest silicic acid, H_4SiO_4, may be regarded as hydrated silica, $SiO_2 \cdot 2H_2O$, or as $Si(OH)_4$; the silicates are salts of the various silicic acids, and the actual chemistry can be unbelievably complex.

We are going to use the following substances as landmarks in reporting pure research in silicon chemistry to the time of Kipping: SiO_2, SiF_4, Si, $SiCl_4$, $SiHCl_3$, SiH_4, $Si(OC_2H_5)_4$, and $Si(C_2H_5)_4$. This will be our Step 1 (see above) in establishing the tetrahedral habit for silicon.

We have already introduced silica as the parent of the silicones. In it and in the silicates, oxygen and silicon, the two most abundant elements in the earth's crust,* are joined by bonds that are among the most *durable*[†] of covalent bonds between elements. This durability long delayed the preparation of elementary silicon, and led silica to be considered itself an element, and hence incapable of being decomposed further. Antoine Laurent Lavoisier (1743–1794) is recorded as the first to suspect the truth. To establish experimentally in those days that silica is in fact a compound, something was needed to dissolve SiO_2 cleanly and to give a volatile silicon compound easy to isolate. That something turned out to be HF, which became available in 1807, and which generates the volatile SiF_4 when it dissolves silica, as it readily does. The reaction between HF and SiO_2 also produces H_2SiF_6,[‡] which forms K_2SiF_6 on being neutralized with KOH. When in 1823 Berzelius made the first silicon useful for

---

* The earth's crust, assumed to be 10 miles deep, has been estimated to contain 46.46% oxygen and 27.61% silicon by Frank W. Clarke, *The Data of Geochemistry,* Bulletin 770, U.S. Geological Survey, 1924, p. 36.

[†] *Durable* is used to describe for the general reader a chemical bond that is at once thermodynamically stable and slow to react with ordinary substances under ordinary conditions. There is no generally valid correlation between *thermodynamics,* which tells what reactions are possible under stated conditions, and *kinetics,* which says how fast the possible reactions occur. When solids react, the surface presents special problems; these are particularly important for silica—comment later.

[‡] Note that silicon does not show the tetrahedral habit in H_2SiF_6, in K_2SiF_6, or in the ion SiF_6^{2-}. Discussion later.

further research, it was by the action of potassium metal on K_2SiF_6. Soon thereafter, he ignited silicon thus prepared in a stream of chlorine gas to make $SiCl_4$. As we shall see, the *Kipping route* to silicones begins with $SiCl_4$; the *Rochow route,* with silicon.

By 1863 Friedrich Wöhler had made the important compounds SiH_4 (silane) and $SiHCl_3$ (trichlorosilane) [17]. In that year he published results on a material [18] that he called *silicon [sic].** The substance deserves to be remembered because Wöhler said that it could be regarded as constructed under the rules of organic chemistry with silicon replacing carbon, implying that there might be a chemistry of silicon analogous to that of carbon.

Also in 1863, C. Friedel and J. M. Crafts [19] announced that they had made $Si(C_2H_5)_4$, this being the *first time* that silicon was bonded to the carbon atom in an organic group. In 1868 Friedel and A. Ladenburg [20a], on the basis of further work that yielded Cl_3SiSH, a derivative of both SiH_4 and H_2S, agreed generally with Wöhler's conclusion about the analogy between the chemistry of silicon and that of carbon, but they did inject a word of caution, namely, that the conclusion must not be extended to imply that analogous compounds of the two elements necessarily react alike. They also extended M. Ebelmen's work [15] on silicon compounds containing ($—OC_2H_5$) groups,† and this work was carried still further by Ladenburg alone. By 1872 Ladenburg had prepared a large number of silicon compounds; in all of these, the silicon was bonded to four other elements or groups [20b]. A representative series listed by him is $SiCl_4$, $Si(OC_2H_5)Cl_3$, $Si(OC_2H_5)_2Cl_2$, $Si(OC_2H_5)_3Cl$, $Si(OC_2H_5)_4$, which the reader may wish to try naming as derivatives of SiH_4, silane; in these names all atoms and groups except hydrogen are included, e.g., dichloromonoethoxysilane (not given above). Ladenburg completed Step 1 in the process of showing the importance of the tetrahedral habit in silicon chemistry.

Many of the silicon compounds mentioned above were prepared by the use of compounds such as $Zn(C_2H_5)_2$ by methods not important here because they are less useful than either the Kipping or the Rochow route to silicones.

In taking Step 2 on the way to establishing the tetrahedral habit for silicon, Kipping founded silicone chemistry. It is time we met him.

---

*Wöhler's material was called *silicone* as late as 1966 in an authoritative inorganic chemistry text. Humpty Dumpty is stubborn!

† Mendelejeff in 1860 first recognized the true nature of this colorless liquid made by Ebelmen 14 years earlier [5b, p. 1424].

F. S. KIPPING (1863–1949)

Frederick Stanley Kipping, the founder of silicone chemistry, was born into an upper-middle-class English family and was well educated as an undergraduate. In his youth a friend of the family, who was a public analyst, interested him in chemistry. In 1886 Kipping entered von Baeyer's laboratory at the University of Munich, where he undertook research on cyclic carbon compounds as the first student of W. H. Perkin, whose like-named father was already a famous chemist. Kipping was granted the PhD *summa cum laude,* and his association with Perkin continued until the latter's death in 1929.

After earning a DSc from London University in 1887, Kipping began his academic career under Perkin at Edinburgh. He had filled two preliminary academic posts by then and had done further research on cyclic carbon compounds with Pope and Lapworth, prominent English chemists both.

In 1897 Kipping was elected a Fellow of the Royal Society and appointed to the chair of chemistry at University College, Nottingham, where he began several kinds of research, that on silicones being the most extensive, the most difficult, and the most noteworthy. The work was done with few students and on resources (some government funding included) minuscule by modern standards. Nevertheless, when he officially retired at Nottingham in 1936, he had made the department internationally known. His scientific activity there continued until 1939.

Kipping was awarded the Longstreth Model of the Chemical Society in 1909, the Davy Medal of the Royal Society in 1918, and an honorary DSc by the University of Leeds in 1936. In that year he delivered the Bakerian Lecture of the Royal Society [21], in which he summarized his research on silicones. He collaborated with Perkin on three books—one on practical chemistry, another on inorganic chemistry, and the third the *Organic Chemistry* (1894–1895), which Kipping rewrote with his younger son in 1939.

Kipping was an athlete and a sportsman (golf, tennis, billiards). His scientific interests were primarily experimental, and his standards were of the highest. He could be difficult· when he believed these standards breached [22]. The modern scientific world could use more like him!

SILICON. THE TETRAHEDRAL HABIT. STEP 2

In 1899 Kipping began pure research to accomplish Step 2 in silicon chemistry—to see whether optical activity as illustrated in Figure 2-2 exists for silicon as it does for carbon. Almost half a century and 56 papers later [23], he had founded silicone chemistry as a consequence of discoveries that had little or nothing to do with optical activity.

In December 1936 Kipping described how the earlier literature looked to him when he began work, and what he found soon thereafter [24]:

. . . Most of [the more important types of organic silicon compounds], it was believed, were represented by a well known carbon analogue, but there was little in the literature to show to what extent corresponding derivatives of silicon and carbon resembled or differed from one another in their chemical behavior . . . .

Even after a very short experience, it was evident that corresponding derivatives of the two elements in question showed very considerable differences in their chemical properties; it may now be said that the principal if not the only case in which they exhibit a really close resemblance is that of the paraffins and those particular silicohydrocarbons, containing a silicon atom directly united to four alkyl radicals.

But of far greater importance in any general comparison of carbon compounds with the organic derivatives of silicon is the fact that many, if not most, of the more important types of the former are not represented among the latter. Apparently this is not merely a consequence of the insufficient experimental investigation of silicon derivatives but is due to the fundamental differences in the properties of the atoms of silicon and carbon; . . . .

In order words, the analogy between the chemistry of carbon and that of silicon is severely limited, and a more sophisticated interpretation of the periodic table is needed.

The *Kipping route* to silicones [25] was discovered by him in 1904, when he learned to make ethylchlorosilanes* by using a Grignard reagent, a compound of the general formula RMgX made by the action of magnesium on an organic halide, RX, usually with an ether as solvent [26]. In his route the siloxane bridges that characterize silicones result from *condensation reactions that accompany the hydrolysis* of organohalogenosilanes thus prepared. The reactions are easy to carry out, but the route has the drawback—common in silicone chemistry—of yielding mixtures of products that are difficult to separate and that often contain the desired substances in small amounts. Nevertheless, the Kipping route was a great improvement over the prior art, and it is still the most versatile laboratory method for making silicones "tailored to order";

---

* In speaking of chlorosilanes, we shall have to use both names and formulas. The reader will remember the "naming exercise" proposed. In ethylchlorosilanes, a silicon atom is bonded to at least one C_2H_5 group and to at least one chlorine atom.

$SiCl_4$ is a usual starting material. With CCl_4 no analogous route to carbon compounds is possible.

The Kipping route will be illustrated by sketching how he reached his objective of demonstrating optical activity in compounds of silicon [27]. In his first step a Grignard reagent* was used as follows to generate a needed Si-C bond:

$$C_2H_5MgBr + SiCl_4 = C_2H_5SiCl_3 + MgBrCl \qquad (2\text{-}12)$$

Suitable successive analogous reactions gave him Compound A, among many other products:[†]

$$\begin{array}{cccc}
C_3H_7 & C_3H_7 & C_3H_7 & C_3H_7 \\
| & | & | & | \\
C_7H_7\text{—Si—}C_6H_5 & C_7H_7\text{—Si—OH} & C_7H_7\text{—Si—O—Si—}C_7H_7 & (2\text{-}13) \\
| & | & | & | \\
C_2H_5 & C_2H_5 & C_2H_5 \quad C_2H_5 \\
A & B & C
\end{array}$$

Compound A shows that one can attach four different organic radicals to a central silicon atom, which means that a multitude of silicones is possible. Compound B resulted *unexpectedly* when A was treated with H_2SO_4 as the first stage in the long process of resolving (separating) the optically active isomers of A. The formation of B shows the Si—C_6H_5 bond to be the least durable toward H_2SO_4. Compound C formed along with B as the result of a *condensation reaction* that yielded a complete molecule with a silicone bridge, but not a long backbone. Kipping later obtained better results on optical activity than he did at first, and elegant modern work of this kind has followed [28]. Step 2 had thus been accomplished for silicon: Figure 2-2 applies to the chemistry of silicon as well as to that of carbon. But there the analogy stops. The reactions of analogous carbon and silicon compounds can be enormously different. Had Kipping done no more than establish the facts just summarized, he would have made a great contribution.

---

* The Grignard reagent, for the discovery of which Victor Grignard shared the 1912 Nobel Prize, is so widely useful in organic chemistry that thousands of papers relating to it have been published.

[†] Professor Rochow has kindly pointed out to me that early users of the Grignard reagent often made life difficult for themselves by using compounds of more than one halogen [e.g., bromine and chlorine in (2-13)] in their work. The number of reaction products is greatly increased when this is done.

Note that Compound A in (2-13) corresponds to Example 4 in Figure 2-2, which please see.

Quite early in his career, Kipping's interest in organosilicon chemistry began to dominate his interest in optical activity. In all his work he felt driven (as did many organic chemists of his time) to prepare, isolate, and characterize pure compounds, preferably compounds that occurred as beautiful crystals. Organosilicon crystals of any kind seldom come easily. This spurred him on, but it also led him to regard *polymers* as *omnipresent nuisances.* He encountered ("collided with" seems more suitable) these messy *persona non grata* immediately after he began work in 1899. Kipping and L. L. Lloyd [29], after distilling $(C_6H_5)_3SiOH$ from the crude *by-product* formed in the preparation of $(C_6H_5)_4Si$, could get nothing but a "sticky mass" that "melts, not sharply, at about 109°, and would seem to be diphenyl silicone." This material they could not "obtain in a *crystalline condition,* so that its nature is *not definitely established*" [italics mine: no crystal, no definitive conclusion]. In 1909 G. Martin and Kipping [30] found that $(C_2H_5)_2SiCl_2$ poured into cold water gave "an oil of about the same consistency as glycerol" with molecular weight 694. The formula of this oil we write as $[(C_2H_5)_2SiO]_n$ today, with $n = 7$, approximately. When searching for a "definite polymeric form" thereof, they once got a "yellow, transparent jelly." Kipping and A. G. Murray [31] made complex products of molecular weight 3900 ($n = 19$ or 20) by the action of *hot alkali* on $(C_6H_5)_2Si(OH)_2$; the gelatinous material contained 14.2% Si, or 0.1% less than $(C_6H_5)_2SiO$, which Kipping and Murray took it to be. Note that materials such as this, for which Kipping did not use the subscript n, and which he then regarded as *oxides, not as polymers,* were actually complex polymeric mixtures.

The examples just cited show what Kipping had to contend with throughout his work in organosilicon chemistry. As the degree of condensation increased, oils, jellies, glues, and uncrystallizable solids became more common. One can almost feel his distaste for these polymers permeating Kipping's fine, restrained scientific prose.

FURTHER GUIDE TO CHEMISTRY. THE HYDROLYSIS-CONDENSATION SEQUENCE

Kipping's polymeric nuisances (oils, jellies, glues, "sticky masses"), properly transformed and controlled, are today the silicone industry's bread and butter. They are formed by what we shall call the *hydrolysis-condensation sequence.* Kipping had no notion he would meet these substances; he was searching for molecules containing a *single* silicon atom, such as A in (2-13).

The first man—Berzelius, in all likelihood—who brought $SiCl_4$ and

H_2O together must have encountered the hydrolysis-condensation sequence as it occurs in the inorganic chemistry of silicon. Today we know the sequence to be so complex that a simplified introduction thereto will have to suffice here.

Consider the behavior of $SiCl_4$ when led as a gas into an *excess of water*. It reacts violently with the water. So much heat is liberated that the presence of ice is advisable. A white, gelatinous solid (hydrated SiO_2) forms. Were this a tidy reaction, we could describe it thus when H_2O is in excess:

hydrolysis:

$$SiCl_4 + 4H_2O = HO-\underset{\underset{H}{\overset{|}{O}}}{\overset{\overset{H}{\overset{O}{|}}}{Si}}-OH + 4H^+ + 4Cl^- \quad (2\text{-}14)$$

followed by

condensation: $n\ Si(OH)_4 = n\ SiO_2(aq) + 2n\ H_2O \qquad (2\text{-}15)$

to give the sum

hydrolysis- $n\ SiCl_4 + 2n\ H_2O = n\ SiO_2(aq) \qquad (2\text{-}16)$
condensation: $+\ 4nH^+ + 4nCl^-$

All H_2O represented by (aq) comes from the excess present. As this excess disappears, HCl gas is produced, and the liberation of heat decreases; if an *excess of $SiCl_4$* is present at the start, HCl gas forms and heat is *absorbed*. We may say that we have formed a polymer, hydrated silica, $(SiO_2)_n \cdot (H_2O)_m$, by the hydrolysis-condensation sequence; when m is zero, the product is polysiloxane (silica). But this sequence is never tidy under ordinary conditions. Once the $SiCl_4$ has been attacked by water (and this attack need not affect all four chlorine atoms at once), condensation begins, that is, the hydrolysis-condensation sequence is initiated by hydrolysis. However, all possible hydrolysis and condensation reactions can then occur together until the sequence is terminated; probably condensation has the last word,* although rearrangement seems possible during any hydrolysis-condensation sequence. Clearly, the preceding equations give no clue to the mechanism of the sequence.

---

* Please see the excellent discussion by Lichtenwalner and Sprung [11].

If we now use organohalogenosilanes (i.e., R_nSiX_{4-n})* instead of $SiCl_4$, the hydrolysis-condensation sequence *will produce silicones*; that of course is what happened to Kipping, and that is what the silicone industry is doing today. To elucidate, let us take advantage of hindsight not available to Kipping and restrict the discussion to organochlorosilanes of formulas $RSiCl_3$, $(R)_2SiCl_2$, and $(R)_3SiCl$.*

As with $SiCl_4$, hydrolysis and condensation reactions are intertwined when these chlorosilanes are put through the hydrolysis-condensation sequence. Each hydrolyzable chlorine atom is a reactive site that could be decorated with asterisks as in (2-6). Because the proportions of the three chlorosilanes and of $SiCl_4$ can be varied in the starting mixture for the hydrolysis-condensation sequence, the backbone-skeleton structure of the resulting silicones can be controlled. By referring to the Bakelite case, the reader can convince him- or herself that the use of $RSiCl_3$ or of $SiCl_4$ produces *cross-links*, that the use of $(R)_2SiCl_2$ builds *backbones*, and that the use of $(R)_3SiCl$ produces *end groups* and thus can lead to finished molecules (no free bonds!) of moderate size. Bakelites, in contrast to silicones, do not include oils and rubbers: Bakelites are too highly cross-linked.

An analogous hydrolysis-condensation sequence is not achievable in carbon chemistry. To be sure, CCl_4 can be hydrolyzed, but far less easily than $SiCl_4$; and complete hydrolysis produces CO_2, which of course is not a polymer. Anyone foolish enough to substitute $SiCl_4$ for CCl_4 when attempting to remove grease spots from clothing will quickly discover another great difference in the chemistries of carbon and silicon.

MORE FROM KIPPING

As Kipping never used X-ray or electron diffraction in his research, he could not take Step 3 in establishing the tetrahedral habit for silicon. Step 3 will be described later.

Silicone Cyclamers. Another discovery by Kipping was that silicone *cyclamers,* the molecules of which are *rings* held together by siloxane *bridges*, exist alongside silicones with linear backbones formed in hydrolysis-condensation sequences.

---

*As always, R represents an organic group, e.g., CH_3 or C_6H_5; and X when used thus represents chlorine, bromine, or iodine, the common halogens.

Linear backbone with end groups

A ring compound
or cyclamer

R R
\ /
Si
/ \
R O O R
\ / \ /
Si   Si
/ \   / \
R O O R
\ /
Si
/ \
R R

$$(2\text{-}17)$$

R R R R
| | | |
HO—Si—O—Si—O—Si—O—Si—OH
| | | |
R R R R

A dihydroxy*tetra*siloxane,
$(HOSiR_2O_{1/2})(SiR_2O)_{4-2}(O_{1/2}R_2SiOH)$

A *tetra*mer, $(SiR_2O)_4$; see (2-10)

Notes

1. R is any organic group of one kind. Si-C bonds are not shown.
2. Kipping's class names: open-end chain, and closed-end chain (for the cyclamers).
3. General formulas: in formulas above, replace 4 by *n*.
4. Naming: Greek prefix indicates number of silicon atoms.

Note that internal condensation of the linear molecule could form the cyclamer by splitting out H_2O: picture a dog chasing and catching its tail. The likelihood of such condensation decreases with the length of the molecule; OH groups react with each other more readily the closer they are together, extreme cases being those with three or four OH groups on a single silicon atom. Cyclamers are of practical importance as starting materials in silicone *rearrangements* and *rearrangement polymerization*, about which there will be more later.

Bonds. Kipping drew from his experiments the following conclusions about bonding in silicones. The Si-C, Si-O, and Si-Si double bonds do not appear in stable compounds; the contrast to carbon chemistry is marked. The Si-Si bond is not durable enough to survive even mild treatment with alkaline reagents—again in marked contrast to carbon chemistry.

Siloxane bridges (Si—O—Si), each built of two siloxane (Si-O) bonds, and each formed by the hydrolysis-condensation sequence, constitute the backbones and skeletons of polymeric silicones. The existence of Si-O bonds is essential to the existence of silicones as polymers. Kipping, being an organic chemist who regarded silicones as *organic* compounds [21], was understandably slow to appreciate this fact, which derives from *inorganic* chemistry. No doubt this slowness resulted in part from a misconception of what the periodic table tells about the analogy between carbon and silicon. It is not surprising that his coinage of "silicones" as a class name was scientifically unsound. To illustrate, in his early work he regarded $(C_6H_5)_2CO$, benzophenone (a true ketone), and $(C_6H_5)_2SiO$ (see above) as analogous compounds; consequently the latter became a sili*cone* even though the two substances did not look or behave alike.

We have said that silicones consist of molecules in which at least one organic group is bonded via Si-C to a siloxane bridge. "Silicone," now firmly entrenched, is as good as any other short, simple name for these materials. Humpty Dumpty approves.

In assessing Kipping's contributions, one must consider the scientific environment in which he worked, and the tools and resources at his disposal. Polymer chemistry was rudimentary, and the instrumental methods indispensable to modern organic chemistry were not available to him. He had to rely on classical methods to establish composition and molecular weight on samples painfully prepared by distillation in crude stills and by crystallization. Without his pure research the silicone industry would be far from where it is today.

A. [E.] STOCK (1876–1946)

Since silicones mark an important confluence of organic and inorganic chemistry, we ought to seek an effective inorganic counterweight to Kipping. Stock qualifies easily and is the only logical choice.

Alfred [Eduard] Stock was drawn to the natural sciences as a boy. His father, an official in an insurance company, encouraged the boy's interest by purchases of equipment and supplies. An excellent high school ("gymnasium") teacher focused the interest on chemistry. Stock matriculated at the University of Berlin in October 1894. Generous scholarship aid proved all the more useful because he lost his father a few months after matriculation. His early university career was strongly influenced by Emil Fischer (Nobel Prize, 1902) and van't Hoff. Oskar Piloty, a son-in-law of von Baeyer, directed his research, and his thesis was one part analytical chemistry and two parts organic. Fischer, who saw the need for reviving inorganic chemistry, eventually sent Stock to Paris for a year with Henri Moissan (Nobel Prize, 1906), the discoverer of fluorine; similarly, Otto Ruff, a close friend of Stock's, went to Wilhelm Ostwald (Nobel Prize, 1909) at Leipzig. The two friends subsequently did all that Fischer could have asked.

Stock's scientific activities were too varied and too extensive to be detailed here. He steadily ascended the academic ladder, beginning as *Privatdozent* (1900–1909) in Berlin after his return from Paris; continuing at Breslau (1909–1916); returning to Berlin (1916–1926), this time to the Kaiser Wilhelm Institut für Chemie; moving to Karlsruhe (1926–1936) as Director of the Chemical Institute; and finally returning to the University of Berlin as Professor Emeritus committed to the investigation of mercury poisoning. After completing this project, he continued scientific work at Berlin until 1943, when World War II led him to migrate westward. Up until 1943 he seems to have been fortunate as regards salary, advancement, and funding for research and for the equipping of new laboratories. The last 3 years of his life were made tragic because of earlier poisoning by mercury, and World War II.

Stock's name is on some 270 articles, naturally not all of first importance, and on 3 books. His honors were many. Primarily an experimentalist, he found most satisfying the preparation and study of new and unusual compounds, particularly of elements then among the poorly known. He was a master at devising methods and equipment for precise but difficult experiments. He wrote much about teaching. He made notable contributions to chemical nomenclature, "borane" and "ligand" being examples outside the field of silicones. Although his work on the hydrides of boron and silicon is generally regarded as his greatest contribution, that on mercury poisoning is not far behind.

Today, when environmental hazards and the like are matters of popular concern, Stock's lonely struggle with mercury poisoning, revealing as it is of his indomitable character, commands added interest as a warning against perhaps the most insidious foe to be found in a chemical laboratory. His contacts with mercury in his boyhood laboratory at home seem to have made him hypersensitive to the element. During his year in Paris, he used a mercury trough to facilitate work with gases, and he introduced this equipment into Germany. Having no notion of what caused his physical and mental difficulties (affected upper air-passages, loss of memory, loss of hearing, inability to concentrate and to calculate), he often worked with large amounts of mercury and took no precautions. He seems often to have had poorly ventilated working quarters that were contaminated with mercury. After 1924, when the source of his troubles

LIBRARY
OF
MOUNT ST. MARY'S
COLLEGE

was accidentally identified, he took precautions, but by then irreversible damage had been done. He bore the cruel burden and continued to work, to lecture, and to write almost until he died [32].

When Stock delivered the 1932 George Fischer Baker Lectures at Cornell University in Ithaca, New York, E. G. Rochow was his assistant, an added reason for recognizing Stock here.

This book hopes to acquaint the reader with the kind of people who engage in science and technology, a topic discussed by Carl Wagner [33] with strong reliance upon Ostwald's *Grosse Männer* [34], which—though published in 1909—makes entertaining and instructive reading even in today's different world. Ostwald believed that research men range from the pronounced romantic to the pronounced classicist, whose personalities lie at the limits of a continuous spectrum. As characterized operationally, the classicist prefers to work in depth on a small number of problems; he is likely to be an introvert. The romantic shifts quickly and easily from problem to problem, his and those of others, in the many fields of interest to him; he is happy to have others do the needed work-in-depth on problems he leaves behind. Ostwald's classification is more useful than one might expect. Kipping and Stock were pronounced classicists. Whitney and Marshall, whom we have met, were pronounced romanticists.

STOCK AND SILICONE CHEMISTRY

A most fortunate accident! Stock felt that he needed to study silicon hydrides so that he could understand hydrides of boron, which were then his principal research interest. To prepare boron hydrides, he allowed "magnesium boride," a complex material made from magnesium and B_2O_3, to react with HCl. As "magnesium boride" always harbored silicon-containing impurities, silicon hydrides and boron hydrides were then produced together. For us a bonus resulted: Stock made an excursion into silicon chemistry during the period 1916–1923, and the results of this excursion appear in 22 papers [35, 36] that complemented Kipping's work and have helped today's silicone industry.

To isolate the silicon hydrides, which he prepared in various ways, Stock had to rely on physical methods (fractional distillation or condensation) carried out on *milligram samples* in *vacuum equipment* that allowed them to come into contact only with glass and mercury. Subsequent investigations of many kinds have benefited from the successful development of this equipment and of the associated techniques, which were experimental marvels in their time.

Among Stock's other contributions were the following:

1. Proof that Si-Si bonds are far less durable than the C-C bonds that hold many polymers together. Stock's discovery was important as it eliminates the possibility of making widely useful polymers with Si—Si backbones or skeletons.

2. Accomplishment of the hydrolysis-condensation sequence with silane derivatives (methyl derivatives included) that could not be studied by Kipping's methods.

3. Observation of the ready interaction of gaseous halogenated silanes with glass surfaces [37] to make them hydrophobic (water repellent).

4. Discovery that hydrides of silicon show chemical behavior identical with that shown by hydrides of boron (not of carbon!). Silicon and boron hydrides are extremely reactive toward oxygen, water, and halogens: Si-H and B-H bonds are *far less durable* than the C-H bond; methane, for example, is relatively inert.

5. Realization that the many pronounced differences in the chemistry of carbon and that of silicon are due to differences in the structures of these atoms.

6. Contribution of "silane" and "siloxane" to silicone nomenclature. (Please compare names and formulas with those of methane and dioxane.)

SILICONE PRECURSORS

Ought we to say that commercial silicone monomers exist? To discuss this deceptively simple question, we shall have need of Adams, Holmes, and Humpty Dumpty.

First, an authoritative definition [12] is in order: "1. *A monomer* is a substance consisting of low molecular weight molecules, capable of reacting with like or unlike molecules to form a polymer." Adams would understand why this definition is so broad.

According to the definition, $(CH_3)_2Si(OH)_2$ is a silicone monomer. It exists and polymerizes [38], and the first stage in the polymerization might be

$$2\ (CH_3)_2Si(OH)_2 = HO-\underset{\underset{CH_3}{|}}{\overset{\overset{CH_3}{|}}{Si}}-O-\underset{\underset{CH_3}{|}}{\overset{\overset{CH_3}{|}}{Si}}-OH + H_2O \qquad (2\text{-}18)$$

with hydroxide groups as active sites. But $(CH_3)_2Si(OH)_2$ is not a *commercial* monomer. The commercial starting material for polydimethylsiloxanes could be $(CH_3)_2SiCl_2$, or it could be a cyclamer, $[(CH_3)_2SiO]_n$, as we shall see later. If $(CH_3)_2SiCl_2$ is considered a monomer, $SiCl_4$ must logically be considered the monomer from which polymeric hydrated silica is made according to a hydrolysis-condensation sequence, (2-16), which is analogous to that needed for making $(CH_3)_2SiCl_2$ into a silicone. This position is not a happy one! The cyclamer $[(CH_3)_2SiO]_n$ is clearly not a monomer.

Contrary to distinguished precedent [39], we shall say that silicones have no commercial monomers—only *precursors*, most commonly the organochlorosilanes. Humpty Dumpty approves once more.

Now, an important scientific question arises. Do polymer units ("mers") such as $(CH_3)_2SiO$ exist as stable species? We have seen that Kipping mistakenly thought they did, which is why he coined the name "silicones." His final word follows: "So far no silicon analogue of a ketone has been obtained" [21]; generalized, this means that stable $Si{=}O$ linkages analogous to the $C{=}O$ linkage in ketones do not exist with silicon in place of carbon. This position is held today. Although $(CH_3)_2Si(OH)_2$ has been isolated [38], $(CH_3)_2SiO$ has not. That Nature will not permit a *stable* Si-O double bond seems a permissible generalization; too great a distortion of tetrahedral bonding would be required.

The story is not over. Beautiful and difficult experimental work led Stock to conclude that H_2SiO, the silicon analogue of formaldehyde, *does* exist, albeit *transitorily*. By our definition H_2SiO is not a silicone, nor are the polymers it forms. Stock called it "prosiloxane," meaning thereby "siloxane predecessor."

Stock's most convincing evidence for the existence of H_2SiO came from the reaction [40]

$$H_2SiCl_2 + H_2O = H_2SiO + 2\ HCl$$

$$(\text{gaseous reactants and products}) \quad (2\text{-}19)$$

When the substances on the left were mixed mole for mole in an 8-liter flask as gases at low pressure and room temperature, approximately the expected pressure increase and approximately the expected amount of HCl for complete reaction were measured. Until Stock's results are controverted or differently explained, we must assume that H_2SiO exists *transitorily* as a gas. It polymerizes quickly, but we do not know how. Stock found the hexamer, $(H_2SiO)_6$, to be a liquid. In our language the Si-O double bond in H_2SiO is not a durable bond, and it readily disappears in favor of the Si—O—Si bridges in $(H_2SiO)_6$.

with hydroxide groups as active sites. But $(CH_3)_2Si(OH)_2$ is not a *commercial* monomer. The commercial starting material for polydimethylsiloxanes could be $(CH_3)_2SiCl_2$, or it could be a cyclamer, $[(CH_3)_2SiO]_n$, as we shall see later. If $(CH_3)_2SiCl_2$ is considered a monomer, $SiCl_4$ must logically be considered the monomer from which polymeric hydrated silica is made according to a hydrolysis-condensation sequence, (2-16), which is analogous to that needed for making $(CH_3)_2SiCl_2$ into a silicone. This position is not a happy one! The cyclamer $[(CH_3)_2SiO]_n$ is clearly not a monomer.

Contrary to distinguished precedent [39], we shall say that silicones have no commercial monomers—only *precursors,* most commonly the organochlorosilanes. Humpty Dumpty approves once more.

Now, an important scientific question arises. Do polymer units ("mers") such as $(CH_3)_2SiO$ exist as stable species? We have seen that Kipping mistakenly thought they did, which is why he coined the name "silicones." His final word follows: "So far no silicon analogue of a ketone has been obtained" [21]; generalized, this means that stable $Si{=}O$ linkages analogous to the $C{=}O$ linkage in ketones do not exist with silicon in place of carbon. This position is held today. Although $(CH_3)_2Si(OH)_2$ has been isolated [38], $(CH_3)_2SiO$ has not. That Nature will not permit a *stable* Si-O double bond seems a permissible generalization; too great a distortion of tetrahedral bonding would be required.

The story is not over. Beautiful and difficult experimental work led Stock to conclude that H_2SiO, the silicon analogue of formaldehyde, *does* exist, albeit *transitorily.* By our definition H_2SiO is not a silicone, nor are the polymers it forms. Stock called it "prosiloxane," meaning thereby "siloxane predecessor."

Stock's most convincing evidence for the existence of H_2SiO came from the reaction [40]

$$H_2SiCl_2 + H_2O = H_2SiO + 2\ HCl$$
$$\text{(gaseous reactants and products)} \quad (2\text{-}19)$$

When the substances on the left were mixed mole for mole in an 8-liter flask as gases at low pressure and room temperature, approximately the expected pressure increase and approximately the expected amount of HCl for complete reaction were measured. Until Stock's results are controverted or differently explained, we must assume that H_2SiO exists *transitorily* as a *gas.* It polymerizes quickly, but we do not know how. Stock found the hexamer, $(H_2SiO)_6$, to be a liquid. In our language the Si-O double bond in H_2SiO is not a durable bond, and it readily disappears in favor of the Si—O—Si bridges in $(H_2SiO)_6$.

Among Stock's other contributions were the following:

1. Proof that Si-Si bonds are far less durable than the C-C bonds that hold many polymers together. Stock's discovery was important as it eliminates the possibility of making widely useful polymers with Si—Si backbones or skeletons.

2. Accomplishment of the hydrolysis-condensation sequence with silane derivatives (methyl derivatives included) that could not be studied by Kipping's methods.

3. Observation of the ready interaction of gaseous halogenated silanes with glass surfaces [37] to make them hydrophobic (water repellent).

4. Discovery that hydrides of silicon show chemical behavior identical with that shown by hydrides of boron (not of carbon!). Silicon and boron hydrides are extremely reactive toward oxygen, water, and halogens: Si-H and B-H bonds are *far less durable* than the C-H bond; methane, for example, is relatively inert.

5. Realization that the many pronounced differences in the chemistry of carbon and that of silicon are due to differences in the structures of these atoms.

6. Contribution of "silane" and "siloxane" to silicone nomenclature. (Please compare names and formulas with those of methane and dioxane.)

SILICONE PRECURSORS

Ought we to say that commercial silicone monomers exist? To discuss this deceptively simple question, we shall have need of Adams, Holmes, and Humpty Dumpty.

First, an authoritative definition [12] is in order: "1. *A monomer* is a substance consisting of low molecular weight molecules, capable of reacting with like or unlike molecules to form a polymer." Adams would understand why this definition is so broad.

According to the definition, $(CH_3)_2Si(OH)_2$ is a silicone monomer. It exists and polymerizes [38], and the first stage in the polymerization might be

$$2\,(CH_3)_2Si(OH)_2 = \begin{matrix} CH_3 & & CH_3 \\ | & & | \\ HO—Si—O—Si—OH \\ | & & | \\ CH_3 & & CH_3 \end{matrix} + H_2O \qquad (2\text{-}18)$$

SILICON. THE TETRAHEDRAL HABIT. STEP 3. THE SILOXANE BRIDGE

The tetrahedral habit and the siloxane bridge are mainstays not only in silicone chemistry, but in the chemistries of silica and of the silicates (organic and inorganic) as well. Elementary silicon itself shows tetrahedral bonding. The completion of Step 3 for the simpler crystals by X-ray diffraction and for vapors by electron diffraction is as important, though accompanied by less fanfare, than establishment of the structure of DNA or of hemoglobin.

Pauling [8, 41, 42] did vital work toward completing Step 3, which entailed making progress in chemical bonding, and X-ray determinations of crystal structure. He built on the covalent electron-pair bond of Gilbert N. Lewis. His X-ray work extended that of W. L. Bragg (Nobel Prize, 1915), who had made a vital contribution to modern structural inorganic chemistry by proving in 1914 that the three C-O bonds in $CaCO_3$ (calcite, limestone) are identical, and by introducing some order into the structure of silicates—a bewildering field.

Kipping, in speaking of $R_2Si(OH)_2$ and $RSi(OH)_3$, said in 1937:

If these hydroxy compounds lose the elements of water only by the formation of Si—O—Si groups [bridges], it seems to be very probable that a similar change would occur in the condensation of orthosilicic acid [$Si(OH)_4$]; and if so, then in all the more complex mineral silicates every silicon atom would be directly combined with four atoms of oxygen [21].

Let us now present an overview, derived largely from Noll [43], of the unification made possible by the completion of Step 3, by the understanding of the siloxane bridge, and by the polymer concept, all used in concert. We shall begin by inventing a *model* regular tetrahedron, **SiO_4,** in which a central silicon atom is bonded to four oxygen atoms.* Let us see how this model must be modified in describing the silicon species that follow.

1.   Silica is the most highly condensed polysiloxane, being constructed entirely of siloxane bridges. Each oxygen at the apex of one tetrahedron is shared with another tetrahedron at one of its apices, where

---

* Please simplify Figure 2-2 by replacing C with Si and pretending that a = b = c = d = O (O = oxygen atom).

$$\mathbf{SiO_4} \quad \text{becomes} \quad \begin{array}{c} O_{1/2} \\ | \\ O_{1/2}-Si-O_{1/2} \\ | \\ O_{1/2} \end{array} \quad \text{(see Figure 2-2)} \quad \text{(2-20)}$$

and four siloxane bridges result. Sharing of any other kind need *never* be considered for silica.

2. In regard to the anion of H_4SiO_4 (orthosilicic acid), all Si-O bonds, though often called covalent, show some ionic character, a situation earlier described as Orwellian. In silicate anions the *second* bond to certain oxygens is so highly *ionic* that it is no longer called covalent and is represented by a minus sign, being completed by cations somewhere in the neighborhood. Here

$\mathbf{SiO_4}$ becomes   (2-21)

$$\begin{array}{c} O^- \\ | \\ O^-\!-Si-O^- \\ | \\ O^- \end{array} \quad \text{when ionic bonding is maximum}$$

In orthosilicic acid and organic silicates, additions to $\mathbf{SiO_4}$ give the tetrahedra

$$\begin{array}{c} H \\ O \\ | \\ HO-Si-OH \\ | \\ O \\ H \end{array} \quad \text{and} \quad \begin{array}{c} R \\ O \\ | \\ RO-Si-OR \\ | \\ O \\ R \end{array} \quad (2\text{-}22)$$

3. In silicones three cases in addition to (2-20) are possible. For the four cases the model $\mathbf{SiO_4}$, now not always exactly tetrahedral, becomes

$$\begin{array}{cccc} \begin{array}{c} O_{1/2} \\ | \\ O_{1/2}-Si-O_{1/2} \\ | \\ O_{1/2} \end{array} & \begin{array}{c} O_{1/2} \\ | \\ R-Si-O_{1/2} \\ | \\ O_{1/2} \end{array} & \begin{array}{c} O_{1/2} \\ | \\ R-Si-O_{1/2} \\ | \\ R \end{array} & \begin{array}{c} R \\ | \\ R-Si-O_{1/2} \\ | \\ R \end{array} & (2\text{-}23) \\ \textbf{Q unit} & \textbf{T unit} & \textbf{D unit} & \textbf{M unit} \end{array}$$

The boldface letters below (2-23) will be explained when we have need of them. Note that a series analogous to (2-23) could have been made to bridge the gap between (2-20) and (2-21).

Now, a most important point must be made. Each apex marked by $O_{1/2}$ is part of a siloxane (Si—O—Si) bridge: in other words, each such apex is evidence of *polymerization*.

An explicit recapitulation is now possible. In silica, silicates, and silicones, tetrahedra with a central silicone atom are united by siloxane bridges. An almost endless polymeric variety, inorganic and organic, results: linear, cyclic, two-dimensional, three-dimensional—all formed by the junction of tetrahedra. Silica can be amorphous, can be a glass, or can have one of several crystal forms. Silicates appear as complex anions; as chains, fibers, and bands (cross-linked chains) with siloxane backbones; as sheets; and as more extended three-dimensional networks with siloxane skeletons.* Silicones can be oils, resins with highly cross-linked siloxane skeletons, gums (long chains with siloxane backbones), or rubbers (specially cross-linked gums); crystals are less common. The variety seems endless. Nevertheless, three concepts—tetrahedral habit, siloxane bridge, and polymerization—bring this bewildering horde underneath one tent. There is no finer example in chemistry of the usefulness of models and of theoretical speculations to produce helpful generalizations, which must of course be viewed with Holmesian skepticism.

Many complexities have been glossed over in this treatment. For example, the tetrahedral bond angle has (or closely approaches) the ideal value only in substances such as silicon, SiH_4, and SiO_2; in others, the tetrahedra are measurably distorted. Distortions result from complexities in chemical bonding that derive ultimately from differences in atomic structure and size. There are limits to how great the distortions in a stable molecule can be.

Silicon does deviate from the tetrahedral habit, but relatively seldom, and in only one instance that we need consider, namely, that of SiF_6^{2-} and the two compounds H_2SiF_6 and K_2SiF_6. Here we must again pay tribute to Berzelius for explaining the chemistry relating H_2O, SiO_2, HF, SiF_4, and H_2SiF_6, which he did when he prepared silicon in 1823 [4]. The modern view of SiF_6^{2-} is this: (1) the six Si-F bonds are identical and directed from the silicon atom to the apices of a regular octahedron; (2) the two negative charges, which are neutralized by cations some distance

---

* In complex silicate minerals (e.g., those containing combined Al_2O_3), octahedra may accompany the silicon-oxygen tetrahedra. Enough cations (e.g., K^+) must always be present to neutralize all negatively charged oxygen.

away, belong to the entire ion; and (3) the six bonds, said to be "hybridized," are complex and—though often called covalent—have a pronounced ionic character.

Among the many facts of silicon chemistry that reveal the strength and limitations of the periodic table as a model of descriptive chemistry, none is more eloquent than the existence of SiF_6^{2-}. It can exist because the relative size of the two ions (F^- being the smallest halide ion) meets the requirements of geometry, because silicon electrons beneath the four valence electrons can come into play, and because fluorine is a highly electronegative element. The carbon analogue, CF_6^{2-}, is impossible. A simple periodic table cannot be a satisfactory model of descriptive chemistry. Adequate modifications are hard to come by because the facts of descriptive chemistry are many and diverse.

SILICA, CARBON DIOXIDE, WATER

The pure research we have described established similarities and differences between the chemistry of silicon and that of carbon. The most noteworthy similarity is a sharing of the tetrahedral habit. The great extent of the differences is brought home most strikingly by a comparison of SiO_2 and CO_2.

One form of SiO_2, α-quartz, is on every beach; free or combined, SiO_2 makes up about 60% of the earth's crust. On the other hand, CO_2, a gas, is in the air around us. Whereas SiO_2 is built of linked tetrahedra (in quartz, these can form either right- or left-handed helices, so that optically active crystals result), CO_2, being a dumbbell-shaped molecule representable as $O{=}C{=}O$, is never tetrahedral and never optically active. The bonds in siloxane bridges rank with the most durable in chemistry; the bonds in CO_2 are so weak that plants accomplish photosynthesis every day. As a consequence, silica and the silicates, in which silicon is in the $+4$ oxidation state, are virtually the only forms in which the element occurs in nature.* By contrast, carbon is found free and forms perhaps a million known compounds, in most of which its oxidation state is lower than in CO_2.

It seems surprising now that a "high-temperature protoplasm," a stuff of life with silicon in place of carbon, could have been seriously suggested as late as 1910 [44]. Evidently the misconception of a close

---

* Naturally occurring silicon carbide is a trivial exception. The Si-C bond in silicones resembles more nearly the C-C bond in organic compounds than it does the bonds in a siloxane bridge, which is why the Si-C bond is not discussed in this book. See Noll [43].

carbon-silicon analogy died hard. Certainly the misconception was still alive in 1916, when Arthur Bygdén wrote his fine thesis, *Silicium als Vertreter des Kohlenstoffs organischer Verbindungen* ("Silicon as the Surrogate of Carbon in Organic Compounds") [45]. Why did it live so long? Boiled down, the chief reason was that chemists were slow to realize that C-O double bonds in CO_2 did not imply similar bonds in SiO_2: witness Kipping's conviction that the Si-O double bond must exist in $(C_6H_5)_2SiO$ because the C-O double bond exists in $(C_6H_5)_2CO$. Figure 2-3 [45, p. 19] shows the slow progress toward the realization that crystalline SiO_2 is a polysiloxane, the simplest possible, of molecular weight limited by the size of the crystal.

Hydrated silica, $SiO_2 \cdot n\ H_2O$, already introduced, gives evidence of a close relationship between SiO_2 and H_2O, two of the world's commonest, most studied, and most interesting substances. Quartz glass, though made at high temperatures, can contain H_2O in the form of OH groups. Opal, a milky gem in one of its many varieties, is partially hydrated SiO_2. A most interesting form of silica is man-made. Sold as Ludox® by E. I. du Pont de Nemours and Company for use as a filler in greases, waxes, and polishes, it consists of tiny, surprisingly uniform, transparent colloidal particles. Of the variety containing 70% water by weight, Alexander* says:

A glass full of "Ludox" looks almost like water. It flows like water, too; yet in an eight-ounce glass there is enough silica to make three pieces of rock the size of ordinary ice cubes. If the surface of all the particles of that silica could be spread out flat, it would cover ten acres—or a city lot one block square. This enormous surface area on such a small amount of solid is one of the features that makes "Ludox" useful and unique.

No trio of simple compounds is of greater importance to mankind than SiO_2, CO_2, and H_2O. Truly, one can see a world of riches in Blake's grain of Sand†—a world characterized by the tetrahedral habit and the

---

* Guy Alexander, *Silica and Me*, Doubleday and Company, Garden City, New York, 1967, p. 22. A delightful book that will acquaint the general reader with applied research in the chemical industry.

† The last word belongs to three intellectual golfers, who say, "Sand . . . . It is neutral, generally inoffensive stuff. Certainly nobody should be scared by sand. What is it, after all, except pulverized rock or minerals, mostly quartz, with smaller amounts of feldspar, garnet, tourmaline, zircon, rutile, anatase, topaz, staurolite, cyanite, andalusite, chlorite, biotite, hornblende, augite, chert, and iron oxide, with traces of olivine, enstatite, tremolite, chromite, and the debris of calcareous shells?" Hollis Alpert, Ira Mothner, and Harold Schonberg, *How To Play Double Bogey Golf*, Quadrangle/The New York Times Book Co., New York, 1975, p. 59.

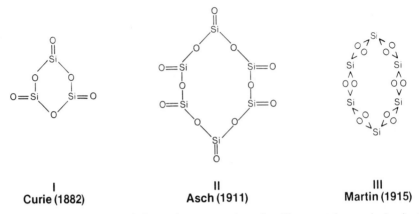

I
Curie (1882)

II
Asch (1911)

III
Martin (1915)

Figure 2-3 Progress toward the modern conception of a silica crystal as a single giant molecule [45]. Note that the Si-O double bond had disappeared by 1915, but not the concept of silica as a relatively small, definite molecule. Today we know, mainly as a result of X-ray diffraction work, that small SiO_2 molecules cannot exist.

The models above show the influence of the benzene ring. But they are not wholly off the track, for they hint that silicones and complex silicate ions will both form cyclamers (ring structures), as in fact they do.

siloxane bond, two features that make silica the logical parent of a diversity of substances, among which silicates and silicones are numbered.

COMMENTS IN CONCLUSION

• As to pure research. The foundations of silicone chemistry were well and truly laid by pure research that antedated the silicone industry. This pure research was marked by misconception and by chance. Elimination of one misconception (the presumed analogy of silicon chemistry to carbon chemistry) made the periodic table more realistic—hence more useful. Chance turned both Kipping and Stock to silicone chemistry. Kipping, an organic chemist, and Stock, an inorganic-physical chemist, complemented each other.

• As to academic and industrial research. The early pure research was largely academic. In laying the foundations of silicone chemistry, academic research performed an indispensable service for the industrial research that followed. The main elements in that foundation were of course the existences of the tetrahedral habit for silicon and of the siloxane bridge. Important practical contributions were the Kipping route to silicones and the inadvertent preparation of polymeric silicones.

Probably neither Kipping nor Stock foresaw a silicone industry. Neither had the industrial point of view. Kipping abhorred his "sticky masses," and Stock's scale of operations did not give him silicones enough to lay his hands on. An anecdote of Kipping's is revealing [46]. During one of his infrequent contacts with von Baeyer, Kipping gave the Herr Geheimrat a new product, which the great man examined under a lens and dismissed with a depreciatory and deprecatory "Ach, Harz," the message being that von Baeyer did not then regard a new "Harz" (resin) as a great gift to mankind. Well, Baekeland changed all that by building an industry on a type of resin discovered by von Baeyer. But it took time—and an industrial point of view!

• As to models. The chapter demonstrates the importance of models in chemical thinking. They help in the interpretation of experimental observations and in the design of new experiments. They exist in the mind before they are committed to paper or constructed in three dimensions. Their usefulness was increased by events such as the explanation of optical activity, the introduction of chemical bonds, and the growth of X-ray diffraction. The chemistries of carbon and of silicon have benefited particularly from the use of models: witness the siloxane bond and the tetrahedral habit. Models facilitate experimentation when understanding is lacking.

• As to nomenclature. Allusions to Humpty Dumpty have been made to point up the difficulties with nomenclature, which are serious hindrances to easy communication. They have two main causes: as knowledge is enlarged, and refined by improved understanding, old names become illogical; concomitantly, increased multiplicity and complexity make exact naming more difficult. In the early days of the General Electric silicone project, to which we turn in the next chapter, the talk was of "polymeric silicone oxides" (instead of "polysiloxanes") and of "methyl silicon chlorides" (instead of "methylchlorosilanes"). How quickly things changed becomes clear from an examination of Rochow's influential *The Chemistry of the Silicones* [47].

"Plastics" is increasingly being used in industry to replace "polymers," even when the polymers are not, nor have ever been, "shaped by flow," as the definition of "plastics" prescribes. Our choice is simple—we use "polymers" because most silicones (notably silicone oils) simply are not plastic.

• As to Chapter 2. Though it received more work than any other, Chapter 2 may impress many readers as the most difficult and least successful in the book. I have reluctantly become converted to Trevor Williams' *disbelief* "that all the mysteries of science can be made clear to the layman if only scientists will take the trouble to explain themselves in very simple terms" [48].

Chapter Three # Applied Research Begins. Methyl Silicones. The Rochow Route to Silicones

In the United States, perhaps, there is a wider nodding acquaintance with industry, but, now I come to think of it, no American novelist of any class has ever been able to assume that his audience had it. He can assume, and only too often does, an acquaintance with a pseudo-feudal society, like the fag-end of the Old South—but not with industrial society. Certainly an English novelist couldn't.

C. P. SNOW, *The Two Cultures: and a Second Look*, p. 31

THE BEGINNING

The General Electric silicone project had a very unusual beginning. During January 1938 the Corning Glass Works was seeking a market for glass (fiber) tape, then a promising new product. Mr. Morrow of Corning

Note. Brackets are used to enclose references and inserted material. For bracketed references, see References and Notes at the end of the book. Numbers in parentheses on the right of pages are often used in the text to identify equations and the like: thus (3-1) means Equation 3-1. Reference to laboratory notebooks is by name and date. CRDA stands for "Chemistry Research Department Archives."

invited representatives of the General Electric Company, a probable purchaser of any new insulating material, for a visit, which was made on January 24, 1938, by A. L. Marshall and W. I. Patnode. At Corning Dr. J. F. Hyde showed the visitors various samples of glass tape, each impregnated with one of three phenyl silicone resins containing no other organic group. He suggested that "aryl and alkyl" groups might be combined in resins not yet prepared.

On January 26, 1938, a Regional Meeting of the American Chemical Society was held at Union College in Schenectady with Marshall, Patnode, and E. G. Rochow in attendance. At this meeting, Marshall told Rochow of the Corning tape impregnated with diphenyl silicone resin; the other two resins were not mentioned. He drew for Rochow a structural formula of the diphenyl silicone polymer* and said it might not be "the last word." He suggested that Rochow begin devoting half his time to silicone research [1].

Seen through Rob Roy McGregor's eyes† [2], the Corning background was this:

The possibility of a hybrid polymer—a cross between the organic polymers (plastics) and the inorganic polymers (glasses)—appealed to some glass manufacturers as a desirable possibility. Dr. E. C. Sullivan, then Director of Research for the Corning Glass Works of Corning, N.Y., engaged Dr. J. F. Hyde, an organic chemist, to investigate this matter. [Hyde reported for work in August 1930. Reference 3, p. 5.]

Hyde was acquainted with the literature describing organosilicon research [note the work of Kipping as described in Chapter 2, especially his "sticky masses"], and he had the advantage of the rapidly accumulating knowledge of large polymers. By using knowledge from both these fields he was able to prepare large polymers containing both organic and inorganic constituents.

At this point there arose one of those coincidences that often direct a course of action. The Corning Glass Works had just begun the development of glass fibers and was on the lookout for appropriate markets. [Hyde was assigned to work on this fiber development in 1936. Reference 3, p. 6.] One of the most promising outlets for this product appeared to be as a woven tape for use in electrical insulation. Cotton, impregnated with a resinous dielectric, had been used but it would char at elevated temperature. This difficulty could be overcome by the use of glass tape in place of the cotton, but it was found that the resin impregnant

* Formula: —[(C$_6$H$_5$)$_2$SiO]$_n$—, with no consideration of end group E. See Chapter 2.
† Dr. McGregor for many years headed the important Dow Corning Fellowship at Mellon Institute. No one was better qualified by his contacts and experience to sketch the Corning background.
The quotation is from *Silicones and Their Uses* by R. R. McGregor. Copyright 1954, by the McGraw-Hill Book Company, New York. Used with permission of the copyright owner.

would stand only slightly more heat than the cotton would. Thus there was but little advantage in using the glass. To realize the full value of the glass tape there was needed a resinous dielectric that was considerably more heat stable than the organic materials in common use.

Hyde was able to point out that the organosilicon polymers he had been developing could be made in resinous form and that certain types were unusually heat stable. The work then turned toward resinous compounds that would be of use as a heat-stable dielectric in tapes made of glass fiber. [Hyde returned to his investigation of silicones late in 1937. Reference 3, p. 7.]

As we have seen, polymers often satisfy needs of the electrical industry for insulating materials with a "property profile" drawn from Table 1-1. As tapes,* such materials are, for example, wound around the electrical conductors in underground cable. No matter how used, the tapes must retain their integrity in service.

In an unimpregnated tape there are voids filled with air. In an electric field the molecules in air lose electrons, which are accelerated by the field. As the electrons gain energy in this way, they begin to knock other electrons out of molecules upon collision. The process, if continued, leads to *breakdown* of the air with *corona* and *arcing* as visible evidence of high currents. In the language of Table 1-1, the *electric strength* is too low. Air at atmospheric pressure has an electric strength near 3000 V/mm, which is far below that of any good solid insulator. That is the reason why the voids in a tape must be filled with a solid, as in impregnation of the tape with a resin. The resin, initially in solution, can be made to fill all voids upon careful evaporation of the solvent. Resins make good impregnants.†

---

* Muslin impregnated with tallow was a forerunner of silicone-impregnated glass tape. The earlier composite insulating material was used in the first underground cable ever installed. In light of Cornell University's contribution to the silicone project, it is particularly appropriate that this first installation should have been made on the Ithaca campus. We shall introduce Cornell's contribution here by mentioning in chronological order eight men active in the silicone project who had previously done research at Cornell: H. H. Race, W. I. Patnode, E. G. Rochow, W. F. Gilliam, A. E. Newkirk, D. T. Hurd, J. R. Donnalley, and A. M. Bueche.

† By 1938 the use of resins as impregnants had long been of major interest to the General Electric Company and had stimulated research on man-made polymers. Consider a letter in the Willis R. Whitney Collection, now on loan to the Schenectady Archives on Science and Technology, Union College, Schenectady, New York. On March 3, 1923, Whitney wrote to L. A. Hawkins, Executive Engineer, Research Laboratory, in part as follows:

"I feel quite sure that his [Professor Moreu's] new resin from acrolein will be [of] some help, if for nothing better than for a study of a high, non-leaking insulation material (better than amber). They can make an alcohol varnish of it, so I naturally want to try it for paper and cloth (condensers, cables, etc.)."

Whitney wrote from Paris. Had he only visited Kipping!

Applied research on silicones thus began with resins, which are high polymers that elude exact definition. Think of rosin from pine trees, and of shellac and Bakelite; then move on to the alkyds in Glyptal® paints, and to the Formvar® in Formex® wire enamel. As patent counsel might say, resins are—but are not limited to—cross-linked polymeric materials at least initially soluble in organic solvents, and capable of forming useful films or filling objectionable voids upon evaporation of the solvent.

The electrical industry has long been indebted to the Corning Glass Works, especially for insulating materials. The recognition by Corning that silicone-impregnated glass tape could fill an important insulation need is a noteworthy item in this debt. The recognition again exemplifies the point of view vital to successful applied research. A new material, glass tape, becomes available, for which a market must be found. The material has a weakness that must be remedied before it can serve satisfactorily as electrical insulation. Had Dr. Sullivan not been willing to assume the risks associated with initiating applied research on silicones as promising substances to remedy this weakness, and had Dr. Hyde not made significant progress before 1938, the arrival of useful silicones could easily have been delayed past the end of World War II. The wartime stimulant to the growth of the silicone industry and the military contributions by that industry would then both have been lost. Near the end of Chapter 5, we shall see what this would have meant to the General Electric silicone project.

E. G. ROCHOW

Gerade das Silicium bietet ein besonderes gutes Beispiel dafür, wie jung und unerfahren die Chemie wirklich ist.

E. G. ROCHOW [4]*

It was fortunate for the silicone industry that Eugene G. Rochow began the General Electric silicone project in 1938. His personal qualities and his scientific abilities, his career at Cornell, and his early experience in the Company all made it virtually certain that he would do successful applied research on silicones if such success was in the cards. His Company experience, to be described along with the silicone project, will be omitted here.

Rochow was born in Newark, New Jersey, on October 4, 1909. Gene

---

* The sense of the quotation is as follows: "It is precisely silicon that is an exceptionally good example to show how young and inexperienced [read, 'naive'] chemistry really is."

and his older brother Ted, who later became a distinguished chemical microscopist in the American Cyanamid Company, collaborated as youths in an attic laboratory they had built with their father's approval [5]. At Cornell University Gene was soon singled out by Professor L. M. Dennis (the "King") and was permitted as a "timid solitary undergraduate" [6] to attend the graduate seminar conducted by Dennis with the help of A. W. Laubengayer, then an Assistant Professor, who soon became a second strong influence on Rochow. The subject was inorganic chemistry; but, with Dennis's abiding interest in the then little-known elements germanium, gallium, and indium, all of which form organic derivatives, it was an inorganic chemistry that stressed organometallic chemistry to an extent uncommon at the time.

After completing his undergraduate course in 1931, Rochow went on to graduate work at Cornell. His association with Stock, which has already been mentioned, occurred during his first graduate year. In his second, he became a Heckscher Research Assistant. As such, he was required to carry on a thesis project that was independent of work he did under the Heckscher grant; accordingly, he did a thesis on fluorine chemistry and paralleled this with research in organometallic chemistry. Both activities proved useful to him in the Research Laboratory, to which he came from Cornell in 1935.

As Rochow's career in the Research Laboratory drew toward its close, he did the silicone project a unique service by summarizing all silicone work for the instruction of newcomers, by his extensive lecturing, and by writing. Under the date of December 19, 1947, his notebook (XXVII, the last but one) contains a copy of his reply to Professor Kipping's comments on Rochow's paper entitled "The Present State of Organosilicon Chemistry." This correspondence may have been the only direct contact between them. On January 29, 1948, he records: "I distilled the sticky solid from hydrolysis etc. related to or near $[(CF_3)_2C_6H_3]_6Si_2O$. . . . Sent distillate to J. T. Cohen [Patent Counsel]. Signed termination papers at Peek St.,* packed." On January 19, 1948, he had received confirmation of his appointment as Associate Professor of Chemistry at Harvard University. His last notebook entry, made on January 30, 1948, and signed, reads: "I cleaned out my desk, turned in notebooks, said farewells. Time with G. E.: 12 yrs, 6 months, 28 days. Part of me stays here."

In 1948 the erstwhile "timid . . . undergraduate" was an outstanding prospect for a distinguished academic career. Harvard must have known

---

* At this time Rochow was a member of the Atomic Power Division of the Research Laboratory. The Division was administered from Peek Street, across town from Building 37, in which Rochow shared Room 528 with up to eight others before his departure.

A later, improved version of the apparatus used by Rochow on May 10, 1940, to discover the direct process. Here three furnaces are in series, fed with CH_3Cl through the large black tube that appears to the right of Rochow's head in the plate. Reaction products are condensed by ice. Note the relatively large volume of products in the flask above the electrical outlets, and compare it with the May 10, 1940, results described later in the chapter.

this, for the University awarded Rochow an honorary AM degree in that year. His papers and patents are too numerous to list; his name is on at least eight books. In 1970 he elected retirement from teaching at Harvard University, of which he is now Professor of Chemistry Emeritus. Before retirement he lectured at the University of Innsbruck; thereafter, he taught at Karlsruhe, Braunschweig, Vienna, and most recently, in 1972, at the Virginia Polytechnic Institute. Until well into 1974, he acted as consultant to the Silicone Products Department, General Electric Company.

Professor Rochow has received the following honors: the Baekeland Medal, 1949 [5]; the Myer Award in Ceramic Chemistry, 1951; the Perkin Medal, 1962; the American Institute of Chemists Award, 1964; the Kipping Award of the American Chemical Society, 1965; an honorary DSc from Braunschweig, 1966 [4]; the Manufacturing Chemists Association Teaching Award, 1970; and the James Flack Norris Award for "outstanding teaching of chemistry," 1973.

At Harvard, Rochow was a worthy successor to Arthur B. Lamb. Rochow's research in inorganic chemistry ranks him with Dennis, Kipping, and Stock. He went beyond them in doing applied research that led to the founding of a major industry.

NAVIAS AND ROCHOW

From July 1935, when Rochow joined the Research Laboratory, until early in 1938, when he began work on silicone resins, his contacts were mainly with Louis Navias, who was in charge of ceramic research and development, which included work on glasses and glass-to-metal seals [7]. Professor Rochow has often acknowledged that this activity and this association were most helpful to him; he became well acquainted with the General Electric Company and with high-temperature insulation during these first years.

Navias had joined the Research Laboratory in 1924 as a specialist in porcelain, ceramics, and glasses as electrical insulating materials. His notebooks (and there were many over the years) show that it was still possible during his time for an exceptional man to conduct scientific liaison inside and outside the Company, to consult, to do specialized service work for others, and to carry on research as well. Today these activities are largely the tasks of organizationally distinct components. Many General Electric people, inside and outside the Research Laboratory, visited Navias on scientific business during his career, as did representatives of other companies, notably from the Corning Glass Works on the properties of glass, and from the Dow Chemical Company on magnesia, both important insulating materials.

One of Navias's principal concerns was Calrod heating elements [8, p. 55]. These devices began with Whitney and Chester N. Moore in the period 1910–1914. Today's Calrod units are familiar to all who use electric ranges. They consist of a spiral metallic conductor surrounded by magnesia (MgO) as insulation and enclosed in a grounded metallic sheath. This description is deceptively simple: the problems encountered in developing the units were many and difficult.

Rochow worked to improve the electrical properties of the MgO. This oxide had to be fused in an arc furnace, crushed, sieved, freed of magnetic materials, and incorporated into specimen units for electrical tests at 1000°C. Subsequently the units had to be sectioned, so that the sections and the MgO in them could be examined. Acting on his own initiative in the favorable environment of the Research Laboratory, Rochow also carried out pure research on the electrical conductivity of single crystals of quartz (SiO_2), periclase (MgO), and corundum (Al_2O_3) [9].

To a surprising degree, Rochow's practical and theoretical training at Cornell and his early experience in the Research Laboratory prepared him for silicone research. But there was another important factor that attracted him to silicones. In 1933–1934, long before the visit of Marshall and Patnode to the Corning Glass Works, Patnode had been studying reactions such as that between $Si(OC_2H_5)_4$ and $(NH_2)_2(CH_2)_2$, reactions that give resinous products. To be sure, such products are *not* silicones, for they contain no Si-C bonds; but Patnode's Cornell background must have convinced him that organosilicon compounds were promising insulating materials for high temperatures. Patnode wrote to Professor Dennis about this work, and Rochow was shown the letters.* Dennis himself had a lively interest in the differences between the chemistry of silicon and that of germanium, which is just below silicon in the periodic table. Rochow thus had ample incentive and exceptional preparation for entering upon the work of his early *Meisterjahre*.

ROCHOW EXPLORES RESINS AND DISCOVERS METHYL SILICONE

The important Marshall-Patnode visit to Corning was made on January 24, 1938. Marshall discussed prospective silicone research with Rochow

---

* On March 23, 1934, Rochow wrote to Patnode, saying, "Recently the King [Dennis] pointed out that it is never too early to start after that elusive job. . . ." Patnode helped run the job to earth in Schenectady.

on January 26. On January 31 Navias recorded that "Dr. Marshall came in to suggest that we work on plastics [resins] as impregnants for glass tape." At the time, Rochow was moving into a new laboratory (Room 617) and was preparing a talk on high-temperature insulation, which he gave, after several postponements, on March 8 at a Research Laboratory Colloquium. In describing the talk, he wrote, "I demonstrated semiconduction in Cu_2I_2 and it worked." No one knew then that silicon as semiconductor would eventually overshadow silicon as starting material for silicone polymers.

On February 28, 1938, Rochow began experimental work by mixing $Si(OC_2H_5)_4$ with shellac. Such experiments were extended; on March 3, Patnode helped with test-tube experiments in which $Si(OC_2H_5)_4$ was mixed separately with ethylene glycol, triethanol amine, aniline, and urea. Later $SiCl_4$ was added to the list. Nothing useful resulted.

By March 14, 1938, Rochow was doing library work that included looking up the Kipping preparation* of $(C_6H_5)_2SiCl_2$. A week later he made "the first really resinous material" by pouring a few drops of his $(C_6H_5)_2SiCl_2$ into water at 20°; he had for the first time taken the Kipping route to a silicone. By May 31, 1938, Patnode had tried Rochow's diphenyl silicone as an adhesive for mica. An "animated conference" on possible applications, insulation for electrical cable included, was held that morning with Marshall, Navias, Hubert Race (an electrical engineer), J. G. E. Wright, Patnode, and Rochow in attendance.

Such "animated conferences" were not an unmixed blessing for the research man whose work inspired them. The subject here was a promising new material Rochow had just made. The other conferees, all senior to him, no doubt discussed applications that might bring the silicone infant to maturity *provided* Rochow would make it in the needed amounts. What to do? Stop research and go into pilot-plant production of diphenyl silicones when a better material might be just around the corner? Which is the better alternative for him as a General Electric employee? As a scientist? Which is better for the Company? What if someone else makes the needed diphenyl silicone and controls its applications? An important development anywhere in the Research Laboratory quickly set up ripples felt throughout the Laboratory, and these ripples could easily become amplified into waves that spread into the rest of the Company. In a laboratory with almost no formal organization, an "animated conference" could thus complicate a research man's life by

* Kipping's preparation, which relied upon the Grignard reagent, was used by Rochow. See Chapter 2.

forcing him to find a course of action at once advantageous to him and acceptable to the Company.

Rochow continued his research and made rapid progress [10]. He discovered that successors to "the first really resinous material," when fully "cured" (heated to complete polymerization), were brittle and would burn. Thereupon he proceeded to replace some of the hydrogen in the C_6H_5 group with chlorine. As he expected, the chlorinated resin did not burn, but it was still brittle.*

Rochow then seems to have reasoned intuitively as follows. First, methyl silicones, as they have no C-C bonds, might withstand higher temperatures than other silicones. Second, methyl silicones, being of all silicones the nearest to silica in composition, might show the desirable qualities of the latter (e.g., resistance to oxidation) to the greatest extent. Methyl silicones were thus worth a try.

Rochow's reasoning, in which others concurred, might seem naive today when the complexities of chemical bonding are better known, and when extensive data are available about the behavior of silicones in electric fields at different temperatures. We know today that C-C bonds behave differently in different molecules, and that the deterioration of silicones usually involves other bonds as well [11]. Nevertheless, the Rochow reasoning provided a basis for action at a time when methyl silicones were virtually unknown. They are now the most important products of the silicone industry. It is fortunate that Rochow reasoned as he did.

Rochow began work on methyl silicones on August 10, 1938, again following the Kipping route. Navias' notebook has entries beginning August 17 with respect to dimethyl silicone as a *liquid* (an oil), and an entry dated October 5 mentions a *solid* dimethyl silicone that was a "dark yellow resinous material which on heating became hard and brittle."

In September 1938 Rochow had already applied methyl silicone in solution to a glass-asbestos tape; converted the silicone into a soft, sticky resin by heating the tape; and learned from Wright, who had tested Rochow's product, that the silicone was promising as an adhesive to make flexible tapes of mica. Wright thereupon had such a tape prepared and wrapped on a rectangular bar. This bar was heated, first for 84 hours under conditions no previous binder had been able to withstand, and finally (and more severely) for 265 hours at 230°C—after all that, it was

---

* Quick examination of new materials becomes second nature in work with man-made polymers. Here brittleness and flammability (both probably established in a few minutes) made the new material seem sufficiently unpromising that there was no need to study the other properties listed in Table 1-1.

The first solid dimethyl silicone in bottle with original label by Rochow. Delivered by Rochow to Navias on October 5, 1938. Kindly given to H. A. L. by Dr. Navias in 1976 for use here.

judged to be "still O.K."! This result, at hand by the end of October (still in 1938!), exceeded what anyone had expected. Navias described it in his notebook as "a phenomenal occurrence," and Navias never used words lightly. Methyl silicones promised a bright new day.

The meagerness of knowledge about polymers and about silicones in 1938 explains why a liquid (an oil) and a much different solid (a resin) were both described by Navias as "dimethyl silicone." To make progress, the composition of the silicones had to be established by chemical analysis. As will appear later, this was a difficult task, which Rochow successfully accomplished. He proved that solid methyl silicone polymers always had values below 2 for the CH_3/Si ratio.*

Rochow's early resin work resulted in five U.S. patents [12], all issued on October 7, 1941. They covered several different kinds of silicone polymers, the methyl, the ethyl, and the methylphenyl being of particular interest.

THE CONSTITUTION OF SILICONES. SILICONE SHORTHAND

The significance of the R/Si ratio, R always being an organic group, is best illustrated by examining the case of R = C_2H_5.

Rochow's ethyl patent (U.S. 2,258,220) relates his work to that of Martin and Kipping [13] and shows the significance of the R/Si ratio, here denoted by x. To make the chlorosilane precursors needed for the preparation of polymers in which $x = C_2H_5/Si$ is to be explored as a variable, the first reaction in the Kipping route ideally would be

$$SiCl_4 + x\ C_2H_5MgBr = (C_2H_5)_xSiCl_{4-x} + x\ MgBrCl \qquad (3\text{-}1)$$

followed by the hydrolysis-condensation sequence to give a polymer of composition

$$(C_2H_5)_xSiO_{(4-x)/2} \qquad (3\text{-}2)$$

As (3-1) always forms a mixture of silicone precursors, x in (3-2) is only an average. Different mixtures correspond to the same x. Also, (3-2) gives no information about the molecular weights of the silicones produced.

The product with $x = 2$, an oil that poured about like glycerol, had

---

* When Rochow did this important work, he was still devoting much of his time to Calrod units. He had an assistant, Mr. J. R. Womble, for about 6 months beginning in December 1938. I do not believe he ever had another.

been made by Martin and Kipping. Rochow prepared products with x ranging from 0.5 to 1.5; at still lower values of x, condensation was undesirably rapid even at room temperature, and the products were "hard, brittle, insoluble masses of little or no commercial value." For the more interesting values of x, Rochow's patent states:

By suitable adjustment of the proportions of starting reactants thereby to obtain the above stated ethyl-to-silicon ratio in the final product, I have surprisingly found that new and valuable resinous complexes can be obtained. These new resinous materials are, in advanced stages of polymerization, solid or semi-solid bodies or thick, syrupy, almost non-flowing liquids having valuable and characteristic properties that make them particularly useful in industry, for example as electrically insulating materials and in the plastics and coating arts. *They are quite different in their chemical constitution and in their properties from the oily diethyl silicone obtained by Martin and Kipping.*

The quotation implies a correlation of composition with "stages of polymerization" and with physical state: today we interpret the quotation to mean that the silicones increased in molecular weight, in degree of cross-linking, and in approach to solidity as the C_2H_5/Si ratio decreased below 2. We further surmise that the molecules were spontaneously terminated during the hydrolysis-condensation sequence with OH serving as the end group **E**. The analytical results on the resins were not, nor could they have been, precise enough to establish the nature of the end groups.

In 1978 we must guard against a tendency to underrate the significance of this early work. It satisfied the Patent Office, and it indicated that silicone molecules *might be made to order* not only for electrical insulation, but for other purposes as well.

Let us apply the boldface symbols in (2-23), which are further explained in Table 3-1, to Rochow's work.

The silicone shorthand appeared later in the silicone project as a means for circumventing the cumbersomeness of structural formulas. Its beauty lies in its compactness. In the shorthand a cyclamer with only one type of organic radical is exactly represented as $(\mathbf{D})_n$. A linear backbone of the same molecular weight and composition is $—(\mathbf{D})_n—$: *note the free terminal bonds*; the complete linear molecule is $\mathbf{E}—(\mathbf{D})_n—\mathbf{E}$. When the end groups **E** are not **M** units, the shorthand system needs supplementation if complete molecules are to be described.

So far, so good. But there is a more serious limitation common to all ways of representing cross-linked polymer molecules, which arises from their inherent complexity and their size. Let us consider this in two parts.

Table 3-1   Silicone Shorthand*

Symbol[a] of Unit	Meaning	ABC[b,c]	x = R/Si	Kind of Unit[c]
M	Mono	1	3	Terminating
D	Di	2	2	Backbone-building
T	Tri	3	1	Skeleton-building
Q[d]	Tetra	4	0	Skeleton-building

[a] These symbols represent, *not* chemical species, but hypothesized building blocks for the construction of silicone molecules. See (2-23).

[b] ABC is *available bonding capacity*, defined as the number of bonds formable in the hydrolysis-condensation sequence. ABC + x = 4, because of the tetrahedral habit.

[c] Correlation between the third and fifth columns has been explained in Chapter 2. Only **T** and **Q** units are cross-linking agents.

[d] **Q** from the Latin *quattuor* because **T** has been allocated to *tri*.

First, Table 3-1 shows that x = R/Si cannot uniquely establish the *units* of a silicone polymer. Most values of x are realizable as combinations of not more than four different *kinds* of units; the number of such combinations is limitless. Second, for a given combination of **M**, **D**, **T**, and **Q** *units*, there is wide latitude in the positions occupied by the units within the molecule. The complexity of polymer molecules complicates tailoring them to specific values of the properties in Table 1-1 even when knowledge about closely related polymers is available. Predicting such properties for new polymers is not possible today.

To make a point about cross-linking, let us anticipate a little. On April 15, 1940, Rochow recorded this suggestion by Gilliam: Make polymeric silicones by using pure CH_3SiCl_3 and varying amounts of $SiCl_4$ as precursors to see how far one can go "in tying in SiO_2 with the organic compound in the polymers." In the language of Table 3-1, the suggestion proposes the making of polymers in which cross-linking by **T** and **Q** units is systematically varied. As the proportion of **Q** units is increased, precipitation of silica in some form is to be expected. Such precipitation was well known; see, for example, the Rochow ethyl silicone patent quoted above. Silicone polymers with **Q** units are being made today.

In 1938 the state of the art was this: it was plausible, but not proved, that increased cross-linking due to **T** units was responsible for the gradual

---

* Once this section is concluded, the shorthand will be used *only for methyl silicones*; that is, R = CH_3 is then to be understood by the reader without specification.

transition of silicone polymers from oils to resins. Such proof could be obtained by making polymers in which D and T units were systematically varied. There was much work to be done!

From now on, we shall deal mainly with methyl (R = CH_3) and phenyl (R = C_6H_5) silicones. Please remember that the shorthand will henceforth be used *only for methyl silicones.*

GILLIAM ARRIVES. PURE METHYLCHLOROSILANES

On September 2, 1937, some 5 months before the Marshall-Patnode visit to the Corning Glass Works, Navias made the following entry in his notebook: "Dr. Coolidge informed me in the afternoon that he was seriously considering segregating my group from the Chemical [Chemistry] Section and suggested that the Ceramics Group could grow by the addition of one good man at this time." Dr. Coolidge was looking ahead. The short, but sharp, depression of 1937 discouraged recruiting; nevertheless, the "good man" in question, William F. Gilliam, reported to Navias for work with Rochow on August 1, 1939, after completing graduate studies at Cornell. With one short interruption, Gilliam, today in the Silicone Products Department, has been concerned with silicones ever since. These long years of scientific and industrial experience with silicones make his career unique in the Company.

By now, Rochow had prepared methyl silicone polymers via the Kipping route in amounts large enough for chemical analysis and for tests as components of actual insulation. Such polymers had previously been made by Stock [14] from $(CH_3)_2SiCl_2$ in amounts limited by Stock's experimental method, that is, in amounts too small for analysis or test. Martin [15] claimed* to have derived them from chlorosilanes that were by-products of the reaction between Si_2Cl_6 and CH_3I. Neither of these two methods is a practical way of preparing methyl silicones. Stock and Martin were interested in them as products of pure research; Rochow had applied research very much in mind.

Shortly after Gilliam arrived, Rochow suggested that they collaborate to make pure $(CH_3)_2SiCl_2$ and pure $(CH_3)SiCl_3$ in amounts large enough so that resins of known D and T unit content could be prepared. They soon moved into a new laboratory, Room 605, in which silicone history was to be made. In the isolation and analysis of CH_3SiCl_3 and $(CH_3)_2SiCl_2$, Gilliam encountered two major difficulties. Isolation had to be achieved via distillation, and distillation of the kind then generally

* So far as I know, Martin's claims have not been confirmed.

available was inadequate to the task. Also, there were no analytical methods that could give the percentages of carbon, hydrogen, silicon, and chlorine individually and directly on the necessarily small amounts of these volatile, easily hydrolyzable, and, in many ways, objectionable materials. All four percentages were needed to establish whether they summed to 100 as they should, and whether each percentage had the expected value.

Both difficulties were overcome [16]. The distillation difficulty seemed ominous at the time. In light of earlier experience, especially Kipping's, with separating the components of chlorosilane mixtures, the difficulty was not unexpected, but its acuteness was a shock. A 50-plate column proved inadequate. An 80-plate column, specially built by Gilliam and Sprung, eventually did the job, but only when the distillation was performed very slowly [17].

The numbers quoted are "theoretical plates," which measure the efficiency of distillation columns: the higher the number, the more effectively the column can (in the simplest case) separate liquids with neighboring boiling points. Separation is better when distillation is slow. The powerful stills available today in the petroleum and chemical industries could easily have accomplished the needed separations, but there were no such stills in 1940.

The distillation difficulty had several causes. A constant-boiling mixture (azeotrope) with an awkward boiling point was found. In general, important boiling points were near each other. There existed a "boiling-point anomaly," that is, certain boiling points predicted to lie within the range fixed by those of $Si(CH_3)_4$ and $SiCl_4$ actually lay outside it. The unraveling of complexities such as these was a major task in the early days, when samples were small and compositions difficult to establish.

So long as chlorosilanes must be isolated by distillation, so long is it certain that the industrial production of methyl silicones will require sophisticated and costly distillation equipment. It is fortunate that Gilliam and Sprung found they could successfully replace platinum packing with stainless steel in such stills. The distillation experience was a forewarning that making silicones industrially would be different and more costly than producing the kind of polymers to which the Company was accustomed.

The collaboration between Rochow and Gilliam placed methyl silicone polymers on a firm experimental foundation. The results, given in two short papers published in 1941 [16, 18], were in accord with expectations. The paper by Rochow and Gilliam [18] contains an acknowledgment, concurred in by Dr. Coolidge as Director, of the General Electric indebtedness to the Corning Glass Works. As methyl silicones are still the industry's most important products, the two papers are required reading for anyone with a serious historical, scientific, or industrial interest in silicones.

As has happened often enough to be taken for granted in analytical chemistry, much of the work just described could be done almost without effort today. Take the matter of CH_3SiCl_3 [19], of which the boiling point is now precisely known: it suffices to take material of this boiling point out of a large industrial still, to hydrolyze it, and to establish its chlorine content by titrating the HCl formed by hydrolysis—a simple process that gives no hint of the earlier ordeal to achieve the same result!

DISCOVERY OF THE DIRECT PROCESS FOR METHYLCHLOROSILANES

The saga of Room 605 continues with the most important discovery made in the course of the silicone project. It soon became clear to Rochow, to Patnode [20], and to Marshall that the Kipping route was unsuited to the industrial production of silicones. Among its shortcomings were high cost, need for large volumes of volatile and flammable (hence hazardous) solvents, poor yields of desirable products, and—yet again—high cost. If the Chemistry Section were to succeed in its dream of founding a large-scale silicone business, the Grignard reagent would have to be replaced in the making of silicone precursors.

Rochow began to look for a better route to silicones long before his work on silicone resins was completed. He tried many things [21] that cannot be mentioned here. On May 10, 1940, he found his answer when he succeeded in bringing CH_3Cl to react with Cu-Si.

As early as January 24, 1939, Rochow had tried to make CH_3Cl react directly with silicon, and with Fe-Si. On October 20, 1939, he mentioned that (gaseous) HCl reacts with "copper silicide" (Cu-Si). Clearly, iron and copper were among the elements that he thought might help form the Si-C bond needed to make methylchlorosilanes from methyl chloride and silicon.

On March 29, 1940, Rochow returned from a memorable 18-day Bermuda vacation during which he thought about his work. On April 2, 1940, he listed new research plans. These included the making of $SiHCl_3$ to be used as an *intermediate compound* for the preparation of chlorosilanes; this preparation would not be a *direct* process. He expected to make $SiHCl_3$ by reacting HCl with Cu-Si according to Booth and Stillwell [22]*. He saw, from their article, that these authors preferred using

* There is more about this important reaction in Chapter 4.

silicon alone. On April 3 he ordered both silicon and Cu-Si for the preparation of $SiHCl_3$, and obtained from the Research Laboratory some ferrosilicon (Fe-Si) for immediate trial, his comment being "this has been used . . . too" [for the preparation of $SiHCl_3$]. On April 5 the trial yielded $SiHCl_3$, $FeCl_3$, and other products. On April 17 Rochow simplified his apparatus to virtually the form shown in Figure 3-1. On that day he

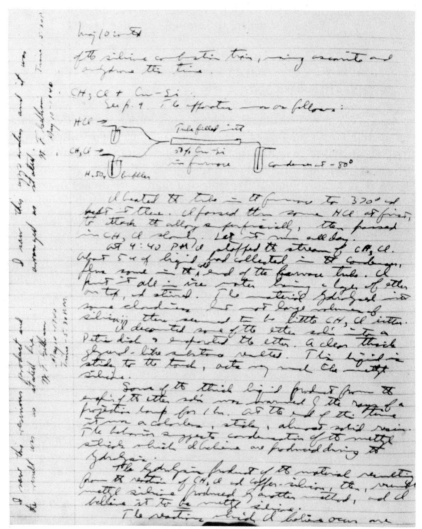

Figure 3-1 Rochow's sketch of apparatus for his experiment of May 10, 1940, in which the direct process was discovered.

received 10 lb of silicon ("Electromet") and about 5 lb of Cu-Si containing 50% Si by weight (Niagara Falls Smelting and Refining Corporation). During this period he was unsuccessfully trying to make methylchlorosilanes according to his plans of April 2 by use of the SiHCl$_3$ he had prepared. On April 30, 1940, he passed CH$_3$Cl and HCl over Fe-Si. A film formed on the walls of a glass flask that had contained the reaction products dissolved in ether. To Rochow this film was visible evidence that a trace of methylchlorosilanes must have been formed by the action, in the presence of HCl, of CH$_3$Cl on ferrosilicon. This was barely a hint, but one from which the direct processes of the silicone industry were to follow. I am reminded here of the discovery of X-rays by Roentgen (Nobel Prize, 1901).

What Rochow did next is best described by quoting from his notebook.

May 9, 1940.

Copper-silicon.

I crushed some of the Niagara Falls Smelting Co. 50% Cu-Si in the jaw-crusher, and packed a Nonex tube with the material (size is about ¼" down to fine powder). I arranged the tube in the furnace and arranged to admit both CH$_3$Cl & HCl. Single CO$_2$ condensing tube on the outlet end.

May 10, cont'd.

I heated the tube in the furnace to 370° and kept it there. I passed thru some HCl at first, to attack the alloy superficially, then passed in CH$_3$Cl slowly. Let it run all day.

At 4:40 P.M. I stopped the stream of CH$_3$Cl. About 5 cc of liquid had collected in the condenser, plus some in the cold end of the furnace tube. I put it all in ice water having a layer of ether on top, and stirred. The material hydrolyzed with some cloudiness, but not large volumes of silica; there seemed to be little CH$_3$Cl either.

I decanted some of the ether solution into a Petri dish and evaporated the ether. A clear thick glycerol-like substance resulted. This liquid is sticky to the touch, acts very much like the methyl silicone.

Some of the thick liquid product from the evaporation of the ether solution was warmed by the rays of the projection lamp for 1 hour. At the end of this time, it was a colorless, sticky, almost solid resin. The behavior suggests condensation of the methyl silicols which I believe are produced during the hydrolysis.

The hydrolysis product of the material resulting from the reaction of CH$_3$Cl and copper-silicon, then, resembles methyl silicone produced by another method, and I believe it to be methyl silicone.

The reactions which I believe occur are as follows. First HCl is passed through the tube:

$$Si + 3\ HCl \rightarrow SiHCl_1 + H_2$$

Only a small amount of HCl is passed through, and this is done principally to etch the surface of the alloy. Small amounts are later mixed with the CH_3Cl, in the ratio of perhaps 1 part to 50 parts of CH_3Cl. The CH_3Cl reacts in this way:

$$3\ CH_3Cl + Si \rightarrow CH_3SiCl_3 + C_2H_6$$

$$2\ CH_3Cl + Si \rightarrow (CH_3)_2SiCl_2$$

and, to a much smaller extent, this might occur:

$$CH_3Cl + 2\ HCl + Si - > CH_3SiCl_3 + H_2$$

The liquid products, which I believe are methyl silicon chlorides, condense in the cooler portions of the tube containing the alloy and are also distilled out into the condensing tube kept at $-80°$. The colorless liquid so collected (in the condensing tube) does not bubble much when warmed to room temperature, hence does not contain much CH_3Cl.

Upon hydrolysis of the combined liquid products,

$$CH_3SiCl_3 + 3\ H_2O \rightarrow CH_3Si(OH)_3 + 3\ HCl$$

$$(CH_3)_2SiCl_2 + 2\ H_2O \rightarrow (CH_3)_2Si(OH)_2 + 2\ HCl$$

and a small amount of

$$SiHCl_3 + 3\ H_2O \rightarrow HSi(OH)_3 + 3\ HCl$$

The methyl silicols undergo partial condensation immediately to form the viscous intermediate products:

$$2\ (CH_3)_2Si(OH)_2 = \underset{\overset{\displaystyle |}{CH_3}}{\overset{\overset{\displaystyle CH_3}{\displaystyle |}}{HO-Si}}-O-\underset{\overset{\displaystyle |}{CH_3}}{\overset{\overset{\displaystyle CH_3}{\displaystyle |}}{Si}}-OH + H_2O\ \text{etc.}$$

This goes on until sticky liquid products result.

On warming, condensation proceeds further, splitting off more water (which evaporates in part or stays behind in globules). The end result is a clear resinous body which I believe to be the methyl silicone.

/S/ E. G. Rochow
May 10, 1940

The experimental setup and the results were witnessed by W. F. Gilliam at about 5:30 P.M. The most important single experiment and the best single day's work in the history of the silicone industry had been done.

This notebook record is a clear, concise description of experimental results acutely observed. Note that Rochow by now no longer considered $SiHCl_3$ as being necessary to form methylchlorosilanes, and that he assigned to HCl the role of "attacking the Cu-Si"; he now had a *direct* process in mind. Note further that the Cu-Si, in which copper remains as end product, was a much better starting material than Fe-Si, in which the iron is changed to $FeCl_3$. Finally, note that a sili*col* (analogous to alco*hol*) is the old name for an organosilicon compound containing at least one Si-OH bond, and that chlorosilanes were called silicon chlorides.

COMMENTS IN CONCLUSION

• As to management and organization. Rochow's accomplishments during the first 2 years of the silicone project make a historically valuable record of industrial research done when it was still possible for informality to be the rule, to have a minimum of management, and to be unconcerned about organization. Silicone chemistry was a field new to industry, and Adamsian multiplicity and complexity had not yet become an unavoidable burden for industrial research to carry.

Early in 1938 Marshall informally asked Rochow to devote about half his time to research on silicones. Marshall did not tell him what work to drop; it is clear from the record that Rochow continued to be responsible at least for research related to Calrod units. What kind of new work Rochow did, how much he undertook, and when he began it, were matters for him to decide. The risks entailed were his to bear. To begin the silicone project, no "paper work" was needed; no new financial arrangements were necessary; and the "front office" did not have to be informed.

At this time the Company was managed largely from its headquarters in New York City; Company-wide decentralization did not come until after World War II. Company control of the Research Laboratory was mainly financial. The great early success of the Laboratory, which rested mainly upon Coolidge's ductile tungsten and the founding of the electronic era [23], had ensured continuance of the policy established by Whitney when the Research Laboratory was founded, namely, that within the budget the nature and scope of research projects were the responsibility of the director, who had commensurate authority and accountability. To this extent decentralization thus came to the Research Laboratory long before it was announced as a policy for the Company.

So long as a research man was reasonably discreet, financial control touched him lightly in the normal course of his work. Usually he did no more than allocate his time once a month to research accounts ("shop orders") identified by numbers, which he also used to obtain services and equipment. Many scientists used only one such number for all their projects. Formal reports of research at fixed intervals were not required.

Organization then mattered little. When Navias became a Section head at some time* after September 1937, he was no longer responsible to Marshall. I believe that Rochow and Gilliam remained responsible to Navias until the chemical engineers took over† Room 605 in 1942. Rochow and Gilliam then moved into Room 528 on the floor occupied by the Chemistry Section, of which they became de facto (and probably de jure) members. I do not believe that organizational changes had any significant effect on the progress of the silicone project.

The Whitney policy of open doors and free discussion had been continued by Coolidge. This policy and the virtual disregard of organization, two important factors in creating the "Whitney atmosphere," did affect the project. Marshall and Patnode, strong men both, were determined that the Company should found new chemical businesses. In the "Whitney atmosphere" they could move unhindered to do all they considered necessary for founding the first such business. In 2 years the silicone project, largely as a result of Rochow's work, had become their first good opportunity. The following chapters will show that this opportunity was promptly exploited on a broad scale unlikely before World War II for any research laboratory outside the chemical industry.

• As to Rochow's accomplishments. A strong research man could flourish in the "Whitney atmosphere." If he wished to do so, he could, despite stresses and strains, work and develop to the limit of his abilities. There was little need to worry about what "the boss" thought. The

---

* I was unable to establish when Navias was named a Section head. Organization announcements were unknown at the time.

† The chemical engineers needed space on the top floor because "penthouses" had to be built on the roof to provide the headroom their equipment required.

conditions, in my opinion, were better suited to individual research than to complex projects.

We have seen that Rochow did flourish. He began silicone work on February 28, 1938, and *in part of his working time* accomplished what we shall recapitulate below. For about 6 months he had an assistant untrained in chemistry. After August 1, 1939, he also collaborated with Gilliam in the *augmentation** of his discovery of methyl silicone resins, a topic excluded from the ensuing discussion.

Between August 1, 1939, and April 27, 1940, Rochow filed five patent applications on various silicone resins intended primarily to serve as insulating materials. Each patent has a cover sheet on which two figures show an electrical conductor surrounded by silicone-containing insulation thicker than bona-fide wire enamel, but wire enamel was certainly not excluded. All five patents were granted, and all five survived interference proceedings in the United States. Not a bad record!

On May 10, 1940, Rochow discovered the direct process for making methylchlorosilanes, the most important process in the world's silicone industry today.

Rochow's accomplishments during these 2 years are outstanding as examples of successful applied industrial research by an individual. Although such successes defy exact analysis, I regard the following as the most important contributory factors:

1. A field in which extensive pure research, but little applied research, had yet been done, and in which materials had been discovered that gave promise of filling important existing needs.
2. An investigator exceptional in ability, training, and experience.
3. An environment in which such an investigator could make rapid progress.

Professor Rochow's biography shows him to be an outstanding scientist. There is no need to list here the qualities such a person must have, but, for the information of the general reader, I shall mention three that are especially useful in applied research. His work on silicones shows that Professor Rochow has all three to an exceptional degree.

The first is the ability to work on a number of things at the same time. Only Rochow's notebooks can tell the whole story, but this chapter and the next adequately make this point.

The second quality is the ability to draw correct conclusions from elusive evidence. In his experiment with Fe-Si and CH_3Cl on April 30,

---

* Augmentation of discoveries will be explained and discussed in Chapter 5.

1940, Rochow saw a film on the walls of a glass vessel that had contained the reaction products dissolved in ether. These reaction products were formed in amounts so small as to preclude identification. The film was no more than a silicone telltale. Yet Rochow concluded that methyl-chlorosilanes must have been formed as silicone precursors, and this conclusion was a crucial step on the way to the direct process. On May 10, 1940, he says that his reaction products "hydrolyzed with some cloudiness, but not large volumes of silica [which means that $SiCl_4$ could not have been an important reaction product]; there seemed to be little CH_3Cl either [which suggests that CH_3Cl had been transformed readily into methylchlorosilanes]." Many investigators would have failed even to make the observations—much less draw the conclusions.

The third quality is the ability to see early that a line of work must be modified or abandoned. Consider Rochow's exploration of resins. He began, with Patnode, doing experiments based on Patnode's earlier ex-perience.* He shifted quickly to diphenyl silicones, which he found to be flammable. He next replaced hydrogen in the C_6H_5 group with chlorine to reduce flammability; the resins were still brittle. He tried ethyl sili-cones; they seemed unsatisfactory. He moved to methyl silicones as a consequence of (not unexceptionable) reasoning described earlier. In the hope of decreasing the brittleness of methyl silicones without eliminating their advantages, he incorporated C_6H_5 groups into them. The point is that years might have been consumed in studying any one class of resins thoroughly enough to establish the applicable properties listed in Table 1-1, and in making tests of possible applications. One must know how and when to jump!

The importance of the early shift to methyl silicone *resins* is obvious from the fact that methyl silicones *in one form or another* continue to be the principal products of the industry. A less obvious point must also be considered. Methylchlorosilanes are, among the large number of possible silicone precursors, the easiest to make by the direct process. Had Ro-chow been using other organic compounds in place of CH_3Cl, his dis-covery of the direct process would have been delayed considerably be-yond May 10, 1940.

• As to academic and industrial research. Chapter 2 contrasted von Baeyer's "Ach, Harz" with Baekeland's founding of the "phenolics" industry to manufacture products related to resins that von Baeyer had

---

* Note that this beginning diverged from Marshall's original suggestion (not an order!) of how Rochow was to start research on silicones. The trusted research man was free to chart his course.

prepared. Chapter 3 revealed that Sullivan perceived that substances rejected by Kipping as "sticky masses" might be put to use. Baekeland and Sullivan both realized that ugly ducklings become beautiful swans if someone finds important needs they can successfully fill.

• As to Sullivan and Corning Glass Works. The silicone industry was particularly fortunate in that it was Sullivan who took the first step toward its founding, for Sullivan was Director of Research for a company that subsisted on glass and consequently knew SiO_2 well. Intimacy with the parent nourished hope for a silicone offspring!

Chapter Four Applied Research Expands. Major Discoveries Continue

For he who naught dares undertake By right he shall no profit take [make].

JOHN GOWER, *Confessio Amantis*, Book iv

With hindsight to help, we are justified in saying that the two major Rochow discoveries described in Chapter 3 did indeed mark the "end of the beginning" of the silicone project. It was far from certain then, however, that the project would ultimately lead to a successful silicone business. Clearly, if this matter were to be settled within a reasonable time, the project would have to be expanded, and the success of the expansion could not be guaranteed. As it turned out, the decision to expand led to important discoveries that fortunately ensured the eventual success of the project, which could have languished had useful silicones other than resins not been found. As the reader of this chapter travels the high road planned for the expansion, he or she will encounter forks marking attractive side roads that made the research exciting, successful, and extraordinarily gratifying.

Note. Brackets are used to enclose references and inserted material. For bracketed references, see References and Notes at the end of the book. Numbers in parentheses on the right of pages are often used in the text to identify equations and the like: thus (4-1) means Reaction 4-1. Reference to laboratory notebooks is by name and date. CRDA stands for "Chemistry Research Department Archives."

ROCHOW AT WORK

I think that May 10, 1940,* was the most important day in Rochow's career at the Research Laboratory. The reader might expect that he and Gilliam would thenceforth have concentrated on direct processes for making various types of organochlorosilanes. A quick look at Rochow in his laboratory will show why this expectation is unrealistic. Rochow's responsibility for research pertinent to Calrod units could not be abrogated. Here is a digest of his notebook entries for May 10, 1940.

1. An idea by Gilliam for a new way of forming Si-C bonds.

2. A conference with patent counsel to discuss methyl silicones as a binder for the powdered phosphors that are the source of light in fluorescent lamps.

3. An item important enough to warrant direct quotation: "Dr. Coolidge desires that we [Rochow and Gilliam] prepare some UF_6 here in the laboratory for experiments on isotopic separation [1]. . . . I began by collecting literature references, etc. The fluorine cell originally borrowed from Cornell University for work on SF_6 is still here."†

4. $U + HgNa_2$. An experiment in progress. Related to Item 3.

5. $Mg + Si(OEt)_4 + CH_3Cl$. An experiment in progress. Objective: to make a silicone precursor from $Si(OC_2H_5)_4$, or $Si(OEt)_4$ in Rochow's shorthand.

6. Readying of combustion train for silicone analysis (see Chapter 5).

7. The all-day direct-process experiment described in Chapter 3.

Item 7 of course made May 10, 1940, Rochow's "big day," but the other items show that an able chemist could have a rewarding career in

---

* On its first birthday the direct process was enthusiastically toasted by some of us with a soft drink kindly provided by Rochow in his laboratory.

† The quick preparation of UF_6 by Rochow and Gilliam helped make possible the rapid separation of uranium isotopes by Kenneth H. Kingdon and Herbert C. Pollock [2], which was then a matter of national concern. Rochow's interest in SF_6 as a gaseous insulating material had led him to borrow the fluorine cell. He had worked with fluorine at Cornell, and Gilliam had acquired needed vacuum-chain experience there.

In October 1940 Dr. Coolidge received from Dr. Lyman J. Briggs, Director, National Bureau of Standards, and head of the newly formed National Uranium Committee, an urgent request for 150 g of UF_6, then a huge amount. On October 21 Kingdon's notebook records cost data for the first batch shipped; "Rochow, 8 days" was one item. A later wry comment [2a] on accounting policy reads, "Apparently we had not heard of overhead yet," to which one might add "or of the desirability of liquidating applicable research costs previously incurred," among which Gilliam's time was an important item.

the Research Laboratory even if he did not make discoveries important enough to influence a future industry. The atmosphere of *Sturm und Drang* to which Rochow's entries testify permeated much of the Chemistry Section. With so much going on, it was not going to be easy to expand the silicone project.

The assignment from Dr. Coolidge had to take precedence over all other activities by Rochow and Gilliam for the next few weeks. Nevertheless, Rochow managed during that time to confirm and extend his discovery of the direct process. On Monday, May 13, 1940, he repeated the experiment of Friday, May 10, with satisfactory results. On May 16 he made the first run on the direct process in improved equipment. Such experiments continued while the UF_6 work was going on. On June 19 he began to add C_6H_5Cl to the CH_3Cl in the hope of generating concomitantly the precursors to methylphenyl silicone resins.

VIEW FROM THE TOP

If a man look sharply and attentively, he shall see Fortune; for though she is blind, she is not invisible.

FRANCIS BACON, *Essays: Of Fortune*

On the third working day after the direct process was discovered, Rochow wrote in his notebook: "I explained the process and results to Dr. Marshall who urged immediate work on the questions that naturally arose. I wrote a letter of disclosure to Dr. Coolidge [Director], copy to Mr. Kauffman [Patent Counsel]." Marshall now had to make a management decision. What were some of the factors he must have considered?

First, the road to commercial silicones would be long and hard. Making silicones would not be like making Bakelite or Glyptal®. Personnel with advanced and diversified training (particularly in chemistry and chemical engineering) would be needed, as would costly and sophisticated manufacturing equipment new to the Company. One thing was certain: there would be recurring reminders that "Money makes the mare go," and these reminders would become sharper as the project grew.

Second, methyl silicone resins were promising, though unproved, as components of insulation that would permit operation of electrical equipment at higher temperatures. They might even make good wire enamels. To what extent they could meet most of the requirements in Table 1-1 was unknown, since no property profile had been established for them. Only a few silicone resins had been prepared. Methyl silicones seemed the best among them, but others yet unknown might well be better. All silicones promised to be costly, which meant that they would need to offer pronounced advantages over competing products if there were ever to be large silicone markets. Fortunately, favorable experience, such as that with Formex® wire enamel, had made the Company willing to take the risks associated with new materials that gave promise of improving electrical equipment.

Third, the direct process offered the hope of an improvement over the Kipping route to silicones, but the hope had so far only a slim foundation. For a successful silicone industry the process would have to prove itself *economic,* and capable of *scale-up* from drops per hour to barrels in the same time.

Fourth, "quitting while you're ahead," inconceivable for Marshall under most circumstances, would have been intolerable research policy for an effort as promising as the silicone project seemed at this time— opportunity seldom knocks so loudly. New silicones, new ways of making them, and new uses for them might all lie ahead. Such possibilities were attractive to a large and diversified company willing to enter a variety of markets.

Marshall had open to him alternatives that ranged from "steady as you go" to "full steam ahead" within the bounds possible in the Research Laboratory. The second alternative would need "front-office" approval, for it would be a much larger step than starting the silicone project had been. Marshall could not, of course, commit the Company to the man- ufacture of silicones. He could not even commit his entire Section to the silicone project, as the Section had the continuing responsibility for corporate research in all kinds of chemistry. At the time federal funding of research, even if available, would probably not have been sought.

The history of the silicone project shortly after May 10, 1940, shows that Marshall's decision had the following elements. Rochow was to continue working as he had been doing. Patnode and William J. Scheiber, with whom we shall become better acquainted in Chapter 6, were to scale up immediately the Rochow route to methyl silicones, which included the direct process for making methylchlorosilanes and the distillations by which the reaction products had to be separated. Gilliam and Murray M. Sprung, an organic chemist then concerned with polymers related to phenol, were to do applied research on the direct process while it was being scaled up by Patnode and Scheiber.

It was indicated above that Marshall's decision placed the silicone project on a *planned high road* to a silicone business. Many added investigators were to become later travelers on that road. As the project grew, so did the complexity of the interactions; it soon became difficult

This and the following two plates are incomplete snapshots that illustrate the journey from silicone research to silicone business. Much of the early research was done by those above. They are, from left: Scheiber, Rochow, Sauer, Marshall, Gilliam, Patnode, Sprung.

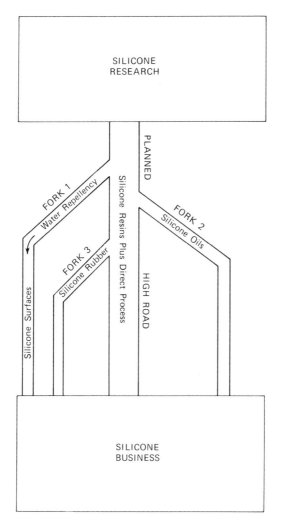

Visualization of the silicone journey: Marshall's planned high road has three unexpected forks. Chapter 7 tells where the journey led.

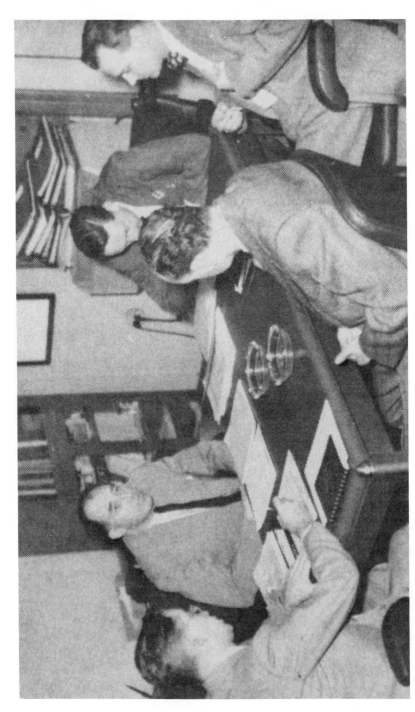

The silicone business comes of age with an organization of its own. A management meeting in Waterford's early days. Clockwise from left: Williams (finance), Reed (General Manager), Donnalley (manufacturing), Howlett (marketing), Kern (engineering).

to assign credit for even major contributions. Surprises along the high road quickly made it necessary to change plans and direction. The most important of these surprises was the encountering of unexpected *forks* that opened attractive side roads—encounters that did more to maintain the project in the face of competing problems brought on by the war than did progress along the planned high road itself.

BASIS FOR SCALE-UP. SiHCl$_3$

One object of this chapter is to show the reader how the important silicone discoveries were made. Accordingly, we shall look further into the basis of the most important discovery of them all—that of the direct process* for methylchlorosilanes. This basis, as recorded in the chemical literature, was significant for both discovery and scale-up.

In setting down his research plans on April 2, 1940, Rochow recorded

$$\text{``CH}_3\text{Cl} + \text{SiHCl}_3 \rightarrow \text{CH}_3\text{SiCl}_3 + \text{HCl?''} \qquad (4\text{-}1)$$

as a reaction he wished to try. Reading Booth and Stillwell [3] revealed to him that the SiHCl$_3$ he needed could be made by carrying out the reaction

$$3\ \text{HCl(gas)} + \text{Si} = \text{SiHCl}_3 + \text{H}_2 \qquad (4\text{-}2)$$

with and without added copper in accordance with the earlier literature they cited. Substitute 2CH$_3$Cl for 3HCl, and you can write a tidy but unrealistic equation for a direct way of making (CH$_3$)$_2$SiCl$_2$, the most important among silicone precursors.

In 1857 Buff and Wöhler [4] had passed HCl(gas) over silicon held below visible red heat and obtained a mixture of products from which they distilled impure SiHCl$_3$. They studied its chemical behavior but did not find its correct formula. They observed that raising the reaction temperature favors the formation of SiCl$_4$. Friedel and Ladenburg [5] repeated their work, isolated pure SiHCl$_3$, and established its formula.

The early work led to an investigation of utmost significance for the direct process. In 1891–1892, under the influence of the periodic table, Combes [6] set out to discover whether the silicon analogues of (C$_6$H$_5$)$_3$COH, a compound of interest to the dye industry, were colored also. They are not. But Combes planned to, and did, make these ana-

---

* By "direct process" we mean a reaction in which organochlorosilanes are made by the action of an organic halide on silicon in the presence of an added metal as catalyst. The reader will not be surprised to learn that such reactions sometimes go by other names, or that the name is sometimes used for other reactions.

logues by using $SiHCl_3$, which he wanted in large amounts because he had visions of a silicon-dye industry. He therefore looked for a commercially available source of silicon other than the crystalline element, which he wished to avoid because it was costly. His choice fell on a "copper silicide" containing 20 parts by weight of copper to 100 parts of silicon.* Whoever made this "copper silicide" commercially available was one of the silicone industry's unsung heroes!

Rochow, U.S. Patent 2,380,995, p. 1, refers to Combes' work as follows:

The reaction of hydrogen chloride with silicon also was known. Thus, Combes [*Acad. Sci.*, **122**, 531 (1896)] obtained a mixture of approximately 80% trichlorosilane (silicochloroform) and 20% silicon tetrachloride by passing hydrogen chloride through an iron tube filled with silicon heated to 300° to 440°C.

When Combes led gaseous HCl over his "siliciure de cuivre," he got complete conversion of silicon to chloro derivatives in all experiments, with metallic copper remaining. His best yield, cited above, came when the reaction temperature was that of "diphenylamine vapor"; this compound boils at 302°C. When the temperature was higher, as with the vapor of mercury (boiling point, 357°C) or with that of sulfur (boiling point, 445°C), the yields of $SiHCl_3$ were less, and those of $SiCl_4$ greater.

Combes also deserves well of the silicone industry. He used copper (which is today irreplaceable in methylchlorosilane manufacture) in (4-2) not as a catalyst, but simply because "copper silicide" was an available source of cheap silicon. His entry into the organosilicon field invites comparison with Kipping's [Chapter 2]. Both men were deluded by the misconception that the chemistries of carbon and of silicon were largely analogous, and both wanted to make silicon compounds they consequently expected would have distinctive optical properties. But Combes had applied research in mind. He was after an easy, viable industrial process for making $SiHCl_3$. He intended to use it as an *intermediate* material, as did Rochow for a different purpose; see (4-1). Finally, Combes' work provided a clear warning, foreshadowed in 1857 [4] and confirmed in 1934 [3], that high reaction temperatures would favor the (wasteful) formation of $SiCl_4$, an almost useless reaction product.

* The importance of Combes' work warrants these direct quotations [6b]:

". . . J'ai cherché à rendre cette préparation facile et pour ainsi dire industrielle en augmentant le rendement et en évitant l'emploi du silicium cristallisé"; and . . . "j'ai employé dès l'origine le siliciure de cuivre à 20 pour 100 de silicium préparé au moyen du four électrique de M. Héroult, et que la Société électrométallurgique française a livré au commerce dès 1889."

WINTON I. PATNODE (1904–1977)

Winton I. Patnode was an influential member of the Chemistry Section almost from its beginning. Born June 6, 1904, in Pittsfield, Massachusetts, he experimented as a boy with a chemistry set in his bedroom workshop. Between undergraduate and graduate careers at Cornell University, he worked for a year as chemist in the Eastman Kodak Company. In 1928 he returned to Cornell as Heckscher Fellow, an appointment he held until 1931, when he completed his thesis on gallium triethyl [Ga(C$_2$H$_5$)$_3$] under Professor L. M. Dennis, "the King."

Patnode was the first to bring to the Chemistry Section the Cornell background in chemistry, which, even at the undergraduate level, was intended to develop persons capable of sound work on any chemical problem because they were anxious to continue learning, knew how to use a library, and were at home in the laboratory.*

When Mackay left the Research Laboratory [Chapter 1], Kienle, whose work on Glyptal® was mentioned in Chapter 1, soon followed him. Meanwhile, Patnode had joined the Pittsfield Works Laboratory and participated there in designing, developing, and building the first modern plant for making alkyd resins, a plant in which the time-honored ways of cooking varnishes were judiciously modernized. In 1933 Marshall recruited Patnode to replace Kienle. The results justified Marshall in devoting, as he continued to do, much time, travel, and effort to building the staff of the Chemistry Section.

Patnode and Edward J. Flynn, an engineer,† soon scored a brilliant

---

* Patnode's interests were always broader than even Cornell training implies. They included botany, specialized photography, and—with the help of his wife—digging on weekends for whatever of interest the desert wilderness of the Pacific Northwest could yield. At the time of his death he was preparing a book unrelated to science. At Cornell he was business manager of the Cornell Dramatic Club, of which Franchot Tone (later to become a well-known actor) was a member, a fact worth recording because Tone's father had helped to make commercial silicon available [Introduction].

† In the main, Patnode concentrated on the chemical, and Flynn on the mechanical, aspects of the problem, and they were jointly granted U.S. Patent 2,085,995. Flynn, who was then in the General Engineering Laboratory (Building 5), will reappear in Chapter 5 in a silicone role as a member of the Schenectady Works Laboratory. Such close and informal cooperation by the different laboratories, always helpful, was particularly important to the silicone project.

Generally such cooperation continues long after the product has been transferred to an operating component, and product improvement and lower cost are the usual results. In a case in point, Edward H. Jackson and Ralph W. Hall of the Fort Wayne [Indiana] Works were granted U.S. Patent 2,307,588 for improving Formex® resin and dissolving it in cheaper solvents; they accomplished both objectives. This is a preview of what was to happen with silicones.

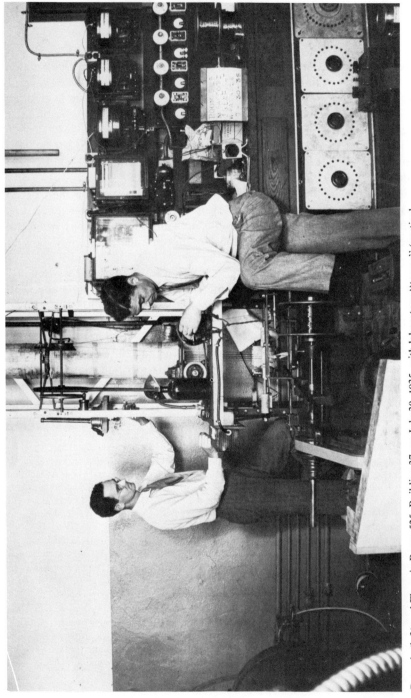

Patnode (left) and Flynn in Room 525, Building 37, on July 30, 1935, with laboratory "tower" (vertical furnace) for application and curing of Formex® wire enamel. Please compare with plate on p. 27.

success, already mentioned, with Formex® wire enamel. Patnode did much to convince Marshall that research on man-made polymers ought to be the main activity of the Chemistry Section, and there was continuing close contact between the two men on all matters relating to these materials, silicones naturally included.

Patnode's contribution to the silicone project, soon to be described, gave evidence of the three qualities of applied research scientists listed at the close of the preceding chapter. His main interest during the project was the making of silicones that could fill useful needs, needs sometimes not obvious to minds less perceptive than his. His successes were ample reward. He seemed not much concerned about publishing his results or about receiving external recognition.

Shortly after scientific and technological liaison* came to be recognized as a separate function in the Research Laboratory in 1945, Patnode became the first Liaison Representative, Chemistry. In 1947 he took charge of the Hanford, Washington, component of the Research Laboratory's Atomic Power Division, which was created after the transfer of the Hanford Works from the Du Pont to the General Electric Company. Subsequently, Patnode became Director of Research of the Weyerhaeuser Company, which he left to embark for himself on various projects, consulting included. After retirement he continued to live in the Pacific Northwest, for which he had acquired a strong preference. He died in Eugene, Oregon, on May 13, 1977.

THE PATNODE-SCHEIBER SCALE-UP

On June 7, 1939, Patnode wrote "for the past week or two, Scheiber and I have been working with Rochow learning how to make methyl silicone with the idea of setting up a small [Grignard] pilot plant." Even before the direct process was discovered, the need for methyl silicones had become so pressing that methylchlorosilanes had to be made by use of the Grignard reagent in amounts greater than Rochow could supply if he were to continue his research. Therefore Patnode and Scheiber went ahead with their Grignard pilot plant, whose short life was ended by Marshall's decision to scale up the direct process, and to learn whether an *economic* scale-up was possible.

Things began to happen on the fifth floor soon after May 10, 1940.

---

* Liaison initially emphasized the role of the Research Laboratory in overseeing, advising, and unifying the complex and extensive laboratory structure within the Company. The scope of the liaison function has been broadened.

On June 4 Patnode wrote "our laboratory has been torn up [for 2 weeks!] . . . to make room for one or two more men. In the meantime I have been . . . setting up . . . [next door] for studying the reaction between methyl chloride and silicon-copper."

Even before this new equipment was used for the first time, Patnode, being Patnode, had begun to think about improving the process. On June 6, after describing the crushing and screening of Cu-Si obtained from Rochow, and after recording calculations to show that he expects to make 100 g of chlorosilanes per hour, Patnode suggests that "a *continuous method* in which *finely powdered silicon* is dropped into a high temperature liquid in the presence of copper and CH_3Cl may be more satisfactory from the standpoint of *temperature control*," [italics mine]. Although methylchlorosilanes are not made in this way, the quotation is noteworthy for the italicized portions, which show that Patnode anticipated three characteristics of the method by which these silicone precursors are now manufactured.

The first run was made on July 1, when Rochow's discovery was not yet 2 months old. On October 1, Patnode summarizes what he had learned:

1. The direct process needs extensive study; microscopic examination of the Cu-Si mass before and after reaction with CH_3Cl is imperative.

2. The first Cu-Si mass used by both Rochow and Patnode contained two solid phases: "copper-colored bands" and "black vitreous crystals."

3. A replacement order for a 100-lb lot of identical material brought a substance of different structure—"all silvery with no copper-colored bands." The 100-lb mass showed much poorer reactivity with CH_3Cl at 350° than had the first, and remelting only made it worse.

4. On July 11 Patnode had mixtures of copper and silicon powders prepared, his idea being that the soft copper could be brought to adhere to the hard silicon during mixing. The mixed powders were made into pellets,* heated, and tested by Rochow, who found them very reactive toward CH_3Cl after they had been treated with HCl.

5. Patnode became convinced that the principal role of the copper was to decompose CH_3Cl, and he concludes: "Today Bob [Robert O.] Sauer came to work and I asked him to look into the decomposition of

---

* No pelleting machine being then available in the Research Laboratory, Marshall found somewhere, and bought, a second-hand pill machine, which was used to get the work under way. Marshall always regarded such instances as presenting personal challenges to be accepted, enjoyed, and met—sometimes in unorthodox ways not quickly forgotten.

CH_3Cl and other chlorides in the presence of Cu and other metals. It may be we can dissociate the halide [CH_3Cl] to produce free radicals at much lower temperature. . . ." Patnode's thoughts about the reaction are in surprisingly good accord with current ideas about its mechanism, to which we shall return in Chapter 5.

The prosaic recital above gives no hint of the drama attending Items 2, 3, and 4. These constitute a "shocker called losing the art" (Professor Rochow's description) and the happy sequel in which the art was restored by the replacement of Cu-Si alloys with properly prepared "contact masses" made from the elements mixed as powders.* Rochow and Patnode collaborated in giving the direct process a firm foundation by the filing of three U.S. patent applications on September 26, 1941. As a result of secrecy orders imposed because of World War II, the patents were not issued until August 7, 1945. They are as follows: Rochow, U.S. 2,380,995, the basic patent on the direct process; Rochow and Patnode, U.S. 2,380,996, which teaches the use of contact masses in the process; and Patnode, U.S. 2,380,997, in which the contact masses and their preparation are described.

A serious obstacle on the planned high road had been overcome, but the troubles were not yet at an end. The specter of an industrial reactor that ought to produce chlorosilanes acceptably, but did not, had receded—not vanished.

While all this was going on, the "big" pilot plant (a slightly tilted, "big" copper reactor, a special still, and associated gear), designed to produce and to distill chlorosilanes in pound lots, was being built in the machine shop of the Research Laboratory. No chemical engineers were on hand to help with the design. There was no time to make engineering drawings. The project had to be superimposed on the normal work load. The machine shop, fortunately accustomed to working from the roughest of sketches, had never built this kind of equipment before. Yet the reactor was built soon enough so that the first large-scale run could be made on

---

* For the benefit of those unacquainted with the patent literature, I quote the following description of a contact mass from Patnode, U.S. Patent 2,380,997:

"In a preferred embodiment of the invention the solid, porous mass initially obtained, as by molding a mixture of comminuted silicon and comminuted copper or other metallic catalyst, is fired in a reducing atmosphere at a temperature sufficiently high and for a period sufficiently long to activate the potentially active catalyst, if initially it is catalytically inactive, or to increase its activity, if initially it is not so catalytically active as may be desired."

Small wonder there is a gulf between Lord Snow's two cultures!

Scheiber in Room 537, Building 37, on July 8, 1940, doing early tests on the "big" reactor in preparation for the "big" run. Please compare with Fig. 3-1, on p. 104.

Monday, November 18, 1940. I doubt whether any other laboratory similarly new to the field could then have done this.

For the first run, pellets (80% Si-20% Cu) had been prepared in the machine shop by mixing and pressing coarse powders, the final treatment being a heating ("firing") in hydrogen for an hour at 1050°C. The new "big" reactor (internal diameter, 3 in.) was charged with 17 lb of these pellets. On that historic Monday morning, CH_3Cl was started through the system, which had previously been shown not to leak this gas at room temperature. Finally, "the heating power was turned on at 9:17 A.M." On Tuesday morning "the 'factory' was still going along as steadily as it was last night." Scheiber recorded the operating data. The run was completed on November 24. It produced 45.75 lb of condensed products from 9.25 of the 12.6 lb of silicon charged into the reactor.

What to do with 45.75 lb of crude chlorosilanes, mixtures of unknown composition and difficult to handle? Distill to separate them—if one has a suitable still! Patnode first did a simple distillation on 100 cc of the product to find out what proportion "is gas and what proportion is useful." The "big" still being built as part of the pilot plant was not yet ready.

Distillation of the products from the "big" run began on December 31, 1940, with 12 lb of crude chlorosilanes in the still, and continued uninterrupted into the new year. By January 16, 1941, 24 fractions ranging from 260 to 1325 cc in volume, and from 33.1 to 69.6°C in boiling point, were on hand. What was in them? Which were significant? Comparison of these numbers with Rochow's 5 cc of May 10, 1940, shows how much the scale of operations and the need for identifying reaction products had grown. By and large, however, Patnode still had to proceed according to Kipping's methods in isolating and identifying reaction products. Patnode's notebooks show that in the use of these methods he was a worthy successor to the older master.

The first "big" run was a brilliant success. The time between May 10, 1940, and January 16, 1941, would have been short for this accomplishment under the existing conditions even had there been no other activity in the interim. Puzzling and discouraging failures were to follow. These brusque comments in CRDA tell some of the story.

Sintered powder work [production of chlorosilanes as in the run described above] was going very well. All of a sudden, couldn't make process work. Found old batch of copper had been used up. Reordered copper but was no good. They found that they had two kinds of copper powder in Carbon Brush Department [Company source of copper for this work], a copper prepared by sintering copper oxide and reducing in hydrogen and an electrolytic grade. The sintered copper was OK. Gave good products but the electrolytic was useless. Finally [came to know] how to control process and were able to make intermediates for resins, [for] rubbers, [and for] oil work. [Process] came close to being abandoned before this.

THE FIRST FORK. WATER REPELLENCY. SILICONE SURFACES

On September 4, 1940, during the hurly-burly of the scale-up, Patnode wrote "during the past few weeks I have noticed that the glassware which I have been using [in the silicone work] acquires a film that is not wet by water." Rochow, it will be recalled, had already used the formation of such a film as evidence that he had made methylchlorosilanes in an experiment with CH_3Cl and Fe-Si. On September 3 Patnode had done simple but convincing experiments to prove that the films were formed by the action of mixed methylchlorosilane vapors on the glass, and that the films were tough and adherent enough to be of practical value in making water-repellent (hydrophobic) surfaces. On September 19 he prepared water-repellent glass plates to simulate solid electrical insulators and submitted them to the General Engineering Laboratory for test; the results were encouraging. He then listed other possible useful applications of the methylchlorosilanes as waterproofing agents, e.g., for automobile windows.

In the anxious interim between the end of the first "big" run and the beginning of the first "big" distillation, Patnode recorded two items that show how he thought:

1. November 28, 1940. I don't remember whether I noted down the idea of treating cigarette paper to waterproof it or not. Anyway, today I found that a cigarette treated with mixed silicon chlorides in vapor phase followed by ammonia prevents the disintegration of the paper in the mouth. . . . It is not slippery like a straw . . . but rather more like a cork tip.

2. December 4, 1940. I found that gauze bandage treated with methyl silicon chlorides, wrapped on a glass tube to simulate a wounded extremity, sheds water and appears to be useful in keeping a bandaged wound from becoming water soaked. Also, teapots and cream pitchers that dribble in restaurants might be cheaply treated to prevent this annoying behavior.

It is evident that Patnode had found an attractive fork, the first that warranted the diverting of effort from the planned high road. The earlier pure research on monomolecular films (e.g., films of oleic acid on water) by Irving Langmuir and Katharine Blodgett made Patnode's observations easier to understand (see Figure 4-1) and gave assurance that the new fork in the road would not be a dead end. The creation of silicone surfaces, which are capable of easily releasing or of repelling materials unrelated to silicones, was to become ramified and extensive.

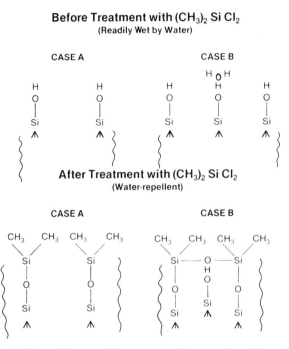

Figure 4-1 Symbolic model to explain water repellency produced by surface treatment with methylchlorosilanes [7].

Notes

1. *Case A:* Two neighboring silicon atoms on the surface of a quartz crystal. *Case B:* Three such atoms. A molecule of adsorbed H_2O is held by hydrogen bonding above the middle atom. The "tripods" represent three Si—O—Si bridges inside the crystal.

2. The treatment consists in allowing $(CH_3)_2SiCl_2$ vapor to react with the surface. A hydrolysis-condensation reaction with the surface occurs. Si—O—Si bridges are formed with the elimination of HCl. In Case B one bridge is exposed more than the others.

3. The surface after treatment is largely covered by CH_3 groups, which resemble saturated hydrocarbons in that they repel water: the surface has been "paraffinized," probably more thoroughly in Case A than in Case B.

4. The figure has symbolic value only. The actual surface will be more complex than the figure indicates, and the nature of the surface will depend on the treatment. Treatments with chlorosilanes have been superseded by others that do not liberate HCl.

The story just told illustrates the discussion in Chapter 2 of the need to regard observations from the point of view of applied research. Stock had seen silicone films on glassware long before Patnode; so had many others, for these films have always been nuisances difficult to ignore. Patnode, however, saw a need these nuisances could fill. Credit him with scoring another success for perspicacity and applied research!

THE SECOND FORK. SILICONE OILS. SILICONE REARRANGEMENTS

In 1872 Ladenburg [8] had discovered that concentrated sulfuric acid could bring about condensation reactions such as

$$2\,[(C_2H_5)_3SiOH] = (C_2H_5)_3Si—O—Si(C_2H_5)_3 + H_2O \qquad (4\text{-}3)$$

a type of reaction we met in Chapter 2. Kipping, and no doubt others before him, made silicone oils via such condensation reactions; but the making of these oils was uncontrolled, and the oils were regarded as liquid polymeric nuisances: *they were not tailor made and not put to use.* All this has changed, and the change began when Patnode discovered the second fork as he was proceeding along the planned high road.

On January 18, 1941, Patnode found that concentrated H_2SO_4 would make viscous products out of oils prepared by hydrolyzing one of the fractions from the "big" run described above. On October 7, 1941, he wrote, "Recently I have been thinking of the possible use of silicones for lubricants again." Two days later Marshall told him of a "possible important application of silicone oils, providing that they lubricate properly." A systematic method of making oils with a range of properties was going to be needed.

On November 24, 1941, Patnode began a crucial, disarmingly simple experiment. He dissolved a little D_4 in concentrated H_2SO_4 at room temperature. I do not know why he did it.* To this *single* liquid layer, he added water, whereupon *two* liquid layers formed: an acid layer below, and an oily layer on top. Upon being heated, the oily layer eventually gave him "a brilliantly clear resin" [more properly a gum, as we shall see later on]. On December 16 he wrote, "It is known that concentrated H_2SO_4 will open the ring of . . . D_4, forming a resin [gum] by a combination of esterification and [subsequent] hydrolysis." Patnode had dis-

---

* Nor did Dr. Patnode in 1975 remember any reason for the experiment. He thinks "the subconscious may have been at work."

covered *rearrangement polymerization* with H_2SO_4 as rearrangement catalyst. He had *broken* siloxane bridges and *re-formed* them; previously it had been assumed that acids and alkalis could function only to *form* such bridges*—that they acted only as catalysts for condensation polymerization, as they do for Bakelite.

Patnode's simple experiment was good news and bad for the silicone project. First, the bad news. Siloxane bridges in silicones, like those in silica, had been regarded as virtually indestructible under all service conditions likely to be met. Now it was clear they could be broken by acid attack at room temperature that left Si-C and C-H bonds untouched. An important fact of silicone chemistry had been discovered.

Now, the good news. On April 17, 1942, Patnode recorded as follows the thinking that led him to a second crucial experiment:

Last year I prepared some of the compounds of the class MD_xM . . . it looks as though these . . . might be very stable, low-temperature oils. These might also be prepared by treating a mixture of D_x and MM with conc. H_2SO_4 . . . which will open Si-O linkages and close them again.

After several days of preliminary work, he did the following definitive experiment on May 5:

An attempt to co-condense D_4 with MM on a somewhat larger scale than formerly was made today. 162.3 g [1 mole] MM and 74.1 g [0.25 mole] D_4 were placed in a bottle together. 10 cc of conc. (95.5%) H_2SO_4 was then added and shaken. A rise in temperature of 3° (from 23° to 26°) was noted. This two-phase [two-liquid-layer] system was shaken at room temperature for 4 hours. Then 25 cc of water was added all at once with shaking, the aqueous layer was discarded (34 cc) and the upper layer washed with two 25-cc portions of fresh water. The oil was then placed in a bottle with K_2CO_3 to remove traces of water and acid. It will be distilled tomorrow.

Distillation of the 281 cc of oil took 3 working days and gave the results plotted in Figure 4-2. On such a curve each of the horizontal portions ("flats") represents a pure compound in the absence of constant-boiling mixtures (azeotropes). In this experiment some MM remained; all the D_4 (boiling point, 175°) had disappeared, and MDM, MD_2M, and MD_3M had been formed. With this one experiment Patnode showed that chance could be controlled in making silicone oils. The experiment suggests that

---

* The essential point here is the breaking and re-forming of siloxane bridges, which results in a *rearrangement*. Polymerization may or may not occur. A mixture of products generally results. Such rearrangement in silicones resembles somewhat the cracking and re-forming of hydrocarbons, which is vital to the petroleum industry.

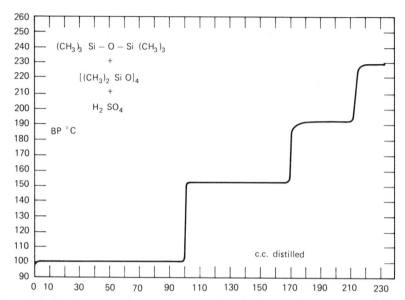

Figure 4-2 The discovery of rearrangement polymerization for the preparation of silicone oils. As the distillation curve shows, an oil made from **MM** and **D$_4$** contained unreacted **MM** (boiling point near 100°C), **MDM** (boiling at 153°C), **MD$_2$M** (boiling at 194°C), and **MD$_3$M** (boiling at 229°C and not identifiable by Patnode when the experiment was done). **D$_4$**, which boils at 174°C, does not appear above because all of it had reacted. The results were "noted & understood" by Robert O. Sauer on May 13, 1942.

one ought to be able to make oils of different, desired viscosities by varying the ratio of **MM** to **D$_4$**, and it suggests also that **T** units be included for further variety. Whether an oil is thick or thin will depend upon the kind of silicone units in it and upon their proportions. Patnode's second fork revealed a side road that would eventually lead to the manufacture of silicone oils as important commercial products.

Patnode's oil work is covered by U.S. Patents 2,469,888 and 2,-469,890, originally applied for on October 29, 1942, and not granted until almost 7 years later, on May 10, 1949. The patents are valuable guides to silicone chemistry, and they can be used to show the usefulness of the shorthand (boldface) nomenclature. Patnode says that his oils

are distinguished from all other known organosilicon compounds by the following criteria:

1. Each silicon atom is joined to at least one other silicon atom through an oxygen atom.

2. Oxygen atoms are found only between silicon atoms.

3. The ratio of the number of R groups to the number of silicon atoms is fixed and is equal to

$$\frac{2a + 2}{a}$$

where a is the number of silicon atoms [and is at least 3].

4. Each terminal silicon atom of the skeletal structure is joined directly to three R groups.

The general formula for such molecules is

$$R_{(2a+2)}Si_aO_{(a-1)} \tag{4-4}*$$

where in the present discussion R = CH_3.

Patnode's *linear* polymers are

$$
\begin{array}{c}
\qquad \text{R} \\
\qquad | \\
R_3Si\!\!-\!\![O\!\!-\!\!Si]_n\!\!-\!\!O\!\!-\!\!SiR_3 \qquad \text{or} \qquad \mathbf{MD_nM} \\
\qquad | \\
\qquad \text{R}
\end{array}
\tag{4-5}\dagger
$$

---

* The reader may wish to use this general formula and the preceding four points to test his knowledge of silicone chemistry. Consider an oil with a silicon atoms in each molecule. Point 1 says that all the silicon atoms are in siloxane bridges. Point 2 says the same thing for all oxygen atoms. Point 3 derives from the tetrahedral habit of silicon in silicones, which requires that the a silicon atoms form $4a$ bonds. These bonds are of two kinds: to oxygen and to R groups. Now, it takes a silicon atoms to form a backbone containing $a - 1$ siloxane bridges, in each of which there is one oxygen atom; the extra silicon atom is needed to complete the last bridge. The $a - 1$ oxygen atoms will form twice as many Si-O bonds. The number of Si-R bonds must therefore be $4a - 2(a - 1)$, or $2a + 2$, which equals the number of R groups in the oil molecule, as each R is joined to silicon by a single bond. The ratio in Point 3 and the general formula for the oil molecule follow from what has been said. Point 4 is implicit in the first three: if the kind of molecules these three specify is ever to end, it will have to end in **M** units, also called "chain stoppers" because their function is to stabilize molecules by inertly ending them.

† A linear polymer with a carbon backbone is representable by E $\wedge\!\!\wedge\!\!\wedge\!\!\wedge$ E [Chapter 2] because there is no difficulty with the end group E. With a siloxane backbone, however, this representation becomes awkward, as is evident from the left-hand side of (4-5); the ends of this molecule are identical, but they look different in the extended formula. The shorthand formula escapes this awkwardness because it makes use of a "split" oxygen atom. The matter is best disregarded by the general reader as a scientific tempest-in-a-teapot. It was implicitly covered when the tetrahedral habit was discussed in Chapter 2.

Let us carry the game a little further by illustrating a point made after Table 3-1, namely,

One of his more complex *branched-chain* polymers is

$$
\begin{array}{c}
R_3Si \qquad\qquad R_3Si \\
| \qquad\qquad\quad | \\
R \quad O \quad R \quad O \\
| \quad | \quad | \quad | \\
R_3Si{-}O{-}Si{-}O{-}Si{-}O{-}Si{-}O{-}Si{-}O{-}SiR_3 \\
| \quad | \quad | \quad | \\
R \quad R \quad O \quad R \\
| \\
RSiR \\
| \\
O \\
| \\
R_3Si
\end{array}
\qquad (4\text{-}6)
$$

$$
\begin{array}{c}
\mathbf{M} \qquad \mathbf{M} \\
\text{or} \qquad \mathbf{M\ D\ T\ T\ T\ M}\ (\text{i.e.,}\ \mathbf{M_5D_2T_3}) \\
\mathbf{D} \\
\mathbf{M}
\end{array}
$$

Patnode's discovery of the second fork made it possible to study such complex silicone molecules systematically, and it foreshadowed the industrial use of cyclic compounds such as $\mathbf{D_4}$ as starting materials in the manufacture of other silicones by rearrangement polymerization.

---

that shorthand formulas for large molecules can correspond to more than one kind of molecule. Note that both the molecules represented below have the formula $\mathbf{D_4T_2}$:

$$
\begin{array}{ccc}
CH_3 & (CH_3)_2 \quad CH_3 \quad H_3C & (CH_3)_2 \\
| & | & \\
(CH_3)_2SiOSiOSi(CH_3)_2 & Si{-}O \qquad\qquad O{-}Si & \\
|\ \ |\ \ | & \diagup \quad \diagdown | \diagup \quad \diagdown & \\
O\ \ O\ \ O & O \qquad Si{-}O{-}Si \qquad O & (4\text{-}7) \\
|\ \ |\ \ | & \diagdown \quad \diagup \quad \diagdown \quad \diagup & \\
(CH_3)_2SiOSiOSi(CH_3)_2 & Si{-}O \qquad\qquad O{-}Si & \\
| & (CH_3)_2 & (CH_3)_2 \\
CH_3 & &
\end{array}
$$

[Each CH_3 group in $(CH_3)_2$ is linked to Si by a bond not shown.]

OILS, RESINS, GUMS, RUBBERS

Before we proceed to the remaining discoveries, let us again use hindsight, this time to compare silicones in a way that would have been enormously helpful to the early investigators who had to feel their way in the dark. *Oils* and *resins* we have met. *Gums* and *rubbers* need to be compared with them. But this is possible here only in the simplified manner shown in Figure 4-3 and explained in the notes thereto.

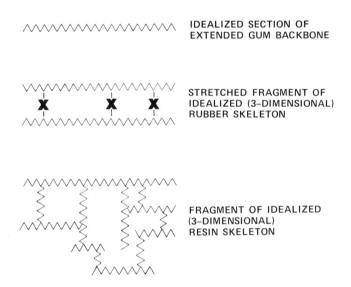

IDEALIZED SECTION OF
EXTENDED GUM BACKBONE

STRETCHED FRAGMENT OF
IDEALIZED (3–DIMENSIONAL)
RUBBER SKELETON

FRAGMENT OF IDEALIZED
(3–DIMENSIONAL)
RESIN SKELETON

Figure 4-3 Comparison in two dimensions of idealized fragments of silicone gums, rubbers, and resins, all with end group omitted. The figure supplements "A Guide to Polymers" in Chapter 2.

Notes

1. The zigzag sections represent joined siloxane bridges, schematically and without regard to number.

2. The relationship of oils to gums, rubbers, and resins is evident from a comparison of the figure with (4-5) and (4-6). The structures of all four materials are determined by siloxane bridges.

3. As to end groups: (*a*) The molecules of oils are terminated by **M** units, which are inert. (*b*) When the molecules become as big and extended as those in gums and rubbers, the influence of end groups on properties decreases. (*c*) When **M** units cannot form, the hydrolysis-condensation sequence often produces molecules ending in OH groups, which are reactive and can promote further condensation, a vital matter in certain products.

The molecules of a gum (end groups disregarded) are longer versions of those in Patnode's linear oils. The resins (end groups disregarded) may be regarded as enormously larger, stiffer, and much more highly branched versions of his branched-chain oils. The rubbers are carefully cross-linked gums with fillers (solid particles such as silica) and other substances admixed.

Rubbers are made in various ways [9], one of which we shall incompletely sketch by way of introduction. Place 100 parts by weight of gum in a dough mixer. Start mixing. As appropriate, add* 100 parts by weight of filler, 10 parts of "process aids or other additives," and 1 part of pigment. After mixing is satisfactory, put the material in a mold at 150°C for about 15 minutes to accomplish the desired cross-linking. If that fails to yield a satisfactory silicone-rubber product in the shape of the mold, improve the rubber by further treatment. Easy, is it not? Well, yes, if you know what more to do.

* For our purposes we may thankfully assume that all the filler is finely divided silica and that benzoyl peroxide is the only material within the quotation marks. Of course, the rubber business is more sophisticated than this.

---

4.   Ideally, gums have linear backbones representable by —$(D)_x$—, where x may be near 10,000.

5.   The rubbers are cross-linked gums further strengthened by the interaction of gum with added filler—preferably SiO_2, finely divided. If the cross-linking is *innate* (considered inadequate today), Si—X—Si is Si—O—Si, a siloxane bridge resulting usually from the presence of T units. If *foreign* cross-linking is produced, as by the action of benzoyl peroxide, Si—X—Si is Si—CH_2—CH_2—Si, whether or not vinyl (H_2C=CH—) groups were present. With benzoyl peroxide, the actual cross-linking occurs between two carbon atoms, each attached to an atom of silicon; see (4-8). For simplicity's sake, we shall use Si—X—Si to cover all cases—even those not listed here.

6.   The idealized model in the figure represents a *stretched* rubber. On release of stress, the rubber would ideally return to its condition before stretching; that is, curl itself up with the ends of the molecules nearer each other and with the rubber occupying its original volume. When this occurs, the "compression set" is zero. In actual rubbers sustained compression often leads to a reduction in the stress-free volume.

7.   Resins are a three-dimensional skeleton built of backbone segments $\wedge\!\wedge\!\wedge\!\wedge$ joined by T or Q units. The backbone fragments are representable as —$(D)_x$—, not to be confused with D_n, the formula of a cyclic compound. Resins have more complex structures than do oils, gums, or rubbers and are the most difficult to prepare reproducibly. Correlation of structure with physical properties is most difficult for resins.

8.   Gums can seem rubbery. Oils can be viscous. Borderline materials of comparable molecular weight can be difficult to classify.

THE THIRD FORK. THE RUBBER SIDE ROAD BEGINS. MAYNARD C. AGENS

The side road revealed by the third fork led, after several major discoveries, to the first usable silicone rubber, a product perhaps more interesting than the resins or the oils. More than one discovery had to be made before the happy situation hinted at in Figure 4-3 could even be approached.

The first of these discoveries, that of the rubber itself, was made by Maynard C. Agens. Agens completed undergraduate work in chemistry at Hamilton College in 1930, began work at General Electric that year in the component responsible for manufacturing alkyd resins, and transferred to the Research Laboratory in 1932, where, after a brief assignment with Marshall, he joined Birger W. Nordlander, whose principal concern was condensation polymers (later called "Permafils") that were related far more closely to alkyds than to silicones.

On January 20, 1941, Agens told Robert O. Sauer that "it would be interesting to react his [Sauer's] hydrolyzed $(CH_3)_2SiCl_2$ with anhydrous $AlCl_3$ to produce a flexible aluminum silicate or synthetic mica." Micas are naturally occurring silicates (e.g., $K_2O\cdot3Al_2O_3\cdot6SiO_2\cdot2H_2O$)* that were formed in the earth at high temperatures. Agens hoped to make similar aluminum-containing flexible materials at moderate temperatures—not a likely prospect. He failed—and discovered silicone rubber.

Agens was then a junior staff member, not assigned to silicones, whose activities were numerous and diverse enough to give his notebooks the flavor of Rochow's [Chapter 3]. How Agens found time for silicone experiments is not easy to see. Two such must be reported. On September 8, 1941, he heated to 191°C in vacuum the hydrolysis products of $(CH_3)_2SiCl_2$ in order to eliminate the smaller molecules present. He next heated the residual liquid with metallic sodium to 180°C and obtained a viscous syrup that gelled at 150°C to form a *rubbery material*. After being heated overnight at 275°C, *the material was still rubbery*.

The other experiment more nearly resembles current silicone-rubber

---

* The formula in the text is that of muscovite, a potash mica, an excellent and useful electrical insulating material, scarce during World War II. Muscovite splits easily into thin sheets. As flakes, and in mica paper, it is used for the insulation of windings in the stators of large electrical machines.

No one has yet made a wholly satisfactory substitute for natural mica, although progress has been made by using fluorine to replace OH in the natural product. Note that the making of mica paper from ground-up muscovite requires a binder, preferably one capable of withstanding high temperatures. This problem thus bears some similarity to the Corning glass-tape problem that led Dr. Sullivan to silicones; see Chapter 3.

technology. On March 13, 1942, Agens made "Siligum" (his designation) as follows:

I did a few experiments with some of the hydrolyzed silicon dichlorides which were in the form of a very viscous syrup [i.e., high in molecular weight]. Poured some of this syrup on a 200°C hot plate and incorporated red iron oxide as filler and then added a small amount of anhyd[rous] AlCl₃ .and after working the material with a spatula it seemed to cure [polymerize by cross-linking] or convert to a tough rubbery material.

The noteworthiness of Agens' work lies not in his products, which were soon to be superseded, but in his demonstration *that rubbery silicones were possible*— that nature's bounty was not yet exhausted. This was no mean achievement for one with no more than undergraduate academic training, who was saddled with the responsibility for supervising the Chemistry Section's stockroom and ancillary activities, and who (in addition to other research) was working mainly on unsaturated polyester resins for aircraft ignition systems.

J. G. E. WRIGHT (1883–1959)

James Gilbert Ernest Wright, the eldest child of Daniel and Mary Andrew Wright, was born in Glasgow, Scotland, on January 16, 1883. His romantic mother believed in the "wee folk," the fairies, and was a dressmaker of unusual artistry. It was she who added the Gilbert and Ernest to the conventional James in her son's name* and gave to him, too, his lifelong interest in line, form, color— indeed, in all the arts [10].

Gilbert Wright attended Anderson's Technical College, Glasgow, on a scholarship and then worked in British industry for some years, during which a laboratory accident cost him an eye. In 1913 he joined the Research Laboratory, where he soon gained distinction as one of those who introduced man-made polymers into the electrical industry. We met him in connection with polyvinyl chloride [Chapter 1] and with the testing of Rochow's early methyl silicone resins [Chapter 3]. When the silicone project began, Gilbert Wright was the senior staff member involved, and he had Curtis S. Oliver and Lester V. Adams to help him with experimental work on his ideas—ideas that were usually unorthodox and seen through a glass darkly. Gilbert Wright was *sui generis*. I think he was

---

* To demonstrate that one of his mother's choices came to be generally accepted, I shall occasionally speak of him as "Gilbert Wright": he was "Gilbert" to almost everyone. Ostwald would have classified him as a pronounced romantic. See Reference 34, Chapter 2.

what William Blake might have been, had the poet deigned to do applied research on man-made polymers. Wright retired on January 31, 1948.

THE THIRD FORK. THE RUBBER SIDE ROAD LENGTHENS

Wright Lengthens the Rubber Backbone.   Gilbert Wright and Agens had a close and sympathetic relationship. They shared thoughts and ideas. In their work they relied heavily on intuition, imagination, and observation—qualities that proved essential to the discovery and improvement of silicone rubber.

Wright discovered that the backbone in gum for the making of methyl silicone rubber could be lengthened through *"controlled hydrolysis"* of $(CH_3)_2SiCl_2$, a term coined by him, which appears in his notebook on August 15, 1942. An added substance present during hydrolysis provided the control. As early as April 22, 1941, he and James Marsden had been trying to make improved methyl silicone resins by adding P_2O_5. On May 22, 1941, Wright listed PCl_3, $POCl_3$, PCl_5, $SbCl_3$, $TiCl_4$, and $SnCl_4$ as other materials to try. He must have thought about this approach for over a year and decided eventually to use added substances that could supply water for hydrolysis. He considered the following classes of substances, which he had been taught to regard as hydrates: acids (e.g., boric acid, the hydrate of B_2O_3); bases (e.g., NaOH, the hydrate of Na_2O); weak bases; crystalline salts (e.g., $FeCl_3 \cdot 6H_2O$); and clays. Lengthy experimental work did lead to an improved gum from which better silicone rubber could be made, $FeCl_3 \cdot 6H_2O$ and $(CH_3)_2SiCl_2$ being the best combination.

On May 28, 1943, Gilbert Wright summarized as follows the results of "controlled hydrolysis" experiments with three phosphates and $(CH_3)_2SiCl_2$. The summary [captions added] tells something of his approach to research.

[Phosphate]	[Immediate Product]	[Final Product]	[Time]
$NaH_2PO_4(H_2O)$	Butterscotch	Maple syrup [then] butter	2 weeks
$Na_2HPO_4(H_2O)$	Molasses	Butter	1 week
$Na_3PO_4(H_2O)$	Syrup	Elastomer	1 day

Would a coldly precise scientific description reveal as vividly what the hydrolysis-condensation of chlorosilanes is really like? Compare Gilbert's summary with the following extract from his U.S. Patent 2,452,416 (applied for, April 26, 1944; issued, October 26, 1948):

The hydrolysis of . . . [$(CH_3)_2SiCl_2$] is most readily carried out by pouring it into water. The product of hydrolysis, polymeric dimethyl silicone, is an oily liquid which floats on the surface of the other product of hydrolysis, hydrochloric acid, and can be separated from the acid by decantation. In general, dimethyl silicone prepared by this procedure contains a large proportion of D_4, boiling at 175 deg. C., along with lesser amounts of D_3, boiling at 134 deg. C., and others. From a third to a half of the total product may be distilled at about 200 deg. C or below, and is thus comparatively volatile.

An object of the present invention is to provide a method of hydrolysis . . . which yields a polymeric dimethyl silicone of high molecular weight and containing a minimum of polymers boiling below 200 deg. C. The product obtained is generally a viscous liquid, but under some circumstances an elastic gum is obtained. In accordance with my invention, these high molecular weight silicones are obtained by employing as the hydrolysis medium a solid inorganic substance containing water of crystallization, water of hydration, or "bound" water. No liquid water is employed.

In general, the materials which are suitable for my hydrolysis medium are members of the class consisting of hydrated inorganic salts in which the anion is unpolymerized, and metal hydrates (hydroxides) other than the hydroxides of the alkali and alkaline earth metals.

In carrying out my invention, it is only necessary to pour the . . . $(CH_3)_2SiCl_2$ onto the solid crystals or powder. As the hydrolysis proceeds, due to the reaction between the chlorosilane and the water, the reacting mass generally separates into two liquid layers with a saturated solution of the salt, or its reaction product with the hydrogen halide below and the viscous oily layer above. On standing, the oily layer is frequently converted into an elastic gum, particularly when the hydrolysis medium comprises compounds of iron, and occasionally when the sulphates of sodium and copper are employed. At the end of the reaction it is only necessary to separate the upper layer of silicone from the lower layer of solution, or excess salt. The product may then be further washed with water, if desirable or necessary, to remove adsorbed portions of the hydrolysis medium.

The dimethyl silicone obtained by the method of hydrolysis described above is of very high molecular weight as compared with that obtained by hydrolysis in water or aqueous solutions and contains only a small quantity of material boiling below 200 deg. C., generally less than 10 per cent.

Figure 4-3 implies that long backbones are needed to make good rubber. Wright found that such backbones form when the hydrolysis-condensation of $(CH_3)_2SiCl_2$ takes place with liquid water absent. Here is a *partial* explanation. The finished backbones end in OH groups (HO $\wedge\wedge\wedge\wedge$ ••• $\wedge\wedge\wedge\wedge$ OH). These groups come from water. By eliminating liquid water, Wright made OH groups less available. Thus $(CH_3)_2SiCl_2$ was forced into making longer backbones if it were to hydrolyze completely, as it has a great tendency to do. Controlled hydrolysis as he saw it was a "fight for [scarce] water."

Wright and Oliver Build the Rubber Skeleton. Wright's notebooks show that he had been familiar with benzoyl peroxide $[(C_6H_5CO)_2O_2]$ as a vulcanizing (cross-linking) agent at least since 1937. Its use for this purpose had been largely confined to (unsaturated) materials with C-C double bonds, bonds toward which the peroxide is highly reactive. Wright had no reason to believe that benzoyl peroxide could cross-link a saturated material. In 1943 Wright and Oliver chose to test its action on methyl silicone gum at temperatures near 150°C, and discovered that effective cross-linking resulted. (See Figure 4-3.) The silicone rubber thus produced had improved tensile strength and greater hardness; see Wright and Oliver, U.S. Patent 2,448,565 (applied for, March 14, 1944; issued, September 7, 1948). The discovery justified the belief that gum produced by controlled hydrolysis could be systematically converted into a useful rubber.

The discovery was a complete and dramatic surprise.* Marsden explained it as cross-linking brought about by "hydrogen abstraction" from methyl groups, a process that may be represented as follows by writing only the minimum number of atoms:

$$2\ SiCH_3 \xrightarrow{150°C} SiCH_2^{\cdot} + SiCH_2^{\cdot} = Si—\overset{\displaystyle H}{\underset{\displaystyle H}{C}}—\overset{\displaystyle H}{\underset{\displaystyle H}{C}}—Si = Si—X—Si \quad (4\text{-}8)$$

(Benzoyl   (Free radicals)   (Cross-link)   (see Figure 4-3)
peroxide)

Shortly after the discovery, an article describing similar cross-linking in saturated hydrocarbon polymers appeared in the literature.

---

* Let Marsden recall the way it was. "After working with a number of compounds to crosslink the . . . filled silicone gum and getting products barely recognizable as elastic, you cannot imagine my amazement when Slim Oliver showed me the product crosslinked with benzoyl peroxide—it really was elastic." Marsden immediately and systematically duplicated the cross-linking experiment—"And my God it really worked; it really vulcanized; the product was really elastic." James Marsden, personal communication to H.A.L., 1975.

Marsden Improves the Skeleton. Today no polymer chemist would be surprised to learn that the skeleton of methyl silicone rubber can be improved by having $H_2C=CH-$ groups present in the gum. One might argue, after the fact, of course, that this improvement was predictable in 1944, when Marsden made his discovery, because the action of benzoyl peroxide on polymers containing double bonds was known. But no one foresaw the discovery, and its history is one more example to show that an investigator working in a broad, complex, new field cannot always take full advantage of prior knowledge. Here is how the discovery was made. A pure substance has a sharp boiling point. When a substance has to be obtained at a practical rate by distillation, as $(CH_3)_2SiCl_2$ must be, it always contains impurities with higher boiling points; generally, the higher the rate, the more of the impurities ("high boilers"). Marshall insisted on higher distillation rates to increase the availability of $(CH_3)_2SiCl_2$ [11]. Different batches of $(CH_3)_2SiCl_2$ gave rubbers with different properties. In 1944 Marsden, for reasons given in Chapter 5, concerned himself with setting specifications, in terms of permissible boiling point ranges, for the $(CH_3)_2SiCl_2$ to be used in making rubber [12]. One then otherwise unidentified "high-boiler," added to $(CH_3)_2SiCl_2$ in small amounts, gave a syrupy liquid at the stage where gum normally appeared. Almost in defiance of common sense, Marsden made a rubber from this liquid. This rubber was unusually hard, a sure indication of increased cross-linking. Systematic further work proved that the crucial "high boiler" was $CH_3(CH_2=CH)SiCl_2$ (methylvinyldichlorosilane), which had to be identified by the analytical chemists (notably Earl H. Winslow) because the forever impetuous Marsden failed: he could not bring himself to wait long enough for analytical chemistry to run its course. Marsden thus made his discovery by a circuitous route. It is no less valuable for all that, and it is one more example of the role of chance in applied research.

Gilbert's Little Accident. Bouncing Putty. I think bouncing putty was Gilbert Wright's favorite discovery. It rivals Marsden's (just described) as an example of the unexpected in research.

Benzoyl peroxide cross-linking gave, as we have seen, a *harder* methyl silicone rubber. On the basis of this result, Wright on July 30, 1943, "conceived the idea of making a silicone rubber ebonite" and asked Oliver to use boron nitride as a filler. By August 4, 1943, the new "hard rubber" silicone had shown good resistance to an electric arc, and it had come to light that the "boron nitride" used as filler contained "maybe 50% tricalcium phosphate and considerable boric acid." Experiments would have to be done to learn whether the "active agent" was $Ca_3(PO_4)_2$, BN, or B_2O_3. When B_2O_3 was added, the product—far from being a hard rubber—was a *putty* that *bounced*, something so undreamed-of that Silverman [13] called it "The Scientists' Goofiest Discovery" and described it as follows in a style that fits his title:

Several years ago, American fathers and mothers began bringing home to their children a doughlike toy which apparently represented one of chemistry's mistakes. Pulled slowly, it could be stretched out like taffy. Hit with a hammer, it shattered like glass. Left alone, it flowed until it was as flat as a pancake. Rolled into a sphere and dropped, it bounced higher than a rubber ball. In addition, as parents soon discovered to their consternation, this peculiar material—known as bouncing putty—could insidiously but permanently seep into rugs, carpets, upholstery, pocket linings, piano keys and children's hair.

Even the chemists who first studied the material were somewhat perplexed. One day one of them stood in a laboratory corridor with a handful of the new substance, rolling it, dropping it on the floor and saying, "golly, look at it bounce!"
"Astounding," observed another scientist. "But what's it really good for?"
"Well," the chemist replied thoughtfully, "you can use it to roll, drop on the floor, and say, 'Golly, look at it bounce!' "

Bouncing putty is now considered to be a methyl borosiloxane that is cross-linked in an unusual way [14]: the boron atoms in the backbones can and do attract lone electron pairs on oxygen atoms in the siloxane bridges of other backbones to form a weak cross-link, representable as follows:

$$
\begin{array}{c}
-O-Si-O-B-O-Si-O- \\
\uparrow \\
-O-Si-\ddot{O}-Si-O-
\end{array}
\qquad (4\text{-}9)^*
$$

Here, only the electron pair involved in the cross-link is shown, and the CH_3 groups are omitted. The boron atoms in the backbones are few and far between.

Bouncing putty, which owes its discovery in part to the stockroom's fortunate deviation from truth-in-labeling, remains an interesting novelty today. It has found limited use in leveling devices; it is sometimes prescribed for use in exercises to strengthen hand and arm muscles; and it continues to delight children.

---

* Hydrogen bonding is more complex. The role of lone electron pairs in condensation polymerization was mentioned in connection with (2-3).

One of the last entries in Wright's final notebook, namely,

$$\begin{array}{c} \text{—Si—O—Si—O—Si—} \\ \diagdown\ \diagup\ \diagdown\ \diagup \\ \text{Ti}\qquad\text{Ti} \\ \diagup\ \diagdown\ \diagup\ \diagdown \\ \text{—Si—O—Si—O—Si—} \end{array} \qquad (4\text{-}10)$$

shows that he was still thinking of incorporating new elements into silicones as he approached retirement [15].

FURTHER PROGRESS ALONG THE PLANNED HIGH ROAD

Exploration of the direct process figured importantly in Marshall's plans. He told Gilliam and Sprung to "throw in the periodic table" in searching for catalysts better than copper; what happened will be told in Chapter 5. Patnode's contact masses have been mentioned. Rochow, soon after May 10, 1940, began to study the suitability of the direct process for making silicone precursors other than methylchlorosilanes. His attempts, already mentioned, to produce mixed precursors for methylphenyl silicones by passing $CH_3Cl\text{-}C_6H_5Cl$ mixtures over Cu-Si were failures.

As the periodic table is large, and the number of possible R's is legion,* complete exploration of the direct process was a formidable assignment. It was necessary to choose, and the phenyl precursors were selected. The choice was logical because phenyl silicones were known, because phenyl is the simplest aromatic group, and because certain physical properties of methyl silicones (notably, brittleness and tendency to crack) made it advisable to discover an economic way of preparing improved alternatives.

The act of May 10, 1940, was not going to be easy to follow. The

---

* It was pointed out in Chapter 3 that Rochow chose to concentrate initially on methyl silicones as resistant to oxidation in air at higher temperatures because these silicones contained no C-C bonds, CH_3 being the simplest aliphatic group, and because they approached silica in composition. The first reason is inapplicable to the phenyls. The phenyl group, because of stabilization attributable to resonance [Chapter 2], resists oxidation very well indeed. Neither is the second reason unimpeachable. Highly cross-linked methyl silicone resins, which are nearer to silica in composition than are those richer in methyl groups, are more susceptible to oxidation by air than when the cross-linking is less and the methyl content greater. The greater susceptibility of resins containing aliphatic groups such as C_2H_5, in which C-C bonds exist, is in full accord with Rochow's position. See the discussion in Reference 7, p. 438 et seq.

discovery of a direct process for phenyl silicones was almost sure to be less dramatic. And so it was. The notebooks make clear that work on the phenyls was less fun.

Differences in the Rochow routes to methyls and to phenyls were anticipated and soon made themselves felt. On the good side, separating the phenyl precursors by distillation was simple. The boiling points of phenyl silicones, unlike those of the methyl precursors, were separated by large temperature differences; there were no serious complications, although the presence of $AlCl_3$ derived from an impurity in the silicon was a nuisance that had to (but could) be dealt with. On the bad side, C_6H_5Cl reacts very sluggishly with silicon, a situation that normally calls for an increased reaction temperature. With phenyl compounds increased temperature means seriously increased risk of undesirable reactions— especially of decomposition that leads eventually to carbon formation, which can be fatal in the direct process. It was necessary to be careful.

Rochow [16] was soon joined by Gilliam [17, 18] in his search for a direct process that used C_6H_5Cl. They found two. Finely divided silicon alone would not react with C_6H_5Cl at any practical temperature. Finely divided silver proved to be a good catalyst. With regard to Cu-Si, the situation was complex. In the reaction with CH_3Cl, Rochow had soon found that HCl could be eliminated—not so with C_6H_5Cl. Here it had to be retained even though its presence increased the chlorine content of the product by promoting the formation of $(C_6H_5)SiCl_3$ (not always desired) and of $SiCl_4$—early experience in the preparation of $SiHCl_3$ [4, 6] repeated itself once more. Another item also has a familiar ring. On December 2, 1941, Rochow wrote that ". . . fresh new 50% Cu-Si alloy does not react with phenyl chloride at all under conditions where the older (oxidized?) alloy converts 15–30% of the phenyl chloride. We [Rochow and Gilliam] have an important fact here." It was facts like this that made a quick penetration of the unknown impossible and necessitated instead a long-drawn-out frontal attack. Though two versions of the direct process (with silver and with copper in the presence of HCl) had now been discovered, much remained to be done before they could be used to produce phenyl silicone precursors on a commercial scale.

The work extended into 1942 largely because a second deviation from truth-in-labeling did not follow the happy precedent set by bouncing putty. I shall spare the reader the long, agonizing, and confusing preliminaries. By February 1942, Batch 1021 of powder labeled Ni-Si, which had originated in the Carbon Products Department, had been shown (along with others) to react well with C_6H_5Cl, but the result was beclouded for various reasons. The denouement and the patent action taken are evident from the following entries in Rochow's notebooks:

February 5, 1942. A sample of the product from Batch 1036 [a batch *known* to contain nickel as the *only* added element] was hydrolyzed and left only some flakes of silica. [The run was a flop!] I shut off the tube and stopped the run.

Found out from Gil [Gilliam] and Balis [analytical chemist] that the trouble with the duplication of Batch 1021 arises from [the] simple fact that 1021 *contains no Ni* [italics mine]. This is Ag-Si instead, a clear case of mis-labeling at Carbon Products dept. We should have looked for such a simple mistake earlier in the work by analyzing everything. . . .

February 9, 1942. Dr. Marshall was up to discuss the preparation of the phenyl Si chlorides by our Ag-Si method, and advised the preparation of a patent docket on this; I spoke to Mr. Rule [Patent Counsel] and he advised filing a continuation in part on the general [Rochow] case.

This action was taken, and the discovery of silver as catalyst is covered by Claim 17 (the last) of Rochow, U.S. Patent 2,380,995: applied for, September 26, 1941; granted, August 7, 1945.

The saga of silver as catalyst contains several points of interest. Silver is useful for the phenylchlorosilanes, but not for the methyls. Copper is useful for both. Why? Furthermore, why should silver be the only catalyst other than copper useful in direct processes? Discussion is easy, but a satisfactory answer is not. Moreover, the deviation from truth-in-labeling is an object lesson that teaches what can happen in research as it is done, and that illustrates once more why analytical chemistry is needed. Finally, the saga includes an interesting problem in patent law that will be explained in a footnote of Chapter 5.

COMMENTS IN CONCLUSION

Discoveries by Rochow's colleagues are now available to enlarge the look at industrial research that closed Chapter 3. The making of discoveries is the most important aspect of any such look, and the most difficult to describe and understand. We are concerned here with industrial research that was predominantly applied in character. It yields valuable evidence about the nature of discovery for the following reasons, some of which have already been mentioned.

1. The evidence is taken mainly from the original laboratory notebooks.
2. It relates to a field, silicone chemistry, then new, and filled with opportunities for unusually attractive and promising research.

3. It is evidence from an industrial laboratory that attempted to give an investigator all the facilities and freedom that could profitably be used.

4. The evidence derives in large part from research that could not have been planned.

5. The evidence is easier to interpret because it comes from a "system" relatively undisturbed, except for the war, by influences from outside the Company. Once begun, the silicone project grew rapidly, but it was largely self-contained as regards discovery.

• As to major silicone discoveries. Ten early major discoveries are summarized in Table 4-1; minor discoveries were too numerous to list. A major discovery is one essential to the survival of the project and/or responsible for giving it a new look.

The applicable criterion for including a discovery in the table is *value at the time*, not *value now*. Consider Discovery 3, Table 4-1. It is of no importance today, quite probably because the violent motion, abrasive action, and efficient mixing characteristic of the "fluidized bed" [Chapter 6] make the direct process insensitive to the kinds of silicon and copper used. Yet, had Discovery 3 not been made in 1940, the silicone project might not have survived.

• As to discoveries generally. Consider the discovery of oxygen, which occurred long ago, in simpler times. "At least three different men have a legitimate claim to it, and several other chemists must, in the early 1770s, have enriched air in a laboratory vessel without knowing it" [19]. A revealing monograph [20] describes the background and origin of the relevant experiments by Lavoisier, who was only one of the three. The example of oxygen teaches us that we are unlikely ever to have the whole truth about any scientific discovery. Even the more modest "nothing but the truth" is often elusive.

The whole truth, needful for complete understanding, demands our awareness of what the discoverer knew and when* he or she knew it, of what he thought and when he thought it, and of what he did and when he did it. Table 4-1 contains 10 discoveries by seven investigators, none of whom operated in a vacuum!

Some further warnings are in order. Important discoveries need not be the most difficult to make. Unimportant discoveries and failures may have more to teach about research than great successes. Under one set

---

* It is not permissible *to assume* in this connection that a discoverer at the time of his or her discovery had knowledge of prior relevant information in whole or in part, published and unpublished—another reason why properly kept notebooks are important. Whether or not a discovery is patentable as an invention is a different question; see Chapter 5.

Table 4-1   Major Discoveries in the Silicone Project Before 1950[a]

No.	Description	Discoverer(s)[b]	Year[c]	Project Size[d]
1	Methyl silicone resins	Rochow	1938	0.5
2	Direct process for methylchlorosilanes	Rochow	1940	3
3	Contact masses for direct process	Patnode and Rochow	1940	3
4	Silicone surfaces	Patnode	1940	3
5	Direct process for phenylchlorosilanes	Rochow and Gilliam	1941	11
6	Controlled preparation of oils	Patnode	1942	16
7	Silicone rubber	Agens	1942	16
8	Preparation of methyl silicone gum	Wright	1943	25
9	Cross-linking of gum	Wright and Oliver	1943	25
10	Cross-linking with vinyl groups	Marsden	1944	34

[a] U.S. patents were granted on all these discoveries. Other discoveries are covered in later chapters. The use of *discovery* instead of *invention* is defended in Chapter 5.
[b] All discoverers had joined the Chemistry Section by August 1, 1939, W. F. Gilliam being the last to arrive. Among the discoverers listed, four of seven had been awarded PhDs.
[c] The choice of year could be argued in some cases.
[d] Project size measured by estimated man-years of research effort during year on left. Data compiled by Arthur E. Newkirk and taken from CRDA.

of circumstances, a discovery may lead to the founding of an industry; under another set the same discovery may come to nothing.

• As to the discoveries in Table 4-1. Discoveries in the silicone project were made in diverse ways. The discovery of bouncing putty has the flavor of Kipling's

> Ah, what avails the classic bent
> And what the cultured word,
> Against the untutored incident
> That actually occurred

> *The Benefactors,* Stanza 1

This example stands in marked contrast to Discoveries 3, 4, and 5, which were made by following direct paths with minimum lost motion. Patnode here displayed an exceptional instinct for the jugular, a useful quality in research as elsewhere.

Discovery 2 is of particular interest and importance. Rochow did not approach it directly. He had planned to prepare $SiHCl_3$ and to use it for making silicone precursors in subsequent reactions such as (4-1). The plan was sound; there are many references to the use of $SiHCl_3$ as an intermediate [21]. The reader will remember that Combes in 1896 prepared it from Cu-Si and HCl and used it as an intermediate (with metallic sodium added) in his vain attempt to produce silicon-containing dyes [6c]. The reaction

$$SiHCl_3 + C_6H_6 = C_6H_5SiCl_3 + H_2 \qquad (4\text{-}11)$$

is of industrial interest today [21]. But what would have happened to the silicone project had Rochow not departed from his research plan?

Richard Müller [22] independently made Discovery 2 a short time after Rochow. The reader will remember a similar close finish between Kipping and Dilthey [Chapter 2]. A criterion resembling that used for Table 4-1 will be applied here. Because of circumstances largely beyond the control of either discoverer, Rochow's work accelerated the founding of today's silicone industry, whereas Müller's did not: ergo, the Rochow route to silicones is appropriately named.

• As to planning in applied research. Unexpected and unplanned forks vital to the silicone project have been described. Without silicone surfaces, oils, and rubbers, the silicone industry could not have grown to its present size and importance. Planning of course becomes increasingly necessary as a project grows and as a field matures, but the plans must always be flexible enough so that unexpected discoveries can be followed up even if planned objectives have to be shelved or abandoned. "You never know where the road is going to fork."*

• As to the frequency of discoveries. Table 4-1 suggests that the law of diminishing returns applies to the making of discoveries in applied research. The man-years of effort per discovery increased sharply as the discoveries multiplied. No discoverer had a monopoly. Limits to discovery cannot be set, but it seems that each successive discovery tends to

---

* Bruce Catton, *Waiting for the Morning Train*, Doubleday, Garden City, New York, 1972, p. 152.

require more effort. Although this conclusion is becoming painfully obvious today, some well-informed scientists would have regarded it as
pessimistic 40 years ago. I believe that the conclusion is valid for the
silicone project even though new, inexperienced PhDs were joining the
Chemistry Section at the high rates suggested by Column 5 of Table 4-1,
and even though emphasis in the project shifted gradually from discovery
to augmentation [Chapter 5].

• As to discoverers. A major discovery did not presuppose the
PhD degree. The three qualities listed at the end of Chapter 3 were
important. Given free communication with PhD colleagues familiar with
the relevant scientific literature, a sensitivity toward industrial needs and
experience in applied research could, as it did with Wright, Agens, and
Oliver, replace graduate training in the making of discoveries. In a favorable research environment it is unnecessary and uneconomical to
have every investigator familiar with all the pertinent literature. Wide
diversity in backgrounds, training, and experience is more desirable.

• As to the usefulness of prior knowledge. Only a subjective estimate is possible. I consider the following prior research achievements to
have been the most useful in connection with Table 4-1: the demonstrated
existence of silicone polymers, the Kipping route, the reactions of $SiHCl_3$
and its preparation by the use of Cu-Si, and the availability of silicon at
a reasonable price. (We take CH_3Cl and C_6H_5Cl for granted.)

The pure research that led to the silicon and silicone chemistry
described in Chapter 2 of course had a pervasive influence, which was
enhanced because Patnode, Rochow, and Gilliam had all been trained at
Cornell.* The silicone project would have moved more rapidly had the
polymer knowledge outlined in Chapter 2 been available in 1938.

As the silicone project grew, it generated knowledge, some of it
peripheral. The Laboratory atmosphere favored such generation of
knowledge to an extent probably unique in applied research at the time—
government sponsorship has made the situation commonplace today.
Applied research, I believe, continues to grow less dependent upon pure
research for subsistence, though perhaps not for discoveries. "Gold is
where you find it."

• As to working conditions. Over the period of Table 4-1, no reg-

---

* It was customary in Patnode's time at Cornell to use silicon whenever possible as a
"stand-in" for the then scarce germanium in work on organic compounds of the latter. An
unusual knowledge of organosilicon chemistry was one result.

ularly scheduled meetings were held and no periodic reports were written. Marshall was away much of the time. Many men carried on more than one investigation at once. Each man (Wright excepted) normally did his own experiments and, to a large extent, his own analytical work. The atmosphere was usually congenial but seldom relaxed. A widespread feeling of urgency arose because the work to be done seemed endless, and the results continued to be interesting. The following conclusions seem warranted. The most comfortable and convenient working conditions are not necessarily the best for the making of major discoveries. Tight management and extensive formal organization are not needed while a project is small. Adequate technical resources are essential.

• As to champions. Industrial research projects need champions to survive. The champion is always at risk—dry holes in research can be costlier than those in oil fields. Championing a project becomes more difficult when it does not go according to plan; for example, when the original objective has to be abandoned or is so difficult to achieve (as were satisfactory silicone resins) that effort must be diverted to other ends (e.g., silicone rubber), and when the overall effort must be concomitantly expanded. The period covered by Table 4-1 was one of uncertainty relieved by major discoveries. At no time during this period did Marshall as champion seem to have serious misgivings. Waterford, as we shall see in Chapter 7, posed a crueler test.

• As to silicone chemistry. The glad tidings proclaimed by Table 4-1 come through loud and clear. Silicone chemistry was an exceptionally promising field for applied industrial research—one in which significant discoveries came frequently enough to stimulate the research staff, to reward their efforts, and to minimize frustration.

• As to bond ''durability''—a concluding technical comment. In Chapter 2, the ''durability'' of the Si-O bonds, of which Si—O—Si bridges are constructed, was emphasized. Now we have found that these bridges are broken and re-formed rather easily during rearrangement polymerization. What would Justice Holmes think? The root difficulty is that of generalizing about descriptive chemistry—of trying to compress *Gmelin* into a few simple sentences.

The Si—O—Si bridge is durable as regards atmospheric oxidation. In SiO_2 this durability is guaranteed by the periodic table and the tetrahedral habit, and by thermodynamics. The following observations constitute a naive but necessary attempt to explain why this durability does not necessarily extend to other reagents or to silicones.

1. As Figure 4-1 implies, the surface of crystalline SiO_2 is far different from its interior. The *untreated* surfaces in the figure presumably resulted when H_2O vapor acted upon surface atoms that were highly active because, unlike interior atoms, they possessed unsatisfied valences—incomplete bonds. Even after bond formation was completed by the formation of OH groups, the surfaces were still active; for example, they could adsorb H_2O molecules or react with $(CH_3)_2SiCl_2$. In Chapter 2 we might have asterisked each OH group as an active site. The *treated* surfaces, being paraffin-like in character, may be regarded as "protecting" the SiO_2 from chemical attack owing to their inertness.

2. In a perfectly protected SiO_2 crystal, the interior Si—O—Si bridges would be immune to chemical attack; they would seem to be "durable." Suppose, though, that protection is imperfect, and that attack by a reagent capable of breaking Si—O—Si bridges forms products that fail to protect: the "durability" will then vanish. The classic example of this kind is HF, which forms H_2SiF_6 (soluble) and SiF_4 (a gas). Strong bases under proper conditions* are a less spectacular example; in forming silicates, they do not destroy *all* the bridges, as HF can be made to do. Acids other than HF attack SiO_2 very little under most conditions. The "durability" of unprotected ("exposed") Si—O—Si bridges is less than one might think.

3. Patnode, when he discovered rearrangement polymerization, was lucky that his D_4 dissolved in concentrated H_2SO_4 because this "exposed" Si—O—Si bridges to attack. Such attack and the consequent formation of OH groups as active sites expedite rearrangement polymerization of silicones with acid or basic catalysts.

4. For want of a better word, we are justified in speaking of "durable" chemical bonds. But, when we do, we must be ready to answer the challenge, "Durable toward what, and under what conditions?" Our main concerns are stability toward heat and toward atmospheric oxidation, both properties being weighed as components of a suitable property profile; see Table 1-1.

---

* In contact with strongly basic solutions at high temperature and pressure, small crystals of α-quartz will dissolve, and seed crystals can be induced to grow until large, useful α-quartz crystals are formed. Properly cut sections of such crystals can be made to vibrate in an electric field and consequently to serve as frequency (time) standards. Such standards are important in the telephone industry and are becoming popular in wrist watches.

In the OH$^-$ process developed by Bell Telephone Laboratories and used by the Western Electric Company, the α-quartz crystals are grown in contact with 1 molar NaOH at about 350°C under a pressure near 24,000 lb/in.^{2} The "nutrient zone," in which the small crystals dissolve, is about 50° hotter [23]. The process obviously involves the breaking (in the nutrient) of Si—O—Si bridges, and a re-forming of such bridges when crystal growth occurs. The process therefore is a *rearrangement polymerization* catalyzed by a base.

Chapter Five The Augmentation
of Silicone Discoveries

To what purpose should our thoughts be directed to
various kinds of knowledge, unless room be afforded
for putting it into practice, so that public advantage
may be the result?

SIR PHILIP SYDNEY

Public advantage or public disadvantage—which, on balance, has the
scientific revolution wrought? The two cultures, literary intellectual and
scientific, have long been in controversy over this question, and the
controversy became prominent after Lord Snow's Rede Lecture. The
main objective of this book is to help each reader, by providing a look
at industrial research and its consequences, to establish his or her own
position.

Major discoveries make news and bring recognition; but, taken
alone, they did not give a sound basis for judging industrial research. To
form a valid judgment, one must regard such a discovery as the point of

Note. Brackets are used to enclose references and inserted material. For bracketed references, see References and Notes at the end of the book. Numbers in parentheses on the
right of pages are often used in the text to identify equilibria and the like: here (5-1) means
Equilibrium (5-1). Reference to laboratory notebooks is by name and date. CRDA stands
for "Chemistry Research Department Archives."

departure on a journey, usually long and costly, that must be taken "so that public advantage [and private profit] may be the result."

It is not a journey easily described. For silicones, we shall say that it comprises four steps: *discovery, augmentation, implementation,* and *marketing.* Chapters 3 and 4 have dealt with discoveries, and there will be more to follow. Here we must deal mainly with the augmentation of certain discoveries, and with their implementation during World War II.

The four steps defy satisfactory definition. "Step" itself is misleading, for it implies an orderly progression. In the early stages, when the silicone business was being established, a four-ring circus would have been a better metaphor—but it would have to be a circus in which the acts influence each other and undergo continuous change to the discomfort and occasional discomfiture of the ringmasters. As the business succeeds and matures, the circus grows less bewildering because the acts become stabilized.

Seldom, when a discovery is made, can anyone make a sound estimate of its promise. The "major" discoveries in Table 4-1 were chosen by hindsight. We shall use *augmentation* to include whatever is done to establish whether a discovery is worth implementing. *Implementation* includes whatever is necessary to make products for the market, for example, building and operating a plant. *Marketing* will not be a principal concern until we reach Chapter 7.

The most important kind of augmentation is *scientific augmentation,* mainly scientific research undertaken as a consequence of prior discoveries, which in its turn can yield discoveries of its own. Other kinds of augmentation are best described with reference to the Company organization of 1956 [1]. At that time the General Electric Company included *operating* and *services* components; the former were expected at some time to make a profit, and the latter existed to help the former succeed. If we now regard the silicone project as an operating component, all kinds of augmentation may be regarded as services to the project. Among these we shall take services such as payroll and purchasing for granted. Scientific augmentation, with emphasis on the discoveries in Table 4-1, will be discussed first.

THE DIRECT PROCESS

The orderly sequence *need, discovery, augmentation,* and *implementation* is a simple strategy for industrial research; of course—as in war—the tactical situation is usually muddled. Consider the Patnode-Scheiber scale-up of the direct process for methylchlorosilanes. It was undertaken to supply these materials inside and outside the silicone project. It was to be a start on the test-tube-to-tank-car program, that is, a start on implementation. Unfortunately, the research being implemented was then so far from complete that additional discoveries (those relating to contact masses) were needed to get the show on the road. In their turn these discoveries had to be augmented. Before this augmentation was well under way, other discoveries (some described in Chapter 4 as forks in the planned high road) began to be made, and these had to be augmented in their turn.

Marshall's decision to gamble on the direct process for methylchlorosilanes was on shaky ground even after Rochow had confirmed his result of May 10, 1940. Copper had been used by Combes in 1896 for the unscientific reason that Cu-Si was an available commercial source of silicon for reaction with HCl. He used nothing else. What, one must ask, was the chance that copper had been the best choice for Combes, and was now the best choice for the reaction of silicon with CH_3Cl? With any of the many other organic halides, RCl? Not good. No wonder Marshall instructed Gilliam and Sprung to "throw in the periodic table" in augmenting Rochow's discovery.

Gilliam and Sprung had to go beyond Rochow's work in identifying the products of reaction, but progress at the rate expected was possible only with experiments on a scale much smaller than that of the "big" run made by Patnode and Scheiber. Gilliam and Sprung were faced with a preliminary screening of likely elements followed by intensive study of the best candidates found.

In 1941 Sprung, on Marshall's instruction, abandoned the interesting and important work he had planned to do on phenolic chemistry related to condensation polymers, and began his collaboration with Gilliam. Their runs were made in a quartz reactor; took several days each; needed adjustments morning, noon, and night, weekends included; and yielded about 25 cc of liquid products from which CH_3Cl, HCl, and $(CH_3)_4Si$ had to be removed in a crude still before the analytical distillation that separated the products into portions (fractions) usually ranging in volume from 1 to 5 cc. The separation and the identification of the products were at that time formidable exercises in analytical chemistry, which will be described later. In addition to all of this, it was of course necessary to prepare "contact masses" consisting of silicon and various added elements for trial, a task made easier by the continuing cooperation of Research Laboratory metallurgists.

After almost a year Sprung and Gilliam concluded that copper was the best single element for use with silicon in the direct process for methylchlorosilanes. Copper in 1896; copper on May 10, 1940; and copper today the world over! Truly, in this case *first was best,* but why?

Even now, the augmentation had only begun. The experience of Patnode and Scheiber [Chapter 4] posed many urgent questions about Cu-Si mixtures. What kind of copper? What kind of silicon? In what ratio? How finely divided? How pretreated? What about impurities and additives? Above all, how does the same mixture behave in different reactors at different temperatures and pressures? To spare the reader, only one more question will be asked: What about product yield?

At the time the last question had a grim resemblance to "Is it colder in the mountains or in the winter?" Yield of *what* and *for what?* Which silicones for which needs? Clearly, $SiCl_4$ was an undesirable product [Chapter 4]; it could be made more cheaply from silicon and chlorine. Similarly, CH_3SiCl_3 seemed unpromising for extensive use because good silicones were unlikely to contain many of the **T** units that CH_3SiCl_3 forms upon hydrolysis. On the other hand, $(CH_3)_2SiCl_2$, the source of **D** units, was highly prized because it builds silicone backbones [Chapters 2 and 4]. As the silicone project progressed, $(CH_3)HSiCl_2$ and $(CH_3)_3SiCl$ became desirable: the former to help produce water repellency, and the latter as a source of **M** units for the termination of silicone molecules.

Rochow's most fruitful day, May 10, 1940, required years of augmentation, with much of which Gilliam was associated before he transferred to the Silicone Products Department to continue what in 1978 was the longest silicone career in the General Electric Company.

It is not possible to do justice here to all that Gilliam and his colleagues did in augmenting the discoveries of the direct processes for methylchlorosilanes and phenylchlorosilanes. Fortunately, the reader can form his or her own estimate of the situation from the following brief reviews of six patents on which Gilliam's name appears.

On March 4, 1942, Sprung and Gilliam applied for U.S. Patent 2,-380,999, granted August 7, 1945, which covers the use of an inert gas as diluent (e.g., N_2 in CH_3Cl) to increase the yield of desirable products [e.g., $(CH_3)_2SiCl_2$] in the direct process for making organosilicon halides (i.e., organochlorosilanes).

Also on March 4, 1942, the same inventors applied for U.S. Patent 2,380,998, granted on August 7, 1945, which discloses the addition of hydrogen to the alkyl halide (e.g., CH_3Cl) in the direct process for the purpose of making a monoalkyldichlorosilane (e.g., CH_3SiHCl_2). Copper is the preferred catalyst, but other metals (e.g., silver) are also useful, and satisfactory results may be obtained without an added catalyst.

On July 26, 1943, Rochow and Gilliam applied for U.S. Patent 2,-383,818, granted August 28, 1945, in which the emphasis is on using copper initially in an oxidized form as catalyst for the direct process, and on having the contact mass (initially silicon and oxidized copper) friable and finely divided to make it easier for the gas (e.g., CH_3Cl or C_6H_5Cl) to react with the silicon in the contact mass.

On February 21, 1946, Gilliam and Robert N. Meals applied for U.S. Patent 2,466,412, granted April 5, 1949, in which the main thrust is toward optimum particle size and particle-size distribution in the Cu-Si contact mass, "optimum" usually meaning "best for the production of $(CH_3)_2SiCl_2$." The patent, which contains experimental data in unusual abundance, is worth examining by anyone interested in learning at first hand what augmentation of the direct process was like. Two quotations give an idea:

In the more specific embodiments . . . [of already-patented versions of the direct process], the hydrocarbon halide is caused to react with the silicon component of a contact mass containing a metallic catalyst for the reaction, for instance copper, said contact mass being in the form of a solid, porous mass, e.g., preformed pellets, or a friable, oxidized alloy of the silicon and the metallic catalyst.

Also:

The present invention differs from the invention disclosed and claimed in the aforementioned patents in that our method of preparing organohalosilanes comprises effecting reaction between the hydrocarbon halide and the silicon compound of a powdered mass obtained by comminuting an alloy comprising silicon and a metallic catalyst for the said reaction to a particle size wherein the distribution, by weight, and the particle size, in microns,* are as follows: 50 to 100% of the particles are from 74 to 105 microns in diameter, less than 15% are smaller than 44 microns in diameter, and at least 95% are smaller than 149 microns in diameter.

Another Gilliam and Meals invention, U.S. Patent 2,466,413, closely resembles the patent just described, the principal difference being that it discloses the use of separate silicon and catalyst powders. This means that one must now worry about the distribution of particle sizes in two materials—not just one, as with the alloy.

On February 27, 1947, Gilliam applied for U.S. Patent 2,464,033, granted March 8, 1949, in which he disclosed the use of zinc (zinc chloride) as a promoter in the direct process. A promoter is a substance added in small proportion to a catalyst to make the latter more effective—

* A *micron* is a millionth of a meter, that is, 10^{-4} cm.

here, to produce more $(CH_3)_2SiCl_2$. As zinc will be changed to zinc chloride when the direct process goes on, either the metal or the salt may be used.

Not all that is claimed in these patents has been mentioned above. Nor do the patents themselves hint at more than a minuscule fraction of the work that has been done on the direct process throughout the world up to the present time [2].

The extensive scientific augmentation related to the direct process for phenylchlorosilanes will not be described. Here is the cream of the jest: experience has shown all direct processes to be so complex that conclusions drawn from the results of augmentation experiments do not necessarily apply under changed conditions—a particularly serious matter as regards kind and size of reactor.

Mechanism.   Every chemist wants to understand how his or her reactions occur—what their true mechanisms are. Unfortunately, except in the very simplest cases, such understanding cannot be attained. The best the chemist can hope for is a *model* of the true mechanism, a model that must be regarded with Holmesian skepticism. The chief virtue of mechanisms, especially in applied research, is that they often suggest useful experiments. However, notions about mechanism should never be allowed to bar experiments that seem attractive on other grounds.

To "prove" the mechanism of the direct process is about as formidable an assignment as any scientist could want [3]. Let us look only at methylchlorosilanes. Silicon is covered by an invisible oxide film that must be breached before reaction with CH_3Cl can occur. This film, fortunately, is less perfect (hence, less protective) on commercial (impure) silicon than on pure.* For the moment, ignore copper. The CH_3Cl must reach (diffuse to) and sit down on (be adsorbed on) active regions of the silicon surface. The products formed by reaction with the solid must leave its surface (be desorbed and diffuse away from), as does unreacted CH_3Cl; all are removed by the stream of flowing gas. In general, not all the silicon will be at the same temperature. The temperature will help to determine the amounts and kinds of reaction products. In particular, the higher the temperature, the more likely the formation of carbon—a second, and highly objectionable, solid.

Now, what can a third and *virtually unique* solid—copper, of course—possibly do to help matters along? Patnode [Chapter 4] thought

---

* Cost ignored, pure silicon is far less suitable than commercial as a raw material for the silicon industry. In the semiconductor industry, on the contrary, pure silicon is at a premium.

early that copper might generate free radicals,* and work by Sauer [4] provided confirmatory, albeit complex, evidence that Patnode was on the right track. Not long thereafter, Rochow began planning beautiful experiments to be carried on under a microscope, experiments in which he was joined by Dallas T. Hurd [5]. They adduced a mechanism in which $CuCH_3$ and $SiCl$ play important roles, $CuCH_3$ being an unstable compound that serves as a ready source of methyl radicals. As might have been expected, many studies of mechanism have been done since. As might also have been expected, disagreements about mechanism continue to flourish [2]. Fortunately for the silicone industry, direct processes continue to operate satisfactorily, unruffled by these disagreements. This happy situation has resulted because of long and hard augmentation that in the end relied upon statistical design and an engineering approach. Nevertheless, mechanisms as models guided thinking and suggested experiments. "Without vision, the people perish."

SILICONE CHEMISTRY

The silicone chemistry known in 1938 was inadequate for founding an industry. The literature up to that time tells us virtually nothing about methyl silicones and makes no mention of the direct process for methylchlorosilanes, the mainstays of today's silicone industry. Much had to be learned. For example, as late as July 7, 1941, Gilliam recorded a request by Marshall that work be done to establish whether the equilibrium

$$(CH_3)_2Si(OH)_2 \leftrightharpoons (CH_3)_2SiO + H_2O \qquad (5\text{-}1)$$

exists. Today we know that $(CH_3)_2SiO$ cannot have more than a fleeting existence, if indeed it exists at all.

Post [6] lists about 60 compounds of which the preparation at moderate temperatures is covered by General Electric patents. More than this number were made in the Chemistry Section. For most compounds some information on properties and chemical behavior was obtained.

Much work of this kind was begun by Patnode and carried further

---

* Organic free radicals are highly reactive species whose existences have been proved by showing that they can, for example, cause mirrors of metallic lead to disappear. The methyl radical is often written as $CH_3\cdot$, the dot representing the unpaired electron responsible for the high reactivity. Alternatively, $CH_3\cdot$ may be described as having an uncompleted valence bond.

by members of the group that was soon formed with him as de facto head. What they did is illustrated in part by References 4, 8, 9, 10, and 33. Here we shall report only on D_4, a cyclamer vital to the silicone chemistry.

As $(CH_3)_2SiCl_2$ is the precursor of silicone backbones, its reaction with water is important in silicone chemistry. Patnode began very early [7] to study the products formed when this precursor undergoes the hydrolysis-condensation sequence [Chapter 2]. When hydrolysis occurs in an excess of water, much of the resulting HCl dissolves in the water so that condensation occurs in an acid medium, a situation reminiscent of Discovery 6, Table 4-1. In such an experiment with water, a light, colorless oil forms and separates from the acid aqueous phase. Distillation of the oil (see Figure 4-2) permits separation and subsequent *characterization* of the principal substances that distill.

In one experiment about half of the reaction products appeared as distillate fractions, namely, D_3 (0.5%), D_4 (42%), D_5 (6.7%), and D_6 (1.6%); and a complex nonvolatile residue (49.2%) remained [8]. No D_1 or D_2 was found. Obviously, in studying the hydrolysis-condensation sequence for $(CH_3)_2SiCl_2$, Patnode had uncovered a complexity, foreshadowed by Kipping's earlier work on other organosilanes, that promised to be both a nuisance and a blessing—a nuisance because simplicity is generally desirable, and a blessing because more silicone species meant more possibilities of useful products. The kind of thinking underlying (5-1) was clearly too simplistic. Much further hydrolysis-condensation work was going to be needed [9], not only to enlarge silicone chemistry, but to arrive at useful oils, resins, and rubbers as well; only a little of this work can be mentioned here.

Here is what was done to characterize D_4:

1. On November 3, 1941, Patnode found 294 (average of four determinations) to be the molecular weight of D_4. The theoretical value is 296.6.

2. On November 13, 1941, Patnode measured the melting point of D_4 fractions obtained by distillation. The three middle fractions had melting points within less than 0.01° of 17.55°C.

3. On November 14, 1941, Patnode measured the density of D_4 as 0.9555 g/cc at 20°C. He reported its refractive index as 1.3968 [determination by Sauer at 20°C for the (yellow) sodium lines]. He measured the boiling point as 175°C.

4. On December 19, 1941, Earl W. Balis reported the following analytical data for D_4 to Patnode: %H, 7.93, 8.13; %C, 32.20, 32.59; % Si, 37.71, 38.04. The (then accepted) theoretical values were 8.16%,

32.38%, 37.86%. Balis reported similar values for D_3 at the same time. (All silicone cyclamers have the same ultimate composition.)

I doubt whether this characterization could be significantly improved upon today. That it was achieved as early as 1941 is a tribute to Patnode's foresight and to the skill of those who did the work.

The work described above would have made an elegant piece of pure research in a university laboratory. Yet we classify it as applied research because Patnode did it as a necessary step toward making useful silicones. It shows the importance of knowing the chemistry of industrially useful materials. Consider only the cyclamers D_3 . . . D_n. Are they useful as individuals? How could they result from the hydrolysis of a single, simple precursor, $(CH_3)_2SiCl_2$? If they are not individually useful, can they be changed into something that is? The results from the kind of work described above convey one important message, much more obvious now than in 1941, and made clearer by subsequent work on oils, resins, and rubbers. The message, hinted at earlier, is this: especially when acids, bases, or certain salts are present, and even in their absence when silicones are heated with oxygen absent [10], siloxane bridges are rather easily made and broken, whereupon the average molecular weights of methyl silicones are raised or lowered, and their molecular-weight distributions are changed. This is another way of saying that methyl silicones undergo *siloxane-bond rearrangements,* of which *rearrangement polymerization* [Chapters 2 and 4] is a special case. In Table 4-1, Discoveries 6, 7, and 8 all involve siloxane-bond rearrangements, albeit in molecules of different kinds.

METHYL SILICONE RESINS

With expanded work on the direct process decided upon, further work on methyl silicone resins had to follow. The reasons were simple. These resins were then the only methyl silicone products in sight. They were as yet unsatisfactory for general use. Moreover, silicone precursors for no other resins could then be made by a direct process.

Silicone resins are usually made by carrying mixed chlorosilanes through the hydrolysis-condensation sequence with a solvent present to remove the reaction products from the aqueous layer. The complexity of the hydrolysis-condensation sequence is by now familiar. The solvent also influences the kind of resin that forms. Diethyl ether, which was naturally used by Rochow as he had used it when he discovered methyl silicone resins via the Kipping route, is costly and flammable, and gave resins that cracked easily. Extensive studies in which solvent and hydrolysis-condensation sequence were both varied yielded better resins: for example, Sauer found it advantageous to dissolve mixed methylchlorosilanes in an inert solvent and add the solution to a mixture of water and a higher aliphatic alcohol; Marsden added a solution of the mixed methylchlorosilanes in an unsymmetrical methyl ketone to water [11].

Marshall had entrusted to James Marsden the development of useful products based on methyl silicone resins. Marsden, who had come to the Chemistry Section from Cambridge University in England through the good offices of Irving Langmuir, soon became a de facto project leader, whose energetic and cooperative efforts did much to accomplish the successful augmentation of Gilbert Wright's discoveries related to methyl silicone resins and rubber.

Marsden and Charles E. Welsh built a tower for the continuous coating of asbestos paper with methyl silicone resins, a composite insulating material being the intended product. In mid-1941 the outlook was gloomy. The best resins then available crazed and cracked after a short time at 150°C. "Indeed, it was difficult to get the resin through the curing tower [in good condition]" [12]. Fortunately the discovery of "precondensed" resins by Wright and Marsden lightened the gloom.

In October 1938 tape coated with methyl silicone resin had aroused great enthusiasm because of its resistance to deterioration in air at high temperatures. By 1941, however, the enthusiasm had moderated, as this quotation from a patent shows:

For example, a methyl polysiloxane resin may be heated for several days at 200°C without discoloration. However, when heated to 300°C in air, it disintegrates within 24 hours. Generally the first sign of deterioration is a localized cracking or crazing of the film. On continued heating, the cracks become more numerous until finally the whole film is cracked or crazed [11].

In 1941 Wright and Marsden, with Oliver's help, began to apply to the resin problem some of the observations Wright had made in experiments that led eventually to his "controlled hydrolysis" [Chapter 4]. On June 18, 1941, Gilliam summarized a conference at which Wright described the preliminary work. Gilliam wrote:

At present, methyl phenyl silicone meets all [current] requirements but cannot be produced cheaply enough. Methyl silicone can be produced cheaply enough but is subject to cracking and crazing. Experiments along J. G. E. W. [Wright's] line may improve methyls.

Improvement soon came via the "precondensed" (now called "bodied" resins), which were made as follows. The reaction products from the hydrolysis-condensation sequence (see above), after removal of acid and solvent, were dissolved anew in a "suitable" solvent (e.g., toluene). This second solution was next refluxed (heated near the boiling point) with $SbCl_5$ and $FeCl_3 \cdot 6H_2O$ for 2 hours. After cooling the mixture, water was added, whereupon an aqueous and a "varnish layer" appeared. The "varnish layer" was a solution of "precondensed resin" in toluene. What had occurred became clear after Patnode made Discovery 4, Table 4-1: *precondensation* was a *rearrangement of siloxane bonds* to give a material that could be *further heated* ("cured" or "condensed") to make *greatly improved* methyl silicone resins. Marsden, after putting such resins on glass, cheerfully described them as "the best I had ever seen" [12].

Marsden and Welsh quickly carried out a thorough investigation of the composition and process variables in the making of precondensed methyl silicone resins, whereupon these products moved out of the Chemistry Section to Building 77 for implementation. The greatest immediate value of the research was that it helped Wright and Marsden down the side road [Chapter 4] that led to a useful silicone rubber. Moreover, although precondensed resins are "old hat" today, they gave life to the silicone project in 1941.

WATER REPELLENCY. SILICONE SURFACES

Patnode's realization that water repellency due to silicone surfaces on glass (Table 4-1, Discovery 4) could prove useful called for augmentation in many different directions, two of which occurred to him in late 1940 [Chapter 4]. Subsequent history shows that the hydrolysis of mixed methylchlorosilanes to produce water repellency was the beginning of a much broader field—that of *silicone surfaces;* such surfaces are now produced on a variety of materials, in various ways, for various purposes, and by the use of various kinds of chlorosilanes and silicones. Patnode's initial observation of water repellency could not reveal or guarantee the broad usefulness of silicone surfaces'. As in all applications, many factors had to be considered; important here were mechanical durability of the surface, effective life, ease of application, cost, and (in the case of films) uniformity and freedom from pinholes.* Luckily, one successful military

---

* During World War II attempts were made to replace the brass in shell cases and cartridges by steel, brass being in short supply. The steel needed a protective coating. Silicones failed in this application; silicone coatings were too variable in thickness and had pinholes.

application more than compensated for the many early frustrations and failures [13].

Here is the story. Steatite is a magnesium silicate ceramic, a good insulator at high electrical frequencies, used extensively during World War II on radio equipment in military aircraft. Sudden descent of the aircraft from high altitudes to regions of increased humidity then entailed the risk of condensing water on the insulators and their supports. On surfaces easy to wet, films of condensed water formed and conducted electricity well enough to put the radio out of action. On steatite insulators properly treated with vapors containing methylchlorosilanes, however, the water condensed in discrete droplets because it could not form a continuous film on the water-repellent silicone surfaces. The risk of short-circuiting was thus enormously reduced and usually eliminated.*

The urgency of the aircraft problem made rapid *implementation* of the applied research necessary. Such implementation almost always involves a *transition* from the research component to a component closer to operations: such a transition had occurred when silicone resins moved to Building 77. At this time the Chemistry Section was fortunate in having

---

* A contemporary General Electric advertisement entitled "How to Cure a Flying Radio's LARYNGITIS" is a prose poem from which I must quote.

"There used to be a lot of trouble, every time an American pilot in a dogfight dropped a radio set 20,000 feet. Not crash trouble, for in the cases we're talking about the radio was in the plane and the pilot pulled out of the dive.

"But sometimes the radio lost its voice. For the sudden plunge from cold to warmer air produced condensation of moisture—like the fog that collects on your glasses when you come indoors on a winter's day. A film of moisture formed on the radio's insulators; the film let the electricity leak away; the radio quit dead! And that was bad—since a modern fighting plane depends almost as much on its radio as it does on its wings.

"But not so long ago General Electric scientists found a way around this difficulty. For if a porcelain insulator is exposed, for just a few seconds, to the vapor of a composition called G-E Dri-film—then the whole nature of the insulator's surface is changed. It looks just the same, but moisture doesn't gather any longer in a conducting film. Instead, it collects in isolated droplets that don't bother the radio a bit. The set keeps right on talking.

"Today the voices of most military radios are being safeguarded by treating their insulators with G-E Dri-film." This apt name was coined by Dr. C. Guy Suits in 1942, about 2 years before he succeeded Dr. Coolidge as Director.

The prose poem needs a little more prose. The insulators were steatite, not porcelain. The steatite compositions contained a high proportion of talc (magnesium silicate), which made it easy to produce them in intricate shapes, and rendered them far superior to porcelain as electrical insulators at radio frequency. But they did need protection against liquid films of surface water.

the Schenectady Works Laboratory * (Buildings 4, 7, and 23) almost next door and in enjoying close relationships with many of its people, notably Edward J. Flynn and Thomas J. Rasmussen. The close contacts of this strong Works Laboratory with the operating components of the Company were indispensable to the Chemistry Section. The treatment of steatite parts was a classic example of effective interlaboratory cooperation.

Mixed chlorosilanes, notably the $SiCl_4$-$(CH_3)_3SiCl$ azeotrope,† proved better than the pure compounds. The films were satisfactory, and dirt could be removed from them. Equipment for the continuous treatment of steatite parts was developed in the Chemistry Section (Francis J. Norton, Moyer M. Safford, and others) and installed first in Building 7 and then in Building 23, for operation under Rasmussen's supervision, the methylchlorosilanes being made by Scheiber in the Chemistry Section. Millions of steatite parts were successfully treated.

When the war ended, so did the business. Other ways were found of solving the aircraft problem. Hoped-for large-scale applications of the methylchlorosilane vapor treatment have not appeared to this day, but the wartime success crucially strengthened the silicone project within the Company.

Augmentation carried out in the Chemistry Section laid the groundwork for future applications of silicone surfaces. Patnode [14] found CH_3SiHCl_2 to be useful for cellulose and other materials. Unfortunately, all methods in which silicone surfaces are generated by the direct action of chlorosilanes produce HCl, which attacks many of the materials (paper, cloth, metals) that would otherwise benefit from the treatment. No satisfactory way of dealing with this drawback has been found.

To give a hint of things to come, we shall select from a great amount of work two examples in which HCl formation is dealt with by indirection. Norton [15] used toluene to dissolve an acid-free oil made from CH_3SiHCl_2, and applied the resulting varnish to make cloth water repellent. John R. Elliott and Robert H. Krieble [16] made a gel by the action of water on CH_3SiCl_3, and dissolved the gel in strong NaOH to form a methyl siliconate solution‡ that could be used to produce water repellency by the action of the CO_2 in air upon cloth dipped in the solution

---

* The Schenectady Works Laboratory has become the Materials and Process Laboratory of the Turbine Division, which was being managed by E. J. Flynn when he retired. Its present head is Walter L. Marshall, son of A. L. M.

† F. J. Norton, U.S. Patent 2,412,470. An *azeotropic mixture* is one that distills without changing its composition. We shall meet the $SiCl_4$-$(CH_3)_3SiCl$ azeotrope (i.e., constant-boiling mixture) again.

‡ Such a solution contains $CH_3Si(OH)_2ONa$ and related species.

and hung up to dry. Any other weak acid could replace the CO_2 so long as it generated satisfactorily adsorbed polysiloxanes by action on $CH_3Si(OH)_2ONa$ and the like.

A further contemporary hint of things to come was recalled in 1975 by Patnode [17]:

Dri-film-treated glass repelled water. Somebody, perhaps a nurse, in the Sch'dy works medical clinic wondered if it would make blood-stained glassware easier to clean. Somebody else, maybe me, coated some clinical glassware, pipettes, etc., that were to be used to draw or contain blood. Somebody in the clinic noted that the treatment greatly extended the clotting time of blood in contact with glass. Ergo, silicones were biologically inert. I think I also remember the same observation having been made in a Canadian hospital.

Many kinds of silicone surfaces were later found to be biologically inert, with important consequences for medicine and related fields.

METHYL SILICONE OILS

Discovery 6, Table 4-1, offered a rare opportunity for useful augmentation. Here was a class of new materials that could be tailored to order, containing molecules of substances not found in nature and never made before, for which the characterization customary in chemistry would not suffice. Augmentation had to include also the characterization demanded by the lubrication engineer.

On May 8, 1942, even before the significant distillation of Figure 4-2 had been completed, Patnode wrote, "Measure thermal expansion and viscosity over a temperature range of the following: D_x; MD_xM; TM_3; QM_4; thus gaining information of the *effect of structure on physical properties*. Phenyls in place of methyls might also be studied" [shorthand inserted to replace formulas; italics mine].

The expected stability of the oils at high temperatures made them possible replacements for the petroleum derivatives under such conditions, and this alone required that physical properties be known over wide temperature ranges. The sort of properties in question are densities, viscosities, refractive indices, surface tensions, coefficients of expansion, pour points (which help reveal behavior on melting), vapor pressures, boiling points, temperature stabilities, electrical properties, molecular weight distributions, and compressibilities; see Table 1-1. When such measurements were made, the methyl silicone oils turned out to have certain unexpected and favorable properties not only at high temperatures but *also at low*.

The extensive augmentation done by Patnode and his group can only be hinted at here [18]; a little of it will be discussed later. During the summer of 1942, Charles B. Hurd, of Union College, measured viscosities [19]. On December 3, 1942, Patnode expressed the results in the conventional form, which is

$$\log \eta = \frac{A}{T} + B \qquad (5\text{-}2)*$$

His notebook also shows that he had observed pour points near $-70°C$ on the October 16 preceding.

Augmentation was done with the primary objective of establishing the suitability of the methyl silicone oils as lubricants. The discovery was made that these oils flow almost as readily at low temperatures as at high. This unusual characteristic, which has extended the usefulness of silicones, will be discussed later.

On September 21, 1942, Patnode wrote, "Today Rasmussen [Schenectady Works Laboratory] told me that two silicone oils . . . have been lubricating a motor with bearing temperature 155°C satisfactorily for 3 weeks." Extensive work on silicone oils as lubricants, with and without additives, for different combinations of metals under widely different loadings at various temperatures, gave results that were disappointing to the following extent. Silicone oils were unlikely to qualify as all-purpose lubricants though they might replace hydrocarbons at low and high temperatures. They seemed attractive as brake and hydraulic fluids, and in specialty applications such as pumps, provided that they could be sold at reasonable prices.

SILICONE RUBBER

There is a great gap between a rubbery silicone and a useful silicone rubber. Marsden is largely responsible for bridging the gap in time to make silicone rubber useful to the U.S. Air Force and the U.S. Navy in World War II.

In June 1943, shortly after Discoveries 8 and 9, Table 4-1, had been made, Wright showed his best rubber to a visitor from the Taylor Street, Fort Wayne, Indiana, plant of the Company [20], where there was serious difficulty with the gasket for a stamped-steel supercharger, then a vital

---

* Equation 5-2 says that the logarithm of the viscosity (η) less a constant (B) equals another constant (A) divided by the absolute temperature (T).

component of certain military aircraft. A material was needed that behaved like a rubber, that had long life at high temperature, and for which the compression set* was near zero. None had been found. Wright had a gasket made of his new rubber. Tests at Fort Wayne, which showed the gasket to be far better than any previously tested or used, were quickly followed by orders "for a product which [we are] not in a position to make reproducibly" [20]. Not too long thereafter, orders began to come in also for gaskets to be used in the searchlights of battleships. These gaskets, which became very hot when the searchlight was on, had to function well enough so that the glass in the searchlights could survive the concussions produced when the battleship fired its big guns. Both applications eventually showed silicones at their best: a little silicone working invisibly and unobtrusively ensured the successful operation of formidable equipment under conditions that no other material could meet.

What follows is now mainly of historical importance, but it helps to characterize industrial research. The work in question was done under great pressure. Gaskets of silicone rubber were needed "yesterday": they had never been manufactured anywhere, and they now had to be made on the basis of inadequate research, imperfectly understood.

At the time the best silicone rubber was made as follows:

1. Prepare a good gum by the "controlled hydrolysis" of $(CH_3)_2SiCl_2$ with $FeCl_3 \cdot 6H_2O$, which acted as rearrangement catalyst to produce long gum backbones—backbones probably far removed from the idealized ones of Figure 4-3.

2. On compounding rolls, mix thoroughly into the gum at room temperature a small amount of benzoyl peroxide and a much larger amount of a suitable filler (preferably finely divided SiO_2, although the importance of the filler in strengthening the rubber was not yet known).

3. Under pressure at higher temperatures, cross-link the gum by the action of the benzoyl peroxide.

Among the many difficulties encountered in making a good silicone rubber, only those originating in the use of $FeCl_3 \cdot 6H_2O$ are of interest here. Variable amounts of substances derived from this salt could not be kept from appearing in the finished gasket, where they continued to act as rearrangement catalysts but now caused depolymerization or "reversion," that is, they led to products of low molecular weight. Marsden found an ingenious way to cope with this problem at the price of stiffening

* See Notes, Figure 4-3.

the rubber and making it more like a resin. He added 0.2% CH_3SiCl_3 to the $(CH_3)_2SiCl_2$ from which gum was to be made. The result, of course, was a product cross-linked by T units (Figure 4-3, Note 5), but in the *minimum amount* needed to give a satisfactory rubber. An important point: rubber made in this way was better than rubber made by the controlled hydrolysis of D_4 with $FeCl_3 \cdot 6H_2O$.* Silicone rubber thus made did outstanding service in both kinds of gaskets, but it was far inferior to the silicone rubber we have today, and far more difficult to make and to handle.

Implementation of the applied research on silicone rubber was a more formidable assignment than that which yielded water-repellent steatite. The Schenectady Works Laboratory came through again, however, this time in Building 23 with Flynn in charge. Flynn had begun his career in the Company on Formex® wire enamel [Chapter 1] with Patnode, and his close relationship with the Chemistry Section continued after that assignment was completed. He describes, in connection with silicone rubber, how the Company benefited:

Our principal contact was [with A. L. Marshall] in this and other projects. He was principally responsible for our involvement during the early work with silicones and continued to keep us involved, and to put pressure on us to find [applications for silicones in all their forms] wherever possible. He also made certain that we got the materials, help, and guidance whenever needed [21].

Let it be noted that there were no organizational ties between the two men: officially, Marshall's authority did not extend beyond the Chemistry Section.

---

* Today T groups are anathema in the making of silicone rubber, and better rubbers are the result. This situation has arisen largely because Dr. J. F. Hyde, now of the Dow Corning Corporation, made the important discovery that *solid alkalis* are effective rearrangement catalysts, which, added in minute amounts (e.g., 0.01%), to cyclamers such as D_4, change them in a few hours at 140°C into excellent gums. See J. F. Hyde, U.S. Patent 2,490,357: applied for, April 24, 1946; granted, December 6, 1949. The alkaline rearrangement catalyst can be completely washed out, or it can be neutralized ("deactivated") when the lengthening of the backbones has been completed. No one would dream of using $FeCl_3 \cdot 6H_2O$ as rearrangement catalyst today.

The reader will understand that M units and all other "chain stoppers" must also be excluded in the making of silicone rubber—the Patnode reasoning that led to oils is out of place here.

The purity of D_4 as starting material is thus crucial for silicone rubber, and this is why the characterization of D_4 was described earlier.

A detailed history based on Hyde's notebooks would make a welcome and valuable parallel to Chapters 4 and 5.

Marsden, in a photograph taken November 15, 1944, shows how a silicone rubber gasket (O-ring) will serve at 300°F in a turbosupercharger of a B-29 bomber.

Implementation came at record speed. Satisfactory specifications*
were worked out with the U.S. Navy. A Banbury mixer replaced rolls
for the making of "rubber compound" (= gum + benzoyl peroxide +
filler + other additives). Almost always, Scheiber made the chlorosilanes;
Marsden, the gum; Flynn, the compounds and then the rubber. Several
thousand supercharger gaskets were shipped to Fort Wayne, Indiana,
where they were installed. Several hundred searchlights[†] received gaskets
in Schenectady [21]. All this by augmentation and implementation of
discoveries made in 1942! I do not know of a demanding and urgent
transition that was more quickly made.

To the delayed clotting of blood caused by the presence of silicone
surfaces must be added another breakthrough in the medical field. Somehow the biological inertness of silicones and the discovery of silicone
rubber came to the attention of the Leahy Clinic in Boston; the exact
details are lost. In any case, Gilbert Wright was soon sending Y-shaped
rubber tubing to Boston for insertion, with what success I do not know,
as gall-bladder ducts by Leahy surgeons. Patnode, even though then at
Hanford, did his bit to promote the prosthetic use of silicone rubber. He
referred a neighbor of his in Richland, who was a surgeon, to the Research Laboratory, where someone (almost certainly Gilbert Wright)
made a silicone-rubber tube to the surgeon's specifications. By a pioneering exercise of the surgeon's skill this tube replaced the badly ulcerated
urethra of an itinerant male construction worker. For the year or so
before the patient left Richland to science's loss, the silicone-rubber tube
brought satisfactory and welcome relief [17].

ANALYTICAL CHEMISTRY

Throughout the silicone project there were never enough analytical chemists in the Chemistry Section. Moreover, the liberal Whitney-Coolidge
research policy and the minimum of formal organization had made it
possible for these analytical chemists to adopt "Every man a king" as a
way of life. Routine work was sent out. This attitude could be maintained
because the analytical chemists doubled as consultants on the many

---

* The setting of specifications is an important activity foreign to the usual research man.
When pertinent industrial experience is lacking, and when the manufacturing process is
new and difficult to control—as was the case here—reasonable specifications are particularly difficult to set. Unrealistic specifications can wreck a project and those responsible
for it.
† At about this time radar began to supersede the searchlight as an aid to firing the big guns
of battleships at night.

problems brought to them from inside and outside the Research Laboratory. These problems had a wide range, and some came from Company components outside Schenectady. Frequently research had to be done to solve them. The upshot of all this was that scientists on the silicone project could not initially obtain the analytical service to which they were entitled, although they always received advice on request.

Every project dealing with new materials acutely needs analytical chemistry, which, broadly defined, is the *characterization** and *control* of materials [22]. There is no other way to find out what is happening. The needs of the silicone project were acute indeed. Not only were many new substances made, but often there were no textbook methods for establishing their compositions.

As the silicone project grew, the attitude of the analytical chemists changed: they came to regard themselves more and more as a *services function* for the project. The shift came gradually and without open confrontations; it was wiser to ''manage change'' so as to preserve a little freedom for research by the analytical chemists. Even so, much analytical work continued to be done by those in the project. Noteworthy here was the use of instruments (mass spectrometer by Francis J. Norton and infrared spectrometer by Darwin J. Mead) at that time unfamiliar to most analytical chemists.

The changes just described were portents of what was to happen in analytical chemistry everywhere. Analytical chemists now realize that their discipline, especially in industry, is principally a services function. And they have finally admitted that, ''like it or not, the chemistry is going out of analytical chemistry'' [22].

Characterization of Methylchlorosilanes.   In Chapter 3 we saw that Rochow and Gilliam collaborated in placing silicone resins on a sound foundation by preparing them with known contents of **D** and **T** units. For this, the precursors $(CH_3)_2SiCl_2$ and $(CH_3)SiCl_3$† had to be characterized [23], and the matter was important enough to justify major effort. These precursors to silicones, which today are produced in carload lots at high purity, were—when they had finally been made at the best quality attainable in 1940—on hand only in small amounts; they were precious above rubies. The following items taken from Reference 23 show how

---

* *Characterization* implicitly includes needed *separation* or *isolation*. A pure compound must be on hand before it can be characterized. In the silicone project, distillation was of overriding importance as a method of separation, both in the laboratory and in the industry.
† Usually $(CH_3)SiCl_3$ is written as CH_3SiCl_3. The parentheses will be used here to emphasize the relationship of the methylchlorosilanes to each other.

they were characterized and indicate the kind of contribution analytical chemistry in its services function can make to a research project.

1. Statement of "boiling-point anomaly":

Although organic silicon compounds have been studied for almost eighty years the pure methyl silicon chlorides [methylchlorosilanes] have not been previously prepared, probably because the boiling points of the substances are anomalous; that is, dimethyl silicon dichloride [$(CH_3)_2SiCl_2$] and methyl silicon trichloride [$(CH_3)SiCl_3$], both of which we have made [by using the Grignard reagent], boil near 70°, appreciably above the temperature range (26–57.6°) fixed by the boiling points of silicon tetramethyl and silicon tetrachloride.

One would normally expect a systematic increase in boiling point with chlorine content in a series such as this. The high values for $(CH_3)_2SiCl_2$ and $(CH_3)_2SiCl_3$ seriously masked their identities.

2. Difficulties in developing analytical methods for the necessarily small samples:

Since the methyl silicon chlorides [methylchlorosilanes] are volatile liquids that hydrolyze instantly, it was necessary to handle them in a vacuum system in the complete absence of moisture. Furthermore, some of the products of hydrolysis are themselves volatile; others gel in aqueous solution, making exact chloride titrations difficult. The accurate analysis of small samples of these substances presented difficult problems of technique that were satisfactorily solved only after repeated efforts.

...

Fusion in the bomb also preceded the silicon determinations. . . . The ampoules for these determinations were made of a glass containing not over 0.05% Si (spectroscope), which was prepared in platinum from the oxides of boron, aluminum, and barium, and supplied to us by Dr. Navias of this Laboratory.

3. Selected results in Table 5-1, which includes calculated and presently accepted values (both in parentheses) for comparison.

4. Implicit forewarning of distillation trouble to come: In 1940 the characterization work done on $(CH_3)SiCl_3$ showed (a) a chlorine content significantly low,* (b) a high boiling point,† (c) a low freezing point, and

* These determinations of chlorine were considerably more reliable (precision and accuracy both considered) than those of any other elements in these samples.
† The boiling points are the most direct criteria of the effectiveness of distillation as a means for producing substances of acceptable purity. Note that $(CH_3)_2SiCl_2$ boils only 4.1° *above* $(CH_3)SiCl_3$.
There is additional pertinent evidence from Reference 23. Another preparation of $(CH_3)SiCl_3$ was subjected to painstaking *double* distillation with these results from the *second* distillation: $(CH_3)SiCl_3$ fraction, 60 g (of 115-g total sample); boiling point range, 66–67°C; liquid density, 1.23 g/cc.

Table 5-1 Characterization of Two Methylchlorosilanes

Substance	Composition (%)[a]				
	H	C	Si	Cl	Total
$(CH_3)_2SiCl_2$	5.08 (4.67)	20.32 (18.61)	23.46 (21.76)	54.64 (54.94)	103.50
$(CH_3)SiCl_3$	2.16 (2.02)	8.95 (8.04)	18.43 (18.79)	70.04 (71.15)	99.58

Substance	Physical Property		
	Boiling Point (°C)	Freezing Point (°C)	Density (g/cc)
$(CH_3)_2SiCl_2$	69.0–70.2[b] (70.2 ± 0.1)[c]	−80 (−76.1 ± 0.2)[d]	1.06 at 25°C (1.067 at 27°C)[c]
$(CH_3)SiCl_3$	66.2–67[e] (66.1 ± 0.1)[c]	−90 (−77.5)[d]	1.23 at 26.5°C (1.270 at 27°C)[c]

[a] Theoretical values in parentheses are based on 1971–1972 atomic weights.

[b] Boiling-point range at 744.5 mm. Values need to be raised somewhat for comparison with normal boiling points (at 760 mm) given in parentheses.

[c] These values are from Reference 24.

[d] These melting points are from Reference 25.

[e] Boiling-point range at 765.8 mm. Values need to be lowered slightly for comparison with normal boiling point in parentheses.

(d) a low density. All these observations pointed to the presence of $(CH_3)_2SiCl_2$ where it had no business to be. The authors concluded: "The redistillation thus resulted in no further purification, indicating that the limit of the fractionating column had been reached if no constant-boiling mixture was involved." The conclusion stands today with the last seven words deleted.

The work discussed above was later rounded out by the characterization of $(CH_3)_3SiCl$. "Pure" $(CH_3)SiCl_3$ also was eventually characterized by distillation and chlorine titration [24].

Augmentation in analytical chemistry here attained its minimum objective, which was the characterization of silicone precursors at the quality needed to establish the significance of **D** and **T** units in methyl silicone resins, but failed to reach its maximum objective:* the characterization of "pure" $(CH_3)_2SiCl_2$ and $(CH_3)SiCl_3$. As an unexpected important bonus, analytical chemistry gave a clear warning of distillation difficulties to come.

Analytical Distillation. Distillation has never had a serious rival as the best method for separating the chlorosilanes on an industrial scale. Analytical distillation, as we shall see, was largely superseded by gas chromatography well after the industry became established. Before that time, analytical distillation was an indispensable part of the analytical chemistry without which the silicone project could not have grown into an industry.

When the Kipping route to silicones was replaced by the Rochow route, the importance and difficulty of analytical distillation increased because the direct process gave more complex mixtures than those customarily obtained via the Kipping route. Gilliam and Sprung were made unhappy by this situation during their "throw-in-the-periodic-table" augmentation of the direct process. They finally found it necessary to build a still with a column 56 in. long that was demonstrated to have 88 theoretical plates† [24], probably a record-breaking number at the time.

* A "pure" substance is a dream. Absolutes are beyond the reach even of analytical chemistry.

† Until we reach Chapter 6, I ask the reader to be content with the following. Distillation is a way of using heat to separate materials of different volatilities by condensing their vapors, the temperature of condensation being lower the higher the vapor pressure (volatility) of the material. For liquids the separation can occur in the column of a still. The efficiency of such a still is measured in *theoretical plates*—the more plates, the greater efficiency; longer columns should yield more plates. The efficiency improves as the speed of distillation is reduced. Sometimes separation is impossible because of the formation of constant-boiling mixtures (azeotropes), one of which was mentioned in connection with water repellency. To put distillation on a quantitative basis, the compositions of vapor and solution at equilibrium must be known for all conditions of interest.

Analytical distillation was also needed in evaluating the operation of the large stills used in the production of methylchlorosilanes [24], and in characterizing the reaction products (as in Table 5-1), the properties of which are data needed for efficient industrial operation.

The upshot was that analytical distillation had to become a diversified routine operation inevitably carried out under tensions perhaps resembling those existing in the busier air-traffic control centers. The operator was E. Melvin Hadsell, trained for the post by Gilliam about 1943. After completing undergraduate work in sociology in 1935, Hadsell had been a social worker and then an insurance salesman before he learned evacuation techniques as an employee of the Research Laboratory's glass shop in 1936. Eventually, Hadsell operated eight 6-ft, 1-in.-diameter Stedman stills rated at 100 plates, and four miscellaneous stills as well. In about 20 hours the usual sample of crude chlorosilanes was separated into six fractions, each containing two compounds. Hadsell soon became knowledgeable enough to report unusual results in letters that would have done credit to a PhD, and his name is on several scientific papers.

Working conditions were often hectic, as Hadsell later indicated [26]:

Time put in? Seems like forever now, but I think we put in mostly seven-day weeks for about two years before things calmed down. Except for weekends and occasions when I'd be called in for "trouble-shooting," my hours were normal.

In the usual case a fraction produced by analytical distillation contained only two components, both of which were known substances. Then a single determination, which had to be of the highest precision attainable under practical working conditions, sufficed to complete the characterization. Eventually, this determination came to be a precise measurement of density [27].

Often the analytical chemist's lot is not a happy one. In the early days characterization was done by titrating with sodium hydroxide the hydrochloric acid produced when chlorosilane samples were hydrolyzed. This titration, old and well established, should have been foolproof. Nevertheless, the high precision required led to some unexpected shocks. At the time the atomic weight of silicon was uncertain enough to cause trouble—a difficulty that, thanks to the physicists, is no longer with us today. Then it was found that constant-boiling hydrochloric acid, the most logical reference standard for the sodium hydroxide, could not be relied upon to give the high precision needed. In such situations an analytical chemist begins to wonder whether the sun will rise again.

Fortunately, the story has a happy ending. On August 19, 1954, Mr.

H. N. Wilson of Imperial Chemical Industries lectured in the Research Laboratory on a new analytical method, vapor-phase chromatography (known to the British as GLC, or gas-liquid chromatography). In this method the components of volatile liquids can be characterized without resort to distillation. The great advantages of the method were not lost upon Stuart D. Brewer of the Silicone Products Department, or upon the analytical chemists of the Research Laboratory, where work by Earl H. Winslow soon established that the procedure could be used for the characterization of chlorosilanes and other materials important to the Silicone Products Department, where vapor-phase chromatography has since been highly successful. Colloquia are sometimes useful, and taking the chemistry out of analytical chemistry can be a relief!

Composition of Siloxanes.   To deepen understanding and to meet patent requirements, Rochow needed the compositions of the resins he discovered in 1938. He sent to an outstanding commercial microchemist, for determinations of carbon and hydrogen, samples in which these elements were to be burned to CO_2 and H_2O. The results were sometimes reasonable [28], sometimes misleading. Accordingly, Rochow consulted the analytical chemists among his colleagues [29]. These conclusions emerged: (1) the classical (and revered) microcombustion methods could not cope with all siloxanes; (2) volatilization of low-molecular-weight siloxanes* helped defeat these methods; and (3) carbonaceous material (black)* left in the residue, which should have been pure silica (white), made the situation worse. Ways out of these possible problems were suggested to Rochow, whereupon he used his great experimental skill to develop satisfactory equipment in which not only carbon and hydrogen, but silicon also, could be determined; all three elements were completely oxidized. The fourth element, oxygen, could then be estimated by subtracting the combined percentages of the other three from 100. There is no easy way of determining oxygen directly.

Rochow's samples ranged up to 200 mg, and his method required about 8 hours' elapsed time. Complete oxidation ought to be easier, the smaller the sample. When Earl W. Balis joined the Chemistry Section in 1941, he immediately undertook to devise a microcombustion method on the basis of Rochow's experience. Balis was quickly and completely

---

* These two modes of silicone behavior are of widespread importance. The volatilization was undoubtedly similar to that studied later by Burkhard [10] and encountered industrially today. The blackening probably occurs in the following way. Oxidation of the sample forms protective layers of SiO_2, beneath which the organic radicals (CH_3, C_6H_5), because of a deficiency of oxygen, decompose and form carbon. When the temperature is high enough, SiC forms, and the carbon is irretrievably locked up.

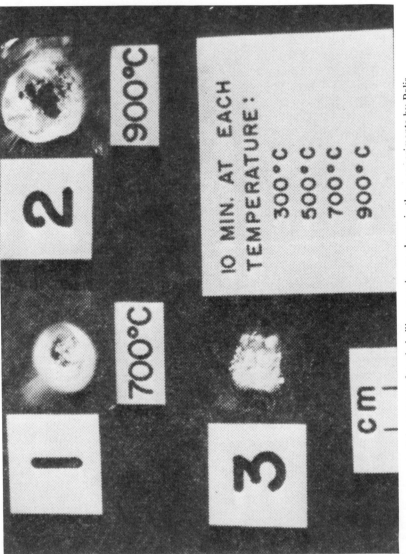

The peculiar "burning" in oxygen of methyl silicone resins as shown in three experiments by Balis. Sudden heating at 700 or 900°C (1 and 2) gives a residue containing black material that will not turn into SiO_2 (white) no matter how long it is heated. The controlled heating in 3 (see schedule) gives complete combustion to SiO_2; no black material could be found when the SiO_2 residue was broken open.

successful [30]; he reduced the sample size to 3 mg and the elapsed time to less than 2 hours; see the admirable results already listed in connection with the characterization of D_4. Furthermore, this approach to microcombustion was successfully extended to materials other than silicones with consequent advantages to analytical chemistry [31, 32].

UNUSUAL PHYSICAL CHEMISTRY

Methyl Silicone Oils. Patnode's discovery that silicone oils could, within limits, be tailor made needed extensive augmentation, of which the measurements of properties have been mentioned. Research in physical chemistry was also needed. Among the questions that presented themselves were these three: Is H_2SO_4 the best rearrangement catalyst? How does such a catalyst function? Will rearrangement ever stop as long as a catalyst is present? The second question is one of reaction mechanism, and we shall dismiss it here. The third deals with equilibration* and is of utmost practical importance.

The kind of thinking that led Patnode to do his experiments with H_2SO_4 pervaded the Chemistry Section. For example, Marsden, on November 22, 1941 (2 days before Patnode's first experiment), planned to discover whether $FeCl_3 \cdot 6H_2O$ might act on cyclamers such as D_4 to open the ring and, if so, whether "when catalyst is removed the rings may form again." As we saw in Chapter 4, Patnode was lucky that D_4 dissolved in H_2SO_4—a single solution makes for a simpler experiment; his luck continued when his second experiment gave linear molecules only, with all the D_4 consumed. The consequence was that his experiment gave him results so simple as to leave no doubt that his thinking of April 17, 1942, had been correct. Naturally, much remained to be done. Figure 4-2 shows that the oil he had made was a mixture, not necessarily an *equilibrated mixture.* Oils that contained *linear, branched,* and *cyclic* polysiloxanes would need to be investigated, as would catalysts other than H_2SO_4.

Donald F. Wilcock [33] and Donald W. Scott [34] provided part of the needed augmentation. They used **M, D,** and **T** units as required to make equilibrated oils of varying complexity, in which linear, branched, and cyclic molecules were present. These oils were equilibrium mixtures,

---

* *Equilibration* is the physical chemist's way of saying that a process is being allowed to run its course until no further change in its rate can be measured. Bringing material in a flask to a desired temperature by immersing the flask in a constant-temperature bath is a simple example.

and their compositions could be established by distilling them after all the H_2SO_4 had been removed. In this way the composition of the *unique* equilibrium mixture that corresponded to a given mixture of starting materials could be established, and oils could be made to order. Also, this composition (called the *equilibrium distribution* of molecules) could, all things considered, be satisfactorily calculated by using mathematical equations based on reasonable assumptions and experimental observations. Wilcock and Scott made physical chemistry give results that were at once elegant and useful.

The study of rearrangement in oils was thus carried far enough to promote an understanding of what happens when any kind of siloxane is treated with a rearrangement catalyst to give other kinds of silicones (e.g., gums) as products. It was unfortunate, though understandable, that the study of the rearrangement catalysts themselves was not taken further than it was. The intent was there; probably too much else had to be done first.

Wright had found that his "controlled hydrolysis" could be accomplished by alkaline (e.g., hydrated Na_2CO_3), as well as by acid, materials. On January 9, 1943, Patnode wrote:

Today I began the study of the effect of HNO_3 on methyl polysiloxanes as a companion study to that with H_2SO_4. The discovery that H_2SO_4 and perhaps also HCl will open the SiOSi bond some time ago suggested that other acids and alkalis deserve study. We have now got the H_2SO_4 process pretty well in hand, and so the next to come is HNO_3.

On July 30, 1943, he observed:

The SiOSi linkage is easily broken by strong acids and alkalis. Acids lead to recondensation with change in molecular size (rearrangement], while alkalis generally lead to salts, as SiONa. [Cf. the work on methyl siliconates reported above.] The exact role of concentration of acids, alkalis, and salts is now being studied more intensively than before . . . under my direction.

Patnode was planning the silicone analogue of Baekeland's study of catalysts for making phenolic resins.

No one can be sure that further work on rearrangement catalysts would have led Patnode or his colleagues to anticipate the important Hyde U.S. Patent 2,490,357 (application, April 29, 1946), mentioned above, according to which a great variety of silicones (not only gums) can be made. Using as *little* as 0.01% of *solid* KOH to effect rearrangement seems to me a considerably more difficult conceptual step than a simple switch from acid to alkali, and a step the likelihood of which could depend more than is usual upon the course that prior related research had taken. The silicone industry will always be indebted to Dr. Hyde. Rearrangement by the use of acid catalysts is also done differently today than it was by Patnode and his colleagues.

Methyl Silicone Gum. At about the time that silicone-rubber gaskets were beginning to be made, a question seriously worrying Marsden and others was why the rubber had so little strength. Polymers then being imperfectly understood, the general belief in the Chemistry Section was that the rubber lacked strength because its molecules were too small. This belief was about to be annihilated.

   Scott, who had joined the Section for silicone work with Marsden in 1942, undertook to test this belief by establishing the range of molecular weights (more precisely, the molecular-weight distribution) to be found in the kind of gum from which methyl silicone rubber was being made. The investigation was not one to be undertaken lightly or with joyful anticipation. The silicone gum had first to be separated painfully into parts ("fractionated"), in each of which molecules of approximately the same size were to be found. These "fractions" had then to be dissolved, and the osmotic pressure of the solutions measured: the larger the molecule, the more difficult the measurement—it is never easy. Nevertheless, Scott [35] carried on. On June 26, 1944, he summarized the results to date as follows:

Fraction	A	B	C	D	E
Percentage of gum sample	16.7	13.5	17.7	27.2	12.2
Nature of fraction	Elastic	Soft and plastic	Soft and plastic	Very plastic	Viscous liquid
Molecular weight[a]	2,500,000	1,400,000	630,000	310,000	50,000[b,c]

[a] These interim results differ a little from the final, published values. They are the average molecular weights for each fraction.
[b] Considered unreliable and not published.
[c] Note the transition from liquid to "elastic" solid with increasing molecular weight. The solid is too weak to be called a rubber.

   The results were from work in progress, but they show beyond all doubt that the molecules in silicone gum are large, and that many of them are huge. How huge is easy to see: one need only divide Scott's numbers by 74, which is the molecular weight of a **D** unit. Although the quotients are only averages, the backbones of some of the molecules in this gum surely contained *thousands* of **D** units. Silicone gums *are not* cluster polymers of the kind Staudinger had rejected as unrealistic:* in this respect polymeric silicones were not a class apart from other polymers. But high molecular weight does not guarantee strength in silicones to the extent that had been taken for granted. It was a shock to learn that "big could be weak." Scott had freed the Chemistry Section of a serious misconception.

With this misconception laid to rest, the belief grew in the Chemistry Section that various physical properties of different silicones, such as the weakness of the gums, the low boiling points of some liquids, and the ease with which many oils flowed at low temperatures as well as high,† pointed to unusually low attractive (cohesive) forces between all kinds of silicone molecules.

X-Ray Evidence.   The diffraction (coherent scattering) of X-rays by crystals has become so sophisticated that it is now possible not only to establish crystal structure in this way, but also to make electron-density patterns for atoms or groups within a molecule that is a building block of the crystal. If such an atom or group undergoes unusual (thermal) motion, its electron-density pattern will be distorted. In favorable cases such distortion can be quantitatively correlated with a kind of motion assumed for the atom or group.

By October 1, 1945, when Walter L. Roth joined the Chemistry Section, Marsden and David Harker had attempted to obtain X-ray diffraction patterns on silicone rubber made crystalline by stretching, a measure known to crystallize organic polymers. Their experiments failed, mainly because their silicone rubbers were too weak.

Roth sensed a paradox. How could the intermolecular attractive forces in *silicones* be weak when their molecules were held together by Si-O bonds, which in *silicates* were known to be highly ionic? "Covalent" bonds of highly ionic character ought to make for electrical forces strong enough to hinder molecular movement in silicones under ordinary conditions. Roth began X-ray diffraction work on silicone rubber in an attempt to resolve the paradox. The results in 1945 were inconclusive.

---

* Please see "A Guide to Polymers" in Chapter 2.
† In scientific language such oils have a low temperature coefficient of viscosity.

Fortunately for Roth, Scott had made and crystallized the silicone "bicycle" [36]

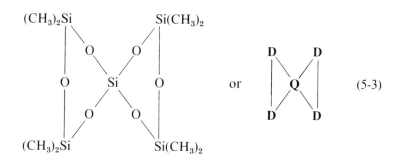

$$(5\text{-}3)$$

using $(CH_3)_2SiCl_2$ and $SiCl_4$ in proper proportion as precursors. On November 19, 1945, Roth decided to make X-ray diffraction measurements on the needlelike crystals about 1 mm long and 0.1 mm in diameter that Scott had given him.

The crystals had to be confined in polymeric envelopes to prevent evaporation during X-ray exposure. The diffraction patterns, excellent for the time, yielded a structure that provided valuable clues to the reason for the weak intermolecular forces.

First, the length of the Si-O bond in silicones agreed with that in silicate minerals, and the Si—O—Si bond angle was large, suggesting easy motion. Ergo, the Si-O bonds in silicones did resemble those in silicates, and presumably did not form the fixed angles characteristic of covalently bonded hydrocarbons, where ionic character is virtually absent. Second, the electron-density map showed that CH_3 groups in silicones precessed with extraordinarily large amplitude and consequently moved so as to sweep out large volumes with umbrella-shaped surfaces around the silicon atoms [37]. The final conclusion [38] was that the $Si(CH_3)_2$ group as a unit underwent large, restricted motions ("librations") in this crystalline *nonpolymeric* molecule. Translated to a *siloxane backbone*, this conclusion suggests that, because of such motions, there exists around the backbone a relatively large volume from which other molecules are excluded. The *effective* attractive forces are consequently reduced: distance makes attraction weaker.

Roth had opened an important field to which nuclear magnetic resonance was soon to contribute [39]. This story will be continued in Chapter 8.

MISCELLANEOUS AUGMENTATION

We shall see only the tip of the augmentation iceberg. By the middle of 1945, some 75 reports had been written.* Work on chlorosilane derivatives, precursors to new silicones, has been given particularly short shrift in this chapter. Sauer alone wrote a series of five reports on the derivatives of only one family of chlorosilanes—the methyls—between July 21, 1941, and November 22, 1942; the series was perhaps half his output for that period of a year plus a few months. Charles A. Burkhard also contributed extensively, as did others.

Brief descriptions of four items of augmentation will follow. These are diversified, outside the main stream of the research, and chosen to give the reader an indication of how comprehensive adequate augmentation can be.

Anomalous Boiling Points. This topic was stressed above as being important and as causing early confusion in the distillation of methylchlorosilanes. Richard N. Lewis and Arthur E. Newkirk [40] showed that the "anomaly" in the boiling points could be predicted from an empirical, but very satisfactory, correlation between the boiling points of many silicon compounds and the natures of the groups attached to the silicon atoms therein.

The Problem of SiCl$_4$. This substance not only is a wasteful product of direct processes, but also complicates the distillation of methylchlorosilanes because of the existence of the constant-boiling mixture (azeotrope), already mentioned, which it forms with $(CH_3)_3SiCl$. The boiling points in question are: $(CH_3)_3SiCl$ and $SiCl_4$, both near 57.6°; azeotrope, near 54.5°. In the early days, when $SiCl_4$ often formed in undesirably large amounts, ways of recovering at least the $(CH_3)_3SiCl$ had to be found. Such recovery is possible by distillation of a mixture in which an additive to the azeotrope has produced a suitable spread in boiling points. Sauer used C_2H_4O in this way; Newkirk, ZnF_2; Reed and Sauer, CH_3N [41, 42, 43].

In the end the $SiCl_4$ difficulty was resolved in the best possible way: the direct process was improved enough so that $SiCl_4$ was no longer significant. As a consequence the research described above entered an all-too-common category, which haunts those who must manage industrial research: admirable, fruitful, but not useful.

* Regular reports were still not required. Many reports would have qualified as PhD theses.

Silicone Nomenclature. We are fortunate because we seldom need to deal with isomers* in this book. Consequently, the formulas of compounds can usually replace their names. For the successful augmentation of silicone chemistry, however, exact names are a necessity; Humpty Dumpty must not run wild. At the beginning of the silicone project, many of the names in the literature were unacceptable. Furthermore, many of the new compounds had complex structures, an added difficulty.

Sauer [44] began before 1944 to look for an improved way of naming organic silicon compounds. Patnode, Rochow, and Sprung were prominent among those who helped him. The American Chemical Society later extended their work and arrived at definitive guides [45].

Two bicyclic compounds will show the reader how difficult naming can be. Consider first

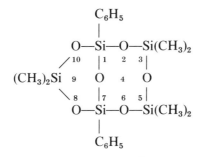

3,3,5,5,9,9-Hexamethyl-1,7-diphenylbicyclo[5.3.1]pentasiloxane

Here the numbers 3, 3, 5, 5, 9, 9 identify silicon atoms to which methyl groups are linked; 1 and 7 likewise serve the phenyls. But the numbers in brackets have an altogether different function: they give the number of bridges (three) in the molecules and also the numbers of (silicon and/ or oxygen) atoms between individual bridgeheads (all of which mercifully are silicon atoms in polysiloxanes): one bridge has five such atoms; another has three; and the third has one.

The second example is the compound prepared by Scott for Roth's X-ray work. It is called octamethylspiro[5.5]pentasiloxane and was shown in (5-3). It has a common bridgehead, the central silicon atom; "spiro" (derived from "spiral") is used to indicate this fact, even though the planes of the two rings are perpendicular.†

* Isomers are compounds identical in composition and molecular weight (hence with identical formulas) but with molecules of different structures. See (4-7).
† At this point *"Dum spiro, spero"* seems a good way to close the subject.

Flammability of the Chlorosilanes. As though their other objectionable properties were not enough, chlorosilanes' can be serious fire hazards.* Investigations of their behavior on ignition [46, 47] revealed many interesting features, of which only two will be mentioned. In the beginning, two explosions of $(CH_3)_2SiCl_2$-air mixtures in liter glass bulbs were so violent that safer apparatus had to be devised. Moreover [46]:

When flame propagation occurs in a darkened room, a dazzling white hemisphere, equator downward, moves up the tube [56 in. long] and reaches the top in several seconds. Incandescent silicalike particles appear to be responsible for the dazzling whiteness. As the flame progresses, the walls in the burned zone become covered with a white deposit. Large white flocs of silicalike material float downward for about a minute after the flame has traveled the full length of the tube.

When combustion was complete, silica particles, less than a millionth of a centimeter in diameter and joined in a loose three-dimensional network, were formed. Such small particles cannot be made by grinding silica. A disclosure letter describing process and product was written, but no patent was obtained.

AUGMENTATION VIA PATENT ACTIVITY

The Patent System. A look at industrial research ought to include at least a layman's glance at the patent system.†

---

* Soon after the Patnode-Scheiber [Chapter 4] scale-up was in full swing, a stockpile of methylchlorosilanes in glass containers began to accumulate in a small room with a single door that terminated the fifth-floor hallway. Robert Palmer, Superintendent of the Research Laboratory, who was responsible for safety measures, became concerned about the risk of fire. To reassure William McGraw, Palmer's representative, Marshall poured some chlorosilanes into a flat disk, held a flame near it, and was rewarded with instant ignition and violent combustion. With his usual aplomb he calmly remarked, "It's a *cold* flame, though." Nevertheless, he immediately directed the analytical chemists to investigate the behavior of the chlorosilanes on ignition: he recognized trouble when he saw it! The first report on the subject is dated September 18, 1943 [CRDA].

Until Marshall did this experiment, the consensus was that these organochlorosilanes contained enough chlorine to make them flame resistant. Another generalization defeated by experiment!

† Inventions and patents are far more important in industrial than in academic research. The U.S. Patent Office seems always to be overloaded. It has been reported that the activity leading to patents was about 25 times greater in 1973 than in 1925, although the number of patents granted to U.S. citizens was about the same in both years. The patent system is so important to industry and to industrial research that Adamsian multiplicity and complexity must not be allowed to choke it.

Article 8 of the U.S. Constitution empowers the Congress "To promote the Progress of Science and useful Arts by securing for limited Times to . . . Inventors the exclusive Right to their . . . Discoveries." In other words, *inventors* make *discoveries* for which letters patent (patents) may be granted. By implication the founding fathers made the two important words synonymous. Humpty Dumpty would not have been pleased!

A chapter could be written on whether it is advisable, or even possible, to distinguish between discoveries and inventions. It was easy in 1783. The *Oxford English Dictionary* quotes H. Blair as then having said: "We invent things that are new; we discover what was before hidden. Galileo invented the telescope; Harvey discovered the circulation of the blood." Well, things are not so simple any more, as anyone who applies these criteria to Table 4-1 will soon discover.

It is 1783 no longer, and it is no longer possible to define "discovery" succinctly—could the founding fathers, in drawing up Article 8, have had a premonition of what lay ahead? Nevertheless, we do need some sort of definition. Building upon Table 4-1, we shall say that a "major discovery" has been made when an experiment justifies a critical change (e.g., expansion, abandonment, new direction) in a research program. And, when we juxtapose "discovery" and "augmentation," we shall normally mean that a "major discovery" is under discussion: after all, most experiments lead to discoveries of one sort or another so that restriction, though necessarily subjective, is desirable. A good rule in writing about industrial research is this: for the same experimental observation, use "discovery" in a scientific article, and "invention" when patent considerations predominate.

Except in this section we shall favor "discovery." Two conceivable, though unlikely, examples show why. Suppose that silicones or related substances were discovered to be so toxic as to force abandonment of the project: a major discovery certainly, but no invention. Suppose that Rochow had discovered that the direct process yields unacceptable amounts of $SiCl_4$, and that augmentation did not improve the situation. The discovery would have been industrially useless, although it would probably have been patented.

Any patent grant is for "an alleged new and useful invention." The second word dominates the quotation. Patents may be struck down in the federal courts; for various reasons almost no patent cases are argued before the Supreme Court. The validity of any patent not tested in court is a fit object of Holmesian skepticism. As intimated in Chapter 1, U.S. Patent 1,082,933, "Tungsten and Method of Making the Same for Use as Filaments of Incandescent Lamps and for other Purposes," was the most important ever granted to the Company, and among the most important ever to issue from the Patent Office. Its subsequent history is revealing.*

To be patentable an invention must, in the opinion of the Patent Office, be new, useful, and not obvious to those "skilled in the art." Reduction to practice (demonstration of usefulness or operability) must

*DUCTILE TUNGSTEN GOES TO COURT: A PATENT DRAMA IN SIX ACTS

Act I

Invention. U.S. Patent 1,082,933; applied for, June 19, 1912, by William D. Coolidge; granted, December 30, 1913.

Act II

Suit. General Electric Co. v. *Independent Wire Co.*; alleged infringement related to lamps. Tried before Judge Hugh M. Morris, sitting on special assignment in District Court, New Jersey District. Decision (see below), June 19, 1920. Reference: 267 F. 824. (F. means "Federal Reporter.")

Act III

Suit. General Electric Co. v. *De Forest Radio Co. et al.*; alleged infringement related to radio tubes. Tried before Judge Hugh M. Morris (*sic!*), sitting at home in District Court, Delaware District. Decision (see below), January 18, 1927. Reference: 27 F. (2d) 590.

Act IV

Appeal by plaintiff from decision of Act III. Circuit Court of Appeals. Third Circuit. Decision (see below) September 18, 1928, by Circuit Judge Victor B. Woolley with Circuit Judge Joseph Buffington dissenting in part. Reference: 28 F. (2d) 641.

Act V

Petition by appellant for rehearing in Circuit Court of Appeals, Third District. Petition denied November 23, 1928. Reference: 28 F. (2d) 650.

Act VI

Petition by G.E. for writ of certiorari to the court of Acts IV and V. Petition made to, and denied by, the United States Supreme Court. The end of the road. Reference: 49 S. Ct. 180, 73 L. Ed.

BOX SCORE FOR ACTS II, III, AND IV

Claim No.	1	2	3	4	5	6	7	8	9	10	11	12	13	14	15	16	17
Nature	p	p	p	p	p	p	p	p	p	p	p	p	p	p	p	p	p
Act II	V	V	V	V	V	V	V	V	V	V	V	V	V	V	V	V	V
Act III	I	I	I	I	I	I	I	O	I	I	I	I	O	I	O	I	I
Act IV	V	V	V	V	V	V	V	O	V	V	V	V	O	I	O	I	V

Claim No.	18	19	20	21	22	23	24	25	26	27	28	29	30	31	32	33	34
Nature	p	p	p	p	p	p	P	P	P	P	P	P	P	P	P	P	P
Act II	V	V	V	V	V	V	V	V	V	V	V	V	V	V	V	V	V
Act III	I	I	I	I	I	I	I	O	I	I	I	O	O	O	O	I	I
Act IV	V	V	V	V	V	V	I	O	I	I	I	O	O	O	O	O*	O*

Symbols. p = process claim; P = product claim; V = claim held valid (or not invalid); I = claim held invalid; O = claim not at issue; O* = claim held by court to be not at issue.

This summary covers all of a story incomplete elsewhere [49]. The court opinions are worth reading for their bearing on Lord Snow's problem of the two cultures: note particularly what Judge Woolley wrote in Act V [28 F. (2d) 650–51].

be done with due diligence. It does not matter to the Office how a discovery was made or whether anyone (discoverer included) understood what was done.

Each key word or phrase in the criteria for patentability necessarily has a governing legal meaning. That alone shows the need for patent counsel. In addition, as the quotations from patents in this book show, a good patent application must be carefully drawn.

The interaction of inventor and patent counsel before the patent application is submitted usually helps them both, especially when the building of a sound patent structure in a new field requires additional research, which can be jointly planned and from which new inventions may result.

In 1939 silicones were such a field. The interaction of Rochow with Harold L. Kauffman, his Patent Counsel, was highly successful, as is shown by the following abbreviated account of the history of Rochow's first silicone patent.

On March 3, 1939, Rochow wrote Dr. Coolidge, "describing methyl silicone and its preparation for patent purposes." On March 21, 1939, his notebook reads, "Literature search for patent application continued" and details follow. Six days later, he noted:

I talked with Mr. Kauffman about the literature search for patent application. Further points to be settled before filing:
(1)   Can other aryl silicones be chlorinated? What ones should be mentioned in process disclosure?
(2)   Can a halogenated aryl silicone be claimed? How about F [fluorine] and I [iodine] on the ring?
(3)   Alkyl silicones probably cannot be claimed. How about claiming methyl silicones and then filing a separate docket on silicones of certain properties made by adjusting composition, etc.? [On April 13, 1939, Rochow describes methyl silicones of varying CH_3/Si ratios.]
(4)   Describe method for prep'n of methyl silicones in full, and desirable properties of various compositions.

After doing the needed experimental work and after discussion of the patent situation, in which Marshall and Patnode participated, Rochow could finally write, on July 31, 1939, that he had signed the papers necessary to speed the application on its way to Washington.

When there is doubt as to who made an invention, a "patent interference" results. Such interference can be internal (within the corporation) or external. Patent counsel is needed in either case. One astute counsel was fond of quoting an obstetrician who said, "It is easier to conceive than to deliver."

Oct. 7, 1941.   E. G. ROCHOW   2,258,218
METHYL SILICONES AND RELATED PRODUCTS
Filed Aug. 1, 1939

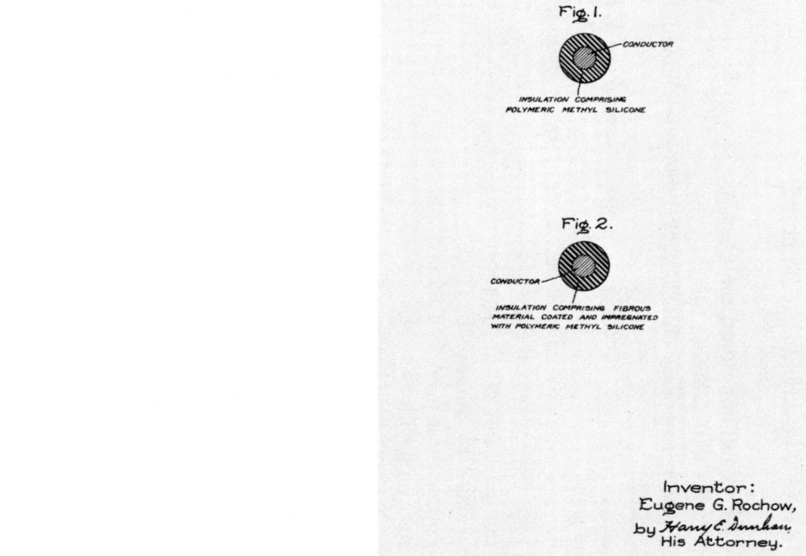

Fig. 1.

CONDUCTOR

INSULATION COMPRISING
POLYMERIC METHYL SILICONE

Fig. 2.

CONDUCTOR

INSULATION COMPRISING FIBROUS
MATERIAL COATED AND IMPREGNATED
WITH POLYMERIC METHYL SILICONE

Inventor:
Eugene G. Rochow,
by Harry E. Dunham
His Attorney.

Patented Oct. 7, 1941   2,258,218

UNITED STATES PATENT OFFICE

2,258,218

METHYL SILICONES AND RELATED PRODUCTS

Eugene G. Rochow, Schenectady, N. Y., assignor to General Electric Company, a corporation of New York

Application August 1, 1939, Serial No. 287,787

29 Claims.   (Cl. 174—121)

This invention relates to new compositions of matter, their preparation and use. More particularly it is concerned with new and useful polymeric bodies comprising chemical compounds of silicon, oxygen and at least one methyl group attached directly to silicon. These polymeric bodies may be defined more specifically as polymers of methyl silicone and of related compounds in which the methyl groups are replaced in part by oxygen.

I have discovered that the heretofore unknown methyl silicone can be obtained in the form of stable polymeric bodies having characteristic properties which make them of particular value in industry. Polymeric di-methyl silicone has a unit structure,

$$
\begin{array}{c}
\mathrm{CH_3} \\
\mid \\
-\mathrm{O}-\mathrm{Si}-\mathrm{O}- \\
\mid \\
\mathrm{CH_3}
\end{array}
$$

which in the high molecular aggregate may have chain endings such as hereinafter described. Related compositions in which the methyl groups are replaced in part by oxygen would give rise to the unit structure

$$
\begin{array}{c}
\mathrm{CH_3} \\
\mid \\
-\mathrm{O}-\mathrm{Si}-\mathrm{O}- \\
\mid \\
\mathrm{O}
\end{array}
$$

which, in its polymeric form, or in combination with the first-named unit structure gives rise to a series of polymeric substances of varying properties.

The polymeric bodies of this invention are unique in that they have no carbon-to-carbon bonds and therefore are free from the type of thermal decomposition which initiates in the rupture of a carbon-to-carbon bond. In marked contrast to previously known silicones, specifically ethyl silicone (Kipping and Martin, Journal of the Chemical Society, 95, 302), propyl silicone (Kipping and Meads, J. C. S., 107, 459), phenyl silicone (Kipping and Martin, J. C. S., 95, 319, 2108, 2121, 3125), benzyl silicone (Kipping and Robison, J. C. S., 101, 2142, 2156) and tolyl silicone (Kipping and Fink, J. C. S., 123, 2830), all have carbon-to-carbon linkages within their structure.

I may use any suitable method for preparing these new compounds of silicon, oxygen and the methyl group. For example, I may hydrolyze a methyl silicon halide, for instance methyl silicon chloride, bromide or iodide, and dehydrate the resulting hydroxy product. I prefer to use methyl silicon chlorides as starting materials for preparing the hydroxy compounds.

In order that those skilled in the art better may understand how the present invention may

be carried into effect, the following illustrative examples are given:

Example I

(A) An ether solution of 1.75 mols of methyl magnesium bromide is added slowly and with rapid stirring to an ether solution containing 1 mol of silicon tetrachloride, the latter solution being cooled to minus 20° C. or lower prior to the addition of the former. The solution of methyl magnesium bromide is added in such rate, and the reaction mixture so cooled, that the temperature of the mixture does not rise above about 0° C. Preferably the reaction mixture is maintained at minus 5° C. or lower. Since the amount of methyl magnesium bromide employed is insufficient to convert all the silicon tetrachloride to the methyl derivative, there is formed a mixture of mono- and di-methyl silicon chlorides, thus:

$$\mathrm{CH_3MgBr + SiCl_4 \rightarrow CH_3SiCl_3 + MgBrCl}$$
$$\mathrm{CH_3MgBr + CH_3SiCl_3 \rightarrow (CH_3)_2SiCl_2 + MgBrCl}$$

Probably a smaller amount of trimethyl silicon chloride also is formed:

$$\mathrm{CH_3MgBr + (CH_3)_2SiCl_2 \rightarrow (CH_3)_3SiCl + MgBrCl}$$

The magnesium salts separate as a granular mass, leaving the silicon derivatives in either solution.

(B) The magnesium salts may be separated, if desired, but it is not necessary to do so. Ordinarily, the entire cold reaction mixture is poured into an equal volume of concentrated hydrochloric acid, or better hydrolyzing the silicon derivatives thus:

$$\mathrm{CH_3SiCl_3 + 3H_2O \rightarrow CH_3Si(OH)_3 + 3HCl}$$
$$\mathrm{(CH_3)_2SiCl_2 + 2H_2O \rightarrow (CH_3)_2Si(OH)_2 + 2HCl}$$
$$\mathrm{(CH_3)_3SiCl + H_2O \rightarrow (CH_3)_3SiOH + HCl}$$

These hydroxy compounds readily condense with loss of water to form initial and intermediate condensation products. These products are either ether-soluble and are collected by separating the ether layer. After washing free of acid, the ether solution is concentrated to obtain a viscous liquid suitable for adhesive and surface-coating applications. This viscous liquid may be further condensed and polymerized to a hard solid by heating at a temperature, for example at 100° to 200° C., to form bodies of desired flexibility and hardness. In many cases it is advantageous to advance the condensation and polymerization of the viscous liquid by heating it in situ, that is, in the position of its ultimate use.

When the liquid reaction product obtained as above described is heated gradually to 200° C. over a period of 24 hours and then kept at 200° C. for 48 hours, the resulting product is a clear, colorless, odorless, horny solid. A typical polymer

Chapter 2 points out that Kipping, in his devotion to pure research, failed to foresee needs that his work might fill. Had he been sensitive to these needs, he might have reduced his discoveries to practice and obtained valuable patents. For example, Kipping's discovery that organochlorosilanes could be made by use of the Grignard reagent would seem to have been a patentable invention.

Stock's case is different, but more interesting. He hydrolyzed methyl silanes (not methylchlorosilanes) to obtain what he inferred to be methyl silicones [48], as he indicated by writing

$$SiH_2(CH_3)_2 + H_2O = [SiO(CH_3)_2]_x + 2 \, H_2 \qquad (5\text{-}5)$$

which we change to read

$$x \, SiH_2(CH_3)_2 + x \, H_2O = D_x + 2x \, H_2 \qquad (5\text{-}6)$$

Is this a patentable discovery of D_x, a methyl silicone? Probably not, for the existence of D_x was inferred from the amount of liberated hydrogen alone: the milligram quantities of siliceous product were never analyzed, otherwise characterized, or proved useful.

The patent literature, court briefs and similar documents included, is a mine of valuable scientific information, but it is not necessarily scientific history. For example, it is possible, though uncommon, for A to make a valid patent claim of a subsequent discovery by B because A anticipated B's discovery as a logical extension of a prior discovery by A.* Also, the patent literature naturally has patents as its objective, and the adversary method dominates litigation. The patent literature is sometimes a better historical source than the scientific, but its nature must be kept in mind.

THE SILICONE PROJECT AND THE CHEMISTRY SECTION

The silicone project was certainly one of the most important in the history of the Company's corporate chemical research; one could argue that it was critical for the Chemistry Section. A comparison of Figures 4-1 and 5-1 makes the case. The combination of a liberal research policy, faith in new materials, military necessity, and an interesting new field of chem-

---

* This example helps to explain the patent action taken in the case of silver as catalyst in the direct process for phenylchlorosilanes (Discovery 5, Table 4-1). In that case Rochow and Gilliam collaborated in the discovery, but Rochow became the sole inventor. On March 12, 1942, Rochow reported that "Mr. Rule [patent counsel] . . . advised filing a continuation in part [claiming silver as Claim 17] on the general [direct process] case." Mr. Rule's advice was aimed at strengthening the patent structure. Patents are not merit awards.

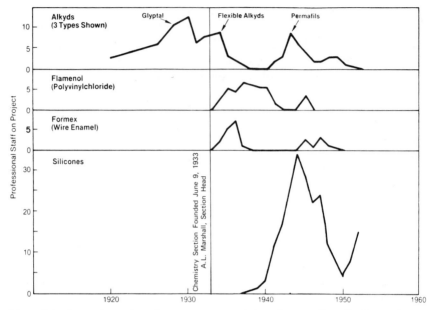

Figure 5-1 Approximate log of corporate research in polymers. Compiled by Arthur E. Newkirk.

istry to explore—all these led, with Marshall as indefatigable champion, to a large and rapid expansion of the Section's staff at a time when management and organization were not yet accorded the emphasis they deserve and receive today. Consequently, the Chemistry Section* during World War II (more precisely, to the peak shown in Figure 5-1 for 1944) could be considered a unique experiment in industrial research—one that shows what happens when young, competent, and enthusiastic scientists are given the opportunity to do research in a new and interesting field with a minimum of management restraint. At that point the Section resembled two small islands (Permafils; analytical chemistry and special problems) awash in a turbulent silicone sea—a comparison not quite accurate because Newkirk was trying to polymerize vinyl fluoride.

Comparison of Polymer Projects.   Figure 5-1 complements general infor-mation given in Chapter 1. The success of the earlier polymer projects

* A reading of L. Thomas, *The Lives of a Cell,* Viking Press, New York, 1974, especially pp. 11–15, 100–102, and 111–120, suggests that the Chemistry Section (and similar orga-nizations) ought to be regarded as living organisms. Experts on management and organi-zation should find the book intriguing.

acted as a powerful springboard for the rapid enlarging of the silicone project. Of course, the greater the enlargement, the greater was the risk that less-than-expected success would jeopardize future polymer projects within the Company. That is the reason why the silicone project was crucial for the Section.

Formex® wire enamel, as we saw in Chapter 1, was a brilliant success. Figure 5-1 shows that the success was gained at low research cost. By 1945 it was clear that silicones would be a far different story. No satisfactory silicone wire enamel was in sight; in fact, it looked as though silicones would need to make their marks in nonelectrical applications if they were to survive.

Silicone-Project Expansion.   Figure 5-1 shows that this expansion was rapid and extensive. For the present we shall not go beyond 1944, when silicones represented over 70% of all corporate research in chemistry. The addition of 31 scientists, most of whom were newly recruited PhDs, to the project during the period 1940–1944 was evidence of Marshall's determination to make the Company an important factor in the chemical industry, of his belief that no better opportunity than the silicone project would soon present itself, and of his willingness to be at risk in championing a cause he thought good.

The staff was increased in part because Marshall took the important step of introducing advanced chemical engineering into the Company. Another factor in the increase was the expected postwar need of the Company for trained personnel with Research Laboratory experience in silicones and in other fields. The rapid expansion would not have been possible, however, had the silicone project not cried out for scientific augmentation, and had the "Whitney atmosphere" not traditionally favored this activity.

Not only did augmentation of the silicone project cost a great deal of money, but also it left very little of the Chemistry Section free to carry out other kinds of corporate chemical activities. As matters turned out, the risk was justified in the situation then existing.

Scientific augmentation brings valuable intangible benefits. It helps recruiting, then a major activity of the Section. It provides entering recruits with worthwhile problems on which to begin their industrial careers. It results in publications that enhance the standing of the Company and of its staff, and that repay the debt owed by the Company to the prior scientific literature; the Company settled this debt handsomely, using its own money. Naturally, these benefits were only by-products of the primary objective of the project, namely, to make silicones profitable and useful. In today's research climate, and in today's Company, augmentation on this scale would be unlikely and unjustified.

About the time the silicone project peaked in 1944, the Chemistry Section began to make, and then to test, boron hydrides as fuels for rockets and jet engines. The new project contrasted with the older in various ways. Stock of course had supplied prior knowledge for both [Chapter 2]. But the boron hydride project was government financed; it aroused serious misgivings in the Section from the start; and it did not result in the making of useful products within the Company. It did lead, however, to a piece of beautiful scientific augmentation by John S. Kasper, Charlys M. Lucht, and David Harker, who in 1948 published X-ray diffraction evidence for a wholly unexpected, basketlike arrangement of boron atoms in crystalline $B_{10}H_{14}$, a compound discovered by Stock and prepared for the first time in relatively large amounts within the Section. An intensive, continuing program of X-ray diffraction work on boron hydrides was thereupon undertaken by William N. Lipscomb (Nobel Prize, 1976) with results that have shown the chemical bonding in these compounds to be far more sophisticated than earlier ideas, notably those of Lewis and Pauling [Chapter 2], had indicated. See the account of Lipscomb's Nobel Prize by Russel N. Grimes in *Science*, **194**, 709 (1976).

Chemistry Section and Organization. I believe that before 1945 the Chemistry Section made spontaneous (i.e., largely unplanned) progress toward a type of organization suitable for industrial research. The belief is based upon Figure 5-1, upon the fragments of organization charts for 1943 and 1945 that constitute Figure 5-2, and upon my knowledge of the Section.

Within the Section there were no formal appointments such as that of Marshall by Coolidge [Chapter 1]; leaders were de facto only, and most of them simply emerged and were accepted. During this period the Section always had a superabundance of interesting problems by no means confined to research. The primitive state of silicone chemistry made it possible to accommodate new staff members without serious dislocation, and the "Whitney atmosphere" helped them to flourish. There was unavoidable crowding.

Figure 5-1 shows that the alkyd project was actually a succession of three subprojects. Within the silicone project also, several subprojects coexisted. Projects and subprojects were born, grew, and died—sometimes to be reborn. Death could follow failure or success, the happier alternative being accompanied by transition of the project to an operating component; troubles in the component might lead to further work and a rebirth.

After Kienle's departure [Chapter 1] Birger W. Nordlander became de facto leader of the alkyd project. Patnode and Marsden have already

1943		1945	
Dr. W. D. Coolidge, Vice President and Director		Dr. C. G. Suits, Vice President and Director	
Dr. C. G. Suits, Assistant to the Director		D. E. Chambers, Executive Engineer	
L. A. Hawkins, Executive Engineer			
		CHEMISTRY	
Dr. A. L. Marshall	Chemical Section	Dr. A. L. Marshall	In Charge of Division
Dr. F. J. Norton	Dielectric Studies	Research Associates:	
Dr. D. J. Mead		Dr. R. Fuoss	Project on Infrared Spectroscopy
Dr. H. H. Race	Electrical Properties of Insulation		
S. I. Reynolds		Dr. J. Marsden	Project on Silicone Elastomers
Miss K. Flickinger			
B. W. Nordlander	Synthetic Resins	B. W. Nordlander	Project on Permafil
M. C. Agens		Dr. F. J. Norton	Project on Special Assignments
Dr. R. E. Burnett			
Dr. W. E. Cass			
Dr. J. A. Loritsch		Dr. W. I. Patnode	Project on Silicone Chemistry
Dr. James Marsden	Silicone Resins	Dr. E. G. Rochow	Project on Organosilicon and Organofluorine Chemistry
Dr. D. W. Scott			
Dr. G. F. Roedel		Dr. M. M. Sprung	Project on Wire Enamels
Dr. W. I. Patnode	Silicone Chemistry	Dr. C. A. Burkhard	
Dr. R. O. Sauer		Dr. R. E. Burnett	
Dr. D. F. Wilcock		Dr. W. E. Cass	
C. P. Haber		Dr. J. R. Elliott	
Dr. E. G. Rochow	Silicone Chemistry	Dr. R. H. Krieble	Organic Chemistry
Dr. W. F. Gilliam		Dr. R. N. Lewis	
Dr. D. T. Hurd		Dr. J. A. Loritsch	
Dr. R. N. Lewis		Dr. G. R. Roedel	

1943		1945	
Dr. A. E. Newkirk		Dr. R. O. Sauer	
E. M. Hadsell		Dr. D. J. Mead	Physical Chemistry
Dr. M. M. Sprung	Silicone Chemistry	Dr. D. W. Scott	
Dr. C. A. Burkhard		R. H. Savage	
Dr. R. H. Krieble	Silicone Chemistry	Dr. W. F. Gilliam	
Dr. J. R. Elliott		Dr. D. T. Hurd	Organometallic Chemistry
J. G. E. Wright	Resin Applications	Dr. A. E. Newkirk	
C. S. Oliver		M. C. Agens	
C. V. Adams		M. M. Safford	
M. M. Safford	Resin Applications	Dr. D. F. Wilcock	Applied Chemical Research
E. G. Tajkowski		J. G. E. Wright	
Dr. C. E. Reed	Chemical Engineering	Dr. H. H. Race	
Dr. A. E. Schubert		S. I. Reynolds	Insulation Measurements
R. K. Eunson		Research Assistants:	
R. C. Faught		L. V. Adams	
W. J. Scheiber	Silicone Pilot Plant	C. P. Haber	
Dr. H. A. Liebhafsky	Inorganic and Analytical	E. M. Hadsell	
Dr. E. W. Balis	Chemistry	I. Hurst	
Dr. E. H. Winslow		E. T. Marx	
A. F. Winslow		E. L. Minscher	
L. B. Bronk		C. S. Oliver	
C. G. Van Brunt	Consultant in Microchemistry	*Chemical Analysis*	
R. H. Savage	Carbon Brushes	Dr. H. A. Liebhafsky	
I. A. Hurst	Mechanical Processing	Dr. E. W. Balis	
		Dr. E. H. Winslow	
		A. F. Winslow	

C. Van Brunt   Analytical Consultant
  Research Assistant:
    L. B. Bronk

Chemical Engineering
  Research Associates:
  Dr. C. E. Reed
    Dr. J. R. Donnalley
    H. A. Fremont
    D. P. Herron
    W. Scheiber   Pilot-Plant Operation
  Dr. A. E. Schubert
    Research Assistants:
      H. P. Blackwell
      R. C. Faught
      B. C. Raynes

Figure 5-2 Partial organization charts for the General Electric Research Laboratory. 1944 chart unavailable. When Dr. Suits took charge, the Chemistry (Chemical) Section became the Chemistry Division. The charts are valuable as early evidence of the gradual evolution of modern research management and organization.

been identified as such leaders of what we have just called silicone subprojects; Gilliam became another such leader as augmentation and implementation of the direct-process discoveries proceeded. Unfortunately, this spontaneous evolution of the subproject organization was always accompanied by an unorganized residue, which disappeared when the entire Section was formally organized after 1945.

Chemical engineering and analytical chemistry also developed into recognizable organizations, but not into *subprojects*—instead they became what we shall call *units*; see Figure 5-2. These units performed functions too complex for description here, but (nominally at least) they were organized around a major discipline—a type of organization that could outlive a project.

Chemistry Section experience relays the message that an unorganized component doing industrial research tends of itself toward an organization in which projects (or subprojects) exist as temporary entities alongside presumably permanent groups (units) in which research is often less important than are other functions, especially services. Exchange of personnel among units and subprojects must be easy and expected if such a "two-tier" organization is to prosper.

The Chemistry Section gave convincing proof of Lord Snow's statement* about the complexity of personal relations in a productive organization. There was, in fact, an added complexity. As our glimpses of Rochow and Patnode at work have shown, members of the Section changed quickly from one kind of investigation to another, and often carried on more than one kind of work at once. No organization chart can represent such a situation: a snapshot cannot replace a motion picture. Corporate research in chemistry was done in 1943 about as it was done in 1945; the reader is invited to compare the two organization charts in Figure 5-2 with each other, and with what Chapters 3, 4, and 5 have told about the Section. Even the best modern organization charts are but idealized models that can only simulate a constantly changing reality.

COMMENTS IN CONCLUSION

• As to the tungsten and silicone projects. I consider Coolidge's ductilization of tungsten the most important research yet undertaken in the General Electric Company. A quick comparison of the tungsten and silicone projects, though necessarily imperfect and incomplete, promotes an understanding of industrial research and indicates how it is changing as the body of prior knowledge grows.

* C. P. Snow, *The Two Cultures: and a Second Look,* p. 31; quoted near the close of Chapter 1.

Clear and urgent needs inspired both projects. At their peaks the tungsten project certainly, and the silicone project probably, were the largest then under way in the Research Laboratory. In the language of Chapter 4, each had a planned high road (in the case of tungsten, the search for a better incandescent-lamp filament) along which appeared important forks (the Coolidge X-ray tube, the thoriated-tungsten thermionic emitter, other uses of ductilized tungsten made possible by the outstanding properties of the element). Tungsten and silicones both benefited from the ''Whitney atmosphere'' in the Laboratory; the interaction of Coolidge and Langmuir was particularly noteworthy [49]. Both projects were helped by chance [49, pp. 10–11]. Both have brought immeasurably more good than harm.

For our look at industrial research, however, the differences between the two projects are more important than the likenesses. Success along the planned high road was less spectacular for the silicones: lamps with filaments of Coolidge's tungsten* came quickly into worldwide use, and the material continues to be manufactured everywhere by substantially his method. Prior knowledge generated by Kipping, Stock, and Combes helped the silicone project; prior knowledge about the ductility of metals could have led Coolidge astray.† The difference most significant for our

---

* Langmuir's brilliant pure research eventually resulted in a new *kind* of incandescent lamp (still with a filament of Coolidge's tungsten) and hastened the use of tungsten in electron tubes. *Industrial* pure research has not contributed comparably to the silicone industry. Note that Langmuir's work was not an augmentation of Coolidge's ductilization of tungsten; it ran a parallel path that Coolidge was unlikely ever to have taken. Yet, without ductile tungsten, Langmuir's career and the history of the electrical industry would have been markedly different.

† Three quotations show how the ductilization of tungsten looked to Coolidge in the light of prior knowledge.

1. ''We have no fear that anyone who has studied the prior art will believe that tungsten is a ductile metal . . .'' [49, p. 57; from a brief asserting the validity of Coolidge's U.S. Patent 1,082,933].

2. ''To a man ignorant of our success, the problem would certainly look more hopeless today than it did [at the beginning]. For since that time millions of tungsten filaments have been produced by widely different methods [other than Coolidge's], and by different men. . . . Yet all of the filaments made have been brittle'' [49, p. 20; quoted from an article by Coolidge].

3. ''Imagine then, . . . a man wishing to open a door locked with a combination lock and bolted on the inside. Assume that he does not know a single number of the combination and has not a chance to open the door until he finds the whole combination, and not a chance to do so even unless the bolt on the inside is open. Also bear in mind that he cannot tell whether a single number of the combination is right until he knows the combination complete. When we started to make tungsten ductile, our situation was like that'' [J. A. Miller, *Yankee Scientist*, Mohawk Development Series, Schenectady, New York, 1963, p. 72; Dr. Coolidge's quoted description].

Fortunately for industrial research, prior knowledge is usually more helpful today.

times was the markedly greater scientific augmentation in the silicone project. This greater augmentation went hand in hand with a greater proportion of PhDs and a greater number of reports and publications. Industrial research was changing.

- As to scientific augmentation in general. *It is easier to augment than to discover.* Augmentation is more readily manageable, can be better planned, and is less at the mercy of chance. It often looks much different to the man at the laboratory bench than it does to his manager, and this can lead to tension between them.

University training predisposes to augmentation. Few who enter industry after completing graduate work would refuse a chance to continue their thesis research. Modern instrumentation—costly, powerful, and necessary—offers tempting opportunities for augmentation: new instruments now appear almost daily; those already on hand invite modification and experiments to test their capabilities.

For the investigator, authorized augmentation is a low-risk activity that can be rewarding. If the investigator can prove that a discovery is not worth implementing, he or she has saved money for the company. If he proves the contrary—so much the better. In either case, understanding will be improved, and worthwhile papers can usually be written if the investigator is anxious to improve his or her scientific reputation.

The manager not only realizes that augmentation is costly, but also knows that it tends to confine research effort to discoveries already made. Thus the incandescent lamp continues to be an attractive object of research the while other light sources are emerging; work on silicones might preclude the discoveries of other materials; prolonged concern with electron tubes might delay a switch to solid-state devices—risks such as these must be balanced against the possibility that major discoveries will be missed because augmentation was prematurely terminated. Also, premature and frequent terminations tend to frustrate those doing the research, particularly if the changes do not bring major discoveries.

For the silicone project Figure 5-1 shows that augmentation declined abruptly after 1944. The decline resulted mainly from a management decision by Marshall. The increase in activity after 1950 will be discussed in Chapter 7.

In company-financed industrial research, Figure 5-1 implies that managers can usually be relied upon to see that augmentation does not get out of hand, and that it does not masquerade as discovery. What of research that is funded by the government? Such funding is so great that the question of discovery versus augmentation ought to concern the average taxpayer, who deserves the best possible assurance that his or her money is being spent to give the public what is needed.

If a small apple can be compared with an enormous orange, one might compare the chronological occurrence of major discoveries in the silicone project (Table 4-1 as a start) and subsequent augmentation (Figure 5-1 and related text) with their counterparts in nuclear fission. As most of us know, the first major discovery in that field was made by Otto Hahn (Nobel Prize, 1944) and Fritz Strassmann late in 1938; and augmentation, long continued on government funding, quite early yielded others. Anyone interested in this matter might well begin by reading the memorable and perceptive review by Don M. Yost of the book, *The Transuranium Elements*, [50], a ponderous work.

Comments on implementation will follow Chapter 6.

Chapter Six Chemical Engineering Enters. A Silicone Business Begins

*To carry on a diversified, growing, and profitable
world-wide manufacturing business in electrical
apparatus, appliances, and supplies, and in related
materials, products, systems, and services for industry,
commerce, agriculture, government, the community,
and the home.*

R. J. CORDINER*

By 1942 it was clear that the commercial manufacture of silicones by the
Rochow route would involve processes resembling those carried out in
diverse industries such as the *mining and metallurgical* (preparation and
processing of Si-Cu);[†] the *chemical* (direct process); *petroleum* (distilla-

---

* The quotation states the first objective of the General Electric Company as given by R.
J. Cordiner, then its President, on p. 119 of his *New Frontiers for Professional Managers,*
McGraw-Hill, New York, 1956.
[†] When used generally from now on, Si-Cu will represent the two elements used in any
form whatever to make silicone precursors in the Rochow route to silicones.
Note. Brackets are used to enclose references and inserted material. For bracketed ref-
erences, see References and Notes at the end of the book. Numbers in parentheses on the
right of pages are often used in the text to identify equations and the like: thus (6-1) means
Equation 6-1. Reference to laboratory notebooks is by name and date. CRDA stands for
"Chemistry Research Department Archives."

tion); again, the *chemical* and *polymer* (hydrolysis-condensation sequence; disposal of HCl formed); and the *rubber industry* (filling, compounding, and cross-linking of silicone rubber). Nor is diversity the end of the matter. Prerequisite to the industrial success of these processes is the satisfying of requirements such as low cost, high efficiency, minimum capital investment, proper choice of materials, suitable instrumentation, prevention of corrosion, and maintenance of adequate safety measures. The list could be continued. Furthermore, the Company would need personnel who could deal expertly with the contractors responsible for building and equipping the plant, and who could turn to universities and to other companies for advice. In short the Company needed chemical engineers with advanced training who could adapt quickly to an industrial environment.

Moreover, such chemical engineers were needed for another crucial reason. Everything hinged on the direct process. Calculations had shown that it would be the best commercial choice *provided* that it could be operated satisfactorily. Augmentation of its discovery had given the encouraging results recorded in Chapter 5, but these results could be implicitly relied upon only for the conditions under which they had been obtained. Their commercial applicability had not been demonstrated; *implementation** was needed. In particular, one could be certain neither that the best reactor for the process had been found, nor that Si-Cu in any of the forms yet studied would be best for the best reactor. Reactor and Si-Cu were related problems, the best solution to which was most likely to be reached via a chemical engineering approach.

Marshall clearly understood all this in 1942, at a time when the Company had not a single chemical engineer with a PhD degree in its employ. Figure 5-2 shows how rapidly he built the chemical engineering unit,† which was headed by Charles E. Reed. The influence of chemical engineering was to go far beyond the silicone project. Even in the beginning the chemical engineers, especially A. E. ("Gene") Schubert, were concerned with Permafil production in Building 234. Today there are many such PhDs in the Company, for training in chemical engineering has proved useful to a degree greater than even Marshall could have foreseen in 1942. Chemical engineering soon gave the silicone project, and eventually much of the Company, a new look. In no instance did Marshall recruit with greater sagacity or to better purpose.

---

* The history of the direct process shows that successful augmentation changes gradually into implementation: exact definitions are illusions.

† J. T. ("Jerry") Coe is not listed in Figure 5-2. As did Reed, he came from MIT. Their collaboration in the Chemistry Section began on May 5, 1942, and continued until Coe reported for navy service in 1943. In 1945 Coe resumed silicone work in Building 77.

CHARLES E. REED

Charles E. Reed was born in Findlay, Ohio, on August 11, 1913, and attended public schools there until his graduation from Findlay Senior High School in 1930. In September of that year, he matriculated in chemical engineering at what was then the Case School of Applied Science in Cleveland. After completing undergraduate work there in June 1934, he enrolled as a graduate student in chemical engineering at the Massachusetts Institute of Technology, where his outstanding performance and abilities earned him the following succession of appointments: National Tau Beta Pi Fellowship, academic year 1934–1935; William Sumner Bolles Fellowship and MIT Fellow, academic year 1935–1936; instructor in chemical engineering, at the opening of the fall term in 1936, even before he received the ScD degree, which was awarded him in 1937. From 1937 until March 23, 1942, when he joined the Chemistry Section under Marshall, he held the rank of Assistant Professor at MIT.

Reed's experience at MIT was excellent preparation for the key position Marshall recruited him to fill. His thesis recorded the results of a collaboration with Professor E. A. Hauser on the structure of bentonite, and on the way in which "solutions" of this clay become quasi-solids (gels). Bentonite, being a silicate and a clay, is related to silicones as we have seen in Chapter 2. Reed's teaching experience at MIT gave him a command of chemical engineering principles then nonexistent in the General Electric Company. His work on a well-received textbook* must also have helped. His close acquaintance with the MIT faculty, notably with Professors W. K. Lewis and E. R. Gilliland, consultants later to the Silicone Products Department, was an added and valuable asset.

Reed's career in the General Electric Company, during which he completed advanced management courses within the Company and at Harvard, mirrors the organizational changes within the Company; and, what is more important, is evidence of the great concurrent increase in the Company's chemical and metallurgical business—an increase that Reed did much to bring about. His career summary follows: Research Associate (in charge of chemical engineering), Chemistry Section, Research Laboratory, 1942–1945; Manager, Chemical Engineering Department, Chemical Division† (Building 77, Schenectady), 1945–1948; Engi-

---

* T. K. Sherwood and C. E. Reed, *Applied Mathematics in Chemical Engineering*, 1st ed., 3rd impression, McGraw-Hill, New York, 1939; and H. S. Mickley, T. K. Sherwood, and C. E. Reed, *ibid.*, 2d ed., McGraw-Hill, New York, 1957.

† The Chemical Division, created on January 1, 1945, with headquarters at Pittsfield, Massachusetts, united all the chemical manufacturing activities of the Company, those of the Resin and Insulating Materials Division (which then ceased to exist) included.

C. E. Reed at about the time he became the first General Manager of the General Electric Silicone Products Department.

neering Manager, Engineering Department, Chemical Division, Pittsfield, 1948–1952; General Manager, Silicone Products Department (Waterford, New York), Chemical and Metallurgical Division,* 1952–1959; General Manager, Metallurgical Products Department, same division, 1959–1962; Vice President and General Manager of that division, 1962–1968; Vice President and Group Executive, Components and Materials Group, 1968–1971; Senior Vice President, 1971 to the present.

Reed began his first notebook on March 24, 1942. Its early pages show that he quickly became conversant with all phases of the silicone project. His influence on the pilot plant, on the Fluidyne, and on distillation—all of which will be described below—was particularly strong. His overall contribution to silicones was recognized as one of the "outstanding developments in the promotion of knowledge of engineering materials" by the American Society for Testing Materials when it invited him to deliver the 1956 Edgar Marburg Lecture, which is quoted at the opening of Chapter 7.

THE METHYLCHLOROSILANE PILOT PLANT

On Monday, July 13, 1942, William J. Scheiber, who had been producing methylchlorosilanes on the fifth floor since the Patnode-Scheiber scale-up described in Chapter 4, took charge under Reed of the new pilot plant in the southeast corner of the sixth floor (Rooms 603 and 605) of Building 37. From that day on, the General Electric Company has been a manufacturer of silicones. Twelve reactors and two stills (distillation columns) were in place. Eventually, a "Stedman" still (see below) was added, which was tall enough to require the building of a penthouse.

One might think that a single source of methylchlorosilanes would have sufficed the Company at this stage in the development of silicones. Not so. The prospect of new polymers with outstanding properties led to methylchlorosilane plants in Pittsfield, Massachusetts, and in Building 77, Schenectady. On March 14, 1942, according to Rochow, Marshall discussed the production planned by those in charge at the last two locations. Pittsfield need not concern us as silicone production there never became significant. The silicone effort of the Resin and Insulating Materials Division in Building 77, however, was the forerunner of the silicone plant at Waterford.

---

* To the probable dismay of many readers, I must report that the Chemical Division was thus renamed when it was given the responsibility for metallurgical products and for Manmade™ diamonds.

Brevity being mandatory, we shall concentrate on Scheiber, and omit description even of Gilliam's production of phenylchlorosilanes in "Gargantua," the reactor that produced these silicone precursors at the rate of 100 lb monthly until Waterford went on stream in 1947.

Scheiber was the key figure in methylchlorosilane production until Waterford took over. In 1920 he joined the group concerned with insulation in the Research Laboratory, and continued taking courses at Union College in Schenectady for several years. His association with Patnode began soon after the latter's transfer from Pittsfield to Schenectady. Operation of the reactors and stills in the new pilot plant required Scheiber to work the usual 44 hours, to be on call during the other 124 hours of the week, and to spend some time at the plant on most Sundays. After about 2 years at Waterford, he transferred to the Knolls Atomic Power Laboratory in 1949 and retired from there in 1967 as Manager, Radiation Facilities, of the West Milton plant. Scheiber prided himself on supplying needed silicone experience to "engineers sitting at desks."

The Reactors. Each of Scheiber's 12 reactors was an 8-ft section of vertical steel pipe, internal diameter 4 in., welded inside a 6-in. steel pipe to form an annular space through which oil was circulated—initially, to heat the reactors to about 250°C, and then to cool them once reaction had begun. The usual Si-Cu was a powder, 10% copper, that had been heated ("fired") in a hydrogen furnace to increase its reactivity. The powder was charged into the reactor from the top, CH_3Cl entered from the bottom, and the gaseous products and the unreacted CH_3Cl left from the top to be condensed and later distilled.

The reactors and the procedure just described had evolved from the earlier experience of Patnode and Scheiber and from experiments by others. Vertical reactors had been found preferable to horizontal or inclined ones; Scheiber had shown that they could be made of steel;[*] and a circulating liquid had come to be accepted as the heat-transfer medium.

When, in July 1944, James R. Donnalley undertook the design of a bank of 12 reactors to serve as replacements, he found no radical changes necessary. In the new bank CH_3Cl was still fed individually to the reactors, but the gases exited into a common header on their way to a brine-cooled condenser. Circulation of the oil was improved to give better control of the reactor temperature.

On Tuesday, March 27, 1945, Donnalley recorded that the new reactors "were put in operation last Thursday," and that it had been

---

* Chapter 4 relates that the Patnode-Scheiber "big" run was made in a copper reactor, and that Combes in 1896 used an iron tube when he reacted HCl with Si-Cu.

necessary to increase the number of Calrod heaters to eight so that the system "could reach the desired temperature of operation [about 250°C] in about 1 hour." Furthermore, "Bill Scheiber tells me that the [12] reactors are producing [a total of] between 5.5 to 6.5 lb per hour," which was a considerable improvement over earlier rates. Donnalley attributed the improvement to the more rapid circulation of the oil. On March 28, 1945, Scheiber told Donnalley that the first 450 lb of crude product had a density of 1.093 g/cc and consisted of the following:

"Low boilers"	(materials boiling below 60°C)*	23.8%
$(CH_3)SiCl_3$		15.8%
$(CH_3)_2SiCl_2$		48.3%
Residue		8.4%

Considerable progress beyond the Patnode-Scheiber "big" run had been made.

The Si-Cu Problem.   A description of all that was done with Si-Cu, fired and unfired, to improve performance and yield in the direct process would make a long, wearying, and confusing tale. On January 6, 1942, Gilliam sadly recorded that Scheiber's runs had been "poor" for 17 weeks after being "good" for 10 months, although the "standard method of making pills was being followed": this was the "copper fiasco" mentioned in Chapter 4. On February 5, 1945, Henry A. Fremont, a chemical engineer, summarized as follows a 2-day discussion of the Si-Cu question with Gilliam and Meals:

These are the facts:

1. unfired powder gives high yields of CH_3HSiCl_2
2. air-cooled alloy gives high yields of CH_3HSiCl_2
3. vacuum-cooled alloy gives low yields of CH_3HSiCl_2
4. fired powder gives low yields of CH_3HSiCl_2

[For yields of $(CH_3)_2SiCl_2$], the above are all reversed. In rate of reaction, they range 1, 4, 3.

...

Now, what causes the common effect between firing powder and making a vacuum alloy as opposed to unfired powder and an air-cast alloy? I can't say.

Neither, it appears, could anyone else.

* Boiling points below 60°C: $SiHCl_3$, 31°; CH_3HSiCl_2, 41°; $(CH_3)_3SiCl$ and $SiCl_4$, near 57.6°; their azeotrope, near 54.5°; and others less important.

Distillation. The importance of distillation in the Rochow route to silicones is clear from this simple statement: the present silicone industry could not exist without distillation to separate the crude chlorosilanes produced by the direct process into the precursors from which the useful silicone products are made. Earlier in the book, the *difficulty* of distillation was exemplified by the problem of obtaining "pure" CH_3SiCl_3 from a crude; the *importance* of distillation was shown by the need to prepare "pure" $(CH_3)_2SiCl_2$ for the manufacture of silicone rubber. Patnode's classical separation of methyl silicones, Figure 4-2, is an example of an easy distillation; the different fractions, each a compound, had well-separated boiling points, and almost all of each compound left the still at the temperature of this boiling point—hence the well-defined "flats" in Figure 4-2.

Figure 6-1 is a distillation primer in which complexities have been swept under the rug. It illustrates the testing of a still by use of a standard mixture, in this case of C_6H_6 (benzene) and $C_2H_4Cl_2$ (ethylene dichloride). It needs to be studied. For distillations of chlorosilane crudes, a far harder task, many of the basic data had to be measured.* Actually, the speed with which a still reaches the $(CH_3)SiCl_3$ "flat," the temperature of the "flat," and the density and/or chlorine content of the product are reliable indications of still performance, and can to some extent replace the testing of stills with known mixtures.

To provide the vapor-liquid interaction needed for efficient distillation, intimate, prolonged contact of liquid and vapor, and large areas of wetted surface, are required, especially for difficult distillations, and there must be no open channels via which vapor can travel toward the top of the still, or liquid toward the bottom; such channels are short circuits. The columns of many stills are packed (filled) to approach these conditions; the packing can be any inert material by use of which distillation is possible at practical rates under conditions where vapor and liquid in contact approach the compositions prescribed by Figure 6-1 for the temperature of contact. Packing not wet is useless because such contact is lacking. When, as often happens, wetting is a sometime thing, stills are exasperatingly temperamental.†

---

* This important work could have been discussed as necessary augmentation in Chapter 5. Typical reports listed in CRDA are the following: C. E. Reed and J. T. Coe, *Vapor-Liquid Equilibrium in Chlorosilane Binaries,* July 30, 1942; and D. J. Mead, *Vapor-Liquid Equilibrium in Binary Systems of Chlorosilanes,* December 22, 1943.

† To produce good wetting in a column, gentle persuasion is often needed. One procedure is as follows. "Flood" the column by increasing the rate of heating at the still pot so that virtually all space in the column is filled with liquid (see Figure 6-1b). Then reduce the rate of heating, and hope for the best. A flooded column has zero theoretical plates.

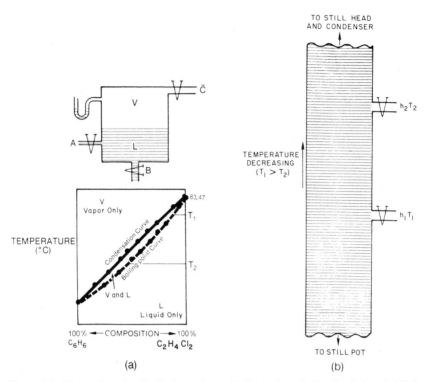

Figure 6-1 Illustrations for distillation primer. (a) Basic data. Refractive index (1.448 for C₂H₄Cl₂; 1.5011 for C₆H₆; both at 20°C) could have been used as abscissa. (b) Testing of section of packed column. Samples for refractive-index measurements are withdrawn at two points (heights, h_1 and h_2; temperatures, T_1 and T_2).

Notes for Distillation Primer

(a)   Basic Data: Condensation and Boiling-Point Curves

1.   *Problem:* Establish condensation and boiling-point curves for C₆H₆-C₂H₄Cl₂ mixtures at total pressure of 1 atm. The pure liquids are completely miscible, and it is known that the refractive index at 20°C reliably establishes composition.

2.   *Known data:* Normal boiling points (boiling points at 1 atm) and refractive indices of C₆H₆ and of C₂H₄Cl₂; refractive indices of all possible mixtures of the two compounds.

3.   *Idealized experiment:* Through *A* introduce liquid mixture into evacuated test cell. Measure temperature *T* at which vapor *V* in equilibrium with liquid *L* has a pressure of 1 atm (note attached manometer). Withdraw vapor sample through *C* and condense. Withdraw liquid sample through *B*. Carry out withdrawals so as not to disturb equilibrium. Measure refractive indices of vapor condensate and of liquid samples. Plot corresponding compositions for each pair of measurements; eight pairs of results are shown above.

4.   The results define the condensation- and boiling-point curves in Figure 6-1a. Between these curves liquid (*L*) and vapor (*V*) are in equilibrium at a total pressure of 1 atm, and the vapor (hence the condensate) is always richer in C₆H₆ (boiling point, 80.1°C) than in the

210

About 1941 the Foster Wheeler Corporation was offering stills with a new packing, invented by Stedman and fabricated from stainless steel wire-cloth. The prediction that ''With these triangular pyramid-type Stedman packings it will be possible to build semicommercial columns having as many as one hundred theoretical plates for use in a laboratory of average height'' rested on convincing work by Bragg [1]. The still he envisioned was exactly what the silicone project needed. Such a still was ordered, and we shall excerpt its early history in Scheiber's pilot plant. The still was to worry Marshall, Reed, Schubert, and Scheiber for a long time, and it drew, especially from the last two, heroic efforts aimed at understanding it and improving its poor performance. All were concerned that some unknown factor peculiar to silicones might complicate the commercial distillations that a silicone business would require and make them more costly.

In scanning the following items, which have their sources in laboratory notebooks identified by initials, the reader should remember that distillation is a physical process, normally free of chemical effects, which was not expected to give trouble. On with the story—but first the follow-

---

(higher-boiling) liquid. This circumstance makes it possible to separate any mixture of C_6H_6 and $C_2H_4Cl_2$ into its components by distillation in a suitable still. Here "suitable" means that the still must be powerful enough (have enough "theoretical plates"; see further on) so that the process described in the idealized experiment can occur for all mixtures of the compounds that boil between 80.1 and 83.47°C. The narrower the space enclosed by the two curves (other things being the same), the greater the number of theoretical plates needed for the separation.

(b)   Testing of Section of Packed Column

1.   To effect separation of C_6H_6 and $C_2H_4Cl_2$, heat is supplied to the still pot below the column and removed along the column and in the condenser. The rates and the manner in which heat is supplied and removed determine whether the column is "flooded" (see text), or whether and how rapidly distillation (as measured by the rate at which vapor escapes from the still) occurs.

2.   In an effective column, boiling and condensation occur all along the column at continually decreasing temperatures as shown above.

3.   *Definition:* A still has *one theoretical plate* when it can deliver a condensate and a liquid, each of the composition fixed by the intersections of a horizontal line with the two curves in Figure 6-1a. The aparatus above the curves is a *one-plate still* if properly used.

4.   *Testing of column:* Wet the packing (see footnote in text). When the column is operating as desired, withdraw samples from it at the two locations shown. Measure the refractive indices of the samples. From the calculated curve of *number of theoretical plates* (n) against *refractive index*, obtain n_1 for h_1 and n_2 for h_2. Then $n_2 - n_1$ is the number of theoretical plates in the section between h_2 and h_1 under the conditions existing. See chemical engineering texts for details.

ing data:

Stedman-still column: height, 15 ft; internal diameter of column, 6.08 in.; stainless steel throughout; expected performance, 70–90 theoretical plates* when fully packed; capacity of still pot, over 1000 lb of methylchlorosilane crude; steam-heated.

    1. W. J. S., Sunday, August 9, 1942:

Returned today from a week's vacation . . . . Dr. Reed has the new . . . column in Room 603 and going through the roof in good working order. It is charged with 210 pounds of crude and has been flooded and allowed to come to equilibrium. . . . Worked all day with Reed on the new still.

    2. W. J. S., Monday, November 16, 1942:

[Still] started up by Reed and Marshall yesterday [Sunday] afternoon. November 18: [Still] is on 1% takeoff† and on a [(CH$_3$)SiCl$_3$] flat. [Product contains] 71.0 to 71.04% chlorine, which is . . . excellent. [It certainly is!] November 21, 1942: We are collecting good [(CH$_3$)$_2$SiCl$_2$] today. Stayed until 10:30 P.M. when still ran out. [The collection of (CH$_3$)$_2$SiCl$_2$ (boiling point, 70°) concluded Scheiber's Run 1F.‡ Everyone must have been happy.] Sunday, November 22, 1942: 83 lb [of] residue was removed from still pot, and the still prepared for the next charge.

    3. W. J. S., November 22–26, 1942: Still charged with 1012 lb methylchlorosilane crude for Run 2F. November 23: "Dr. Reed had flooding trouble." Two days later, the column was working well at reduced heat input. November 25, when CH$_3$SiCl$_3$ should have been distilling: "Hadsell found [only] 69.75% chlorine." Run finished at 11:30 P.M. on November 26, 1942. [Distillation less satisfactory in Run 2F than in Run 1F. Reason unknown.]

    4. W. J. S., July 16, 1943: Still was shut down after Run 45F. About 50,000 lb of methylchlorosilane-containing materials had been put through the still, which had never performed up to expectations—at least not after Run 1F.

---

* The number of theoretical plates achievable in a still will range upward from zero to a maximum realizable under the "best" conditions; see Reference 1. The more rapidly vapor leaves the still, the lower is the expected number of theoretical plates. The numbers given above are my estimates for expected performance under useful conditions.

† At 1% take-off, 99% of the material vaporized in the still pot does not leave the still. The 99% is condensed and works its way down the column; it is "refluxed"—hence the low rate of distillation.

‡ Scheiber's designation (1F, 2F, etc.) will be used to identify distillations made in the still of interest to us.

5. C. E. R., July 22, 1943: "Today we are testing the . . . Foster Wheeler column on a . . . [$(CH_3)_2SiCl_2$-$(CH_3)SiCl_3$] mixture. For detailed description, see . . . Scheiber's notebook." This was the first chapter in what we shall call the "distillation epic." After the close of the chapter, it was deemed necessary to clean the still and to test it with the standard C_6H_6-$C_2H_4Cl_2$ mixture (see Figure 6-1). Accordingly, W. J. S., November 19, 1943:

For the past week, the Foster Wheeler still has been dismantled and cleaned out.

First of all the head . . . cooler coil [were removed]. The metal surfaces were covered with a white powdery deposit which was identified by Dr. Liebhafsky as a silicone. These sections were later treated with alcoholic NaOH followed by a pickling solution for stainless steel. This is a mixture of 25% concentrated nitric acid by volume, 2% concentrated HF by volume, 73% water, temp. 60° approx. They were then washed and dried.

The cleaning procedure gave good results and was therefore used by Earl Balis for the entire column. After other parts of the still and its fittings had been cleaned with HCl and dried, 735 lb of known standard mixture was charged into the still, the column was flooded, and Run 57F was begun. On the next day the number of theoretical plates was 36 at 2 A.M. and had risen to 62 by 3 P.M., mainly because the heating rate had been reduced fourfold to an unacceptably low value. Variations during the rest of the run produced discouraging results: the still did not meet expectations.

6.   Marshall decided on an extremely drastic cleaning of the still. W. J. S., Wednesday, December 1, 1943:

Since Saturday we have been cleaning the column out. On Monday night the whole gang was in and the column was filled with a cleaning solution concocted by Dr. Marshall* . . . .

* Marshall made a memorable return to a laboratory bench in the analytical chemists' bailiwick. While they hovered anxiously nearby, he concocted and tested various awesome cleaning mixtures to establish how much the stainless steel packing could endure. "Marshall's elixir" was the result. "The elixir may be prepared by mixing 1350 cc of chromic-sulfuric acid cleaning solution, 1350 cc of concentrated sulfuric acid, and 150 cc of concentrated hydrofluoric acid in a stainless steel container. . . . This mixture should be good for what ails man or beast, and the cure is sure to be permanent and painless—but painless only after rigor mortis has set in."

The boss's experiments were the curtain raiser to a performance memorable even for the Chemistry Section. The "whole gang" was a carefully selected group composed mainly of trusted analytical chemists and chemical engineers. Better named the "Marshall-elixir bucket brigade," they compounded the elixir in Room 603, Building 37, and passed it by

Run 58F was begun with a C_6H_6-$C_2H_4Cl_2$ test mixture, and the first result, at 4 A.M., was 22 theoretical plates at a fairly high heating rate. After 8 hours this number had increased to 34. W. J. S.: "[This result] is very poor—not up to that obtained before cleaning."

From here on, increased telescoping of the distillation epic is necessary. We shall follow it through 152 pages of W.J.S. Notebook 3789 to September 8, 1944. Marshall, Reed, and Schubert continued as interested observers and frequent protagonists.

    7.   W. J. S., December 1, 1943:

Dr. L. B. Bragg [1] . . . is trying to help us to see what's wrong with the equipment . . . . At 2:30 P.M., we drew a sample of test mixture for Dr. Bragg . . . [to test] in one of his columns.

The mixture was found satisfactory [A. E. S., February 1, 1944]. W. J. S., December 3, 1943:

Today we took the column down. Dr. Bragg found . . . the inside diameter [of the column] variable . . . . [He] said that this poor fit [of packing and column], which allowed liquid to run down the sides [channeling] . . . was responsible for the low number of plates.

    8.   Discouraging results continued. A. E. S., January 31, 1944:

On . . . Jan. 29, . . . I observed Dr. L. B. Bragg . . . re-pack a new set of [their] . . . carbon steel columns with our old . . . Stedman packing . . . . Dr. Bragg

---

hand in suitable receptacles up the penthouse stairs so that it could be poured into the top of the still, the column of which extended 15 ft upward. "The temperature of the elixir was perhaps 50°C, this being the temperature obtained on mixing."

In the first treatment a moderate amount of elixir (accompanied, I believe, by ashes from Marshall's pipe) was dumped into the still and soon thereafter drained out below, whereupon the yet vigorous liquid was adequately diluted and sent down the sewer. All hands descended to the sixth floor for this operation. When they returned to the penthouse, "copious NO_2 fumes were pouring out of the top [of the still]." Marshall ordered immediate and vigorous quenching to kill the reaction, and the menacing red cloud subsided. Undaunted, he bade the bucket brigade continue, and "after this unexpected incident, the rest of the operation went very smoothly . . . . Nearly 400 pounds of [concentrated sulfuric] acid was used. . . . The cleaning with elixir did not improve still efficiency." Nevertheless, "Inspection of the packing showed that the elixir did do an excellent cleaning job. All of the packing seemed to be slightly attacked. . . ."

Balis' shoes and trousers were fatally attacked by the elixir; Marshall kindly arranged reimbursement. After the final curtain the bucket brigade, unscathed but not unshaken, dispersed into the night after having participated in what deserves to be remembered as one of distillation's finest hours.

The quotations are from a letter dated December 6, 1943, H. A. L. to C. E. R.

was well satisfied that the packing was a good job, and that our installation of the columns was mechanically correct.

Testing of the column with C_6H_6-$C_2H_4Cl_2$ was begun. A. E. S., February 1, 1944: "[The] column is performing very poorly." The column should have had about 95 theoretical plates at a heating rate of 29,000 BTU/hour [1]. Schubert found 36 plates at 29,400 BTU/hour, and 47 at the low rate of 4400 BTU/hour.

9. A. E. S., February 7, mentions three conferences with a distillation expert from the petroleum industry (who came back on February 29), and also the design of new distributor plates. (These direct the course of liquid downward from the top of the column.) With new distributor plates, A. E. S., February 18: "The number of theoretical plates in the column was extremely disappointing (16 total)."

10. A. E. S., February 24: "[On February 21] the GE column [Item 8] was removed and a Foster Wheeler column was installed, [being] . . . assembled exactly as it had been [by them]." This column had 19 ft of packed sections. Runs 70F, 71F, and 72F were made to test it with C_6H_6-$C_2H_4Cl_2$. Of the last run, W. J. S., March 6, reports that 74.6 theoretical plates had been measured, the expected value being about 114. There had been flooding troubles. W. J. S., March 6, has a wistful concluding note: "It would [have] been better to run another test . . . . However, Dr. Marshall wouldn't let us take the time . . . and it [the test] is lost forever." Marshall probably felt that enough was enough. The methylchlorosilane inventory must by now have run low.

11. Mercifully, we ring down the curtain with a final quotation. W. J. S., September 8, 1944, Run 117F: Since Item 10, the still had been devoted to methylchlorosilane separations. There had been troubles as the closing observation by W. J. S. indicates: "Once again last night, we had trouble getting up to the [chlorine] value [expected in the distillate]."

So ends our distillation epic—an unexpectedly painful, frustrating, and costly example of augmentation that was necessary at the time, if only to establish expertise in distillation for the silicone plant to be built. The tale should give pause to any who believe that research is a succession of triumphs. In judging the epic, we must remember that here—even more than usual—decisions had to be based on insufficient information, that a research organization was, for the first time in its history, dealing with thousand-pound lots of new materials, and that the distillations here were much more difficult than those associated with the backwoods of Kentucky and Tennessee.

What does the epic teach? Good wetting was needed for efficient still operation ("old hat" even then), but complete wetting seemed practically unattainable with methylchlorosilanes. Distilling methylchlorosilanes industrially was a different problem from distilling them under ideal laboratory conditions [1], and a Stedman packing might not be the best industrial solution. Water had better be rigorously excluded from the still to prevent silicone formation, even though silicones were never proved responsible for reducing distillation efficiency when they did not plug the still. A still operating satisfactorily had best not be disturbed. Finally, any projected silicone plant would need large distillation capacity if distillation were not to limit silicone production.

Modern large-scale distillation has no "epics." Given boiling points and other relevant data, engineers can usually make, install, and operate distillation equipment as a matter of course. This is all the more reason for recording this epic as a reminder that the road to reproducible technology can be rocky!

Conclusion. Of course, the 12 reactors were kept in operation during the distillation troubles. To help in assessing the pilot plant, let us turn to W. J. S., September 2, 1944, when he was planning his Si-Cu needs (10% Cu) up to November 1, 1944. He expected to charge reactors 19 times in the interim with 147 lb of Si-Cu in each charge: a total, in round numbers, of 2800 lb. (Silicon was being ordered a ton at a time, and the performance of each new ton was awaited with misgivings.) The normal procedure was to "hydrogen-fire" Si-Cu in Building 77 to prepare it for the reactors. At this time Building 77, represented by Mr. C. S. Ferguson, could not hydrogen-fire Si-Cu at the requisite rate, and Scheiber was faced with the (then) unwelcome choice of using unfired powder. Scarcely a day without a crisis!

A few remarks should be made about money. The pilot plant was not credited with methylchlorosilanes used by the Chemistry Section for research. Methylchlorosilanes sold by the pilot plant to be used in making products for the government by operating components of the Company were priced as follows: cost of materials plus direct labor plus 80% overhead on direct labor.* No attempt was made to liquidate research expenses. Even on this generous basis, product costs would have frightened commercial customers. W. J. S., October 28, 1942, sold Rasmussen "No. 2 waterproofing fraction" for the treatment of steatite at $60 per kilogram; Rasmussen paid $3850 for 15 gal. Fortunately, very little went

---

* "Direct labor" included no one more highly paid than Scheiber (e.g., Marshall, Reed, Schubert), although such men acted as managers and consultants. The "80% overhead" was low.

a long way. The Research Laboratory offered bargains here that rivaled the UF$_6$ mentioned at the beginning of Chapter 4, but a realistic price would have been so high as to leave no alternative. Methylchlorosilanes would have to be made at lower cost if silicones were to succeed.

The first slump in silicone business came early; on July 16, 1943, Scheiber wrote sadly that he was "back on research and development work" because "there are not many orders" for the pilot plant. The distillation epic began during the business lull.

A last look at the chemical engineering pilot plant is in order. During its life—from 1942 until Waterford went on stream in 1947—it produced some 100,000 lb (probably more) of methylchlorosilanes, from which resins, rubbers, oils, and waterproofed steatite insulators were made. It was the source of methylchlorosilanes needed for research. It had the only large-scale facility in the Company for chlorosilane distillations. It provided on-the-job training and know-how for the men who were to design and operate Waterford. And it made the Company a manufacturer of silicones.

KINDS OF PROCESSES

In the pilot plant only *batch processes* were carried on. These are *intermittent* by nature, as is heating just enough water for a cup of tea. *Continuous processes* are just that—think of heating water centrally for continuous withdrawal at a tap. In general, continuous processes offer lower labor and instrumentation costs. But Mr. Justice Holmes's criticism of generalizations is again in order: any actual situation usually involves a trade-off, the proper making of which is a good engineer's stock-in-trade.

The distillation experience with chlorosilanes pointed to a continuous process. Note that the vapor-liquid experiment in Figure 6-1 shows in principle how such a process might operate in a still with one theoretical plate. Stopcocks A, B, and C would have to be opened, each to an extent required to keep a constant amount of matter in the "still." The liquid to be distilled (the "feed") would enter through A; the liquid would move down through B, and the vapor up through C.

The direct process is more complex than distillation. Let us speak of the solids as a *bed*. We must consider three kinds of reactors: *fixed-bed*, *fluidized-bed*, and *stirred-bed*. We have met the first kind in the pilot plant; the other two appear below. Note that even a fixed bed will be stirred somewhat (especially at the outset) by CH$_3$Cl passing through it; as this stirring does not greatly disturb the bed, it need not disturb us.

Stirred-bed reactors are those in which the bed is moved mechanically by any means or in any way whatever. Which reactor is best?

LIMITATIONS OF THE FIXED-BED REACTOR

In the pilot plant why was there not a single reactor with a cross section equal to those of the 12 reactors combined? (This single reactor would have been about 7 in. in diameter.) To answer this question, let us say first that the principal objective of the pilot plant was to convert silicon into $(CH_3)_2SiCl_2$ (not, Heaven forbid, into $SiCl_4$) quickly and completely, the while consuming a minimum of CH_3Cl. Given the complexity of the direct process, we may take for granted that there is an *optimum temperature* to attain this objective for each pressure at which CH_3Cl enters the reactor on its way through a fixed bed of specified character. The stoichiometric equation that describes the objective is

$$2 \ CH_3Cl + Si = (CH_3)_2SiCl_2 + \text{heat} \qquad (6\text{-}1)$$

Because considerable heat is liberated by (6-1), the reaction bed will become very hot unless heat is efficiently removed through the reactor walls—the gases cannot do the job. Because silicon is a very poor thermal conductor, being only about 1/200 times as effective as steel, heat cannot flow through the fixed bed to the walls unless the temperature is higher at the center of the bed than at the wall.* The temperature of the bed is then nonuniform and will get out of control. All hope of keeping a *fixed* bed at optimum temperature for (6-1) must be abandoned, and trouble expected. Moreover, other things unchanged, the trouble is worse the higher the center bed temperature; this temperature is higher, the wider the reactor. That is the reason why the pilot plant had 12 small reactors when a single large reactor should have sufficed.

In Chapter 4 we saw that Patnode sensed such trouble in 1940. In 1942–1943 Reed and Coe went further: they made temperature control the crux of the reactor problem. They tell why [2]:

With a furnace *only one inch* in internal diameter it is very easy, even at moderate feed rates of methyl chloride, to build up temperatures in excess of 600°C—under which conditions the silicon powder *hardens to a coke-like mass,* removable from

---

* The actual situation is complicated further by the occurrence of "hot spots" in the bed. These form at points where the rate of reaction is high. They can move through the bed. The height of the temperature level possible (600°C is cited above) and the speed (say, a matter of minutes) with which it can be reached are twin invitations to catastrophe.

the furnace only with great difficulty [an understatement], and the product consists largely of the comparatively valueless $SiCl_4$ with some CH_3SiCl_3 and very
little $(CH_3)_2SiCl_2$ [formulas and emphasis mine].

Fixed-bed reactors had to be used in the pilot plant because no other
type could then be relied upon. The horror story quoted above illustrates
why an alternative reactor had to be found for the Waterford plant:
"coke-like masses" on a large scale are bad for business!

THE FLUIDYNE

Soon after Coe joined the Chemistry Section in 1942, Reed and Coe
began work that led to the *Fluidyne*, a name coined by Reed on May 28,
1943, for what Coe had called a "jiggling apparatus," both being appel-
lations for a reactor in which the direct process goes forward in a fluidized
bed* [2, 3]. In a fluidized bed one or more solids, so finely divided that
they remain suspended in a moving gas, form with the gas a mixture that
flows and, for all other practical purposes, acts like a fluid. Such beds
are now used the world over to make chlorosilanes. Consequently, when
Reed and Coe proved that the direct process could advantageously be
carried out in a fluidized bed, they made one of the *most important
discoveries* in the silicone project. Claim 1 of their patent reads:

1.   The method of preparing an organo-silicon compound which comprises introducing a powdered mixture of silicon and a metal catalyst into a hot reaction
zone with a hydrocarbon compound capable of reacting with the silicon to form
an organo-silicon compound [3].

The foundations for the fluidized-bed technique were laid at BASF in
1924 by Clemens Winkler in attempts to fluidize lignite for the manufacture of gases useful for chemical syntheses [4a, p. 15]. About 1950 the
technique began to be used commercially by BASF to roast pyrites and
produce SO_2 needed for making H_2SO_4 [4a, p. 64]. In 1977 government
contracts were awarded in the United States for investigating fluidized
beds of limestone and burning coal suspended in air as a way of reducing
the sulfur oxides emitted by coal-fired power plants [4b].
    When Reed joined the Chemistry Section in 1942, fluidization was
well known in the petroleum industry and at MIT [4c], because of the
following chain of events. High-octane gasoline could advantageously be

* "Bed" must not be taken too literally. Humpty Dumpty once more!

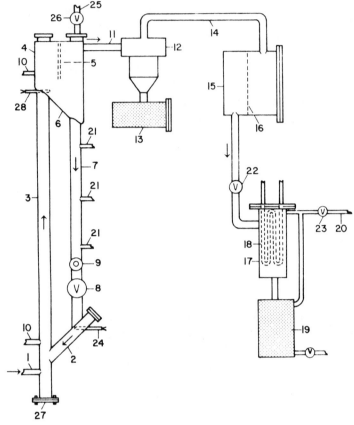

Figure 6-2 Simplest version of the Reed and Coe Fluidyne [3]. Not all the numbers in the figure are used in the abbreviated "Method of Operation." For a full account, see the patent [3]. Also, the need for initial heating to start the reaction is taken for granted.

Method of Operation

"A reactive gas, such as methyl chloride, is introduced into the apparatus through inlet pipe *1*, flows upward and meets a stream of silicon powder flowing down inclined leg *2*. The gas fluidizes the powder and the resultant mixture ascends through upright reactor . . . [*3* to separator chamber *4*] having a larger cross-sectional area than reactor *3*. In separatory chamber *4*, most of the unreacted silicon powder is disengaged from the unreacted methyl chloride and the gaseous products of reaction *3* (and to some extent in separatory chamber *4*) since the gas velocity in chamber *4* is almost negligible as compared with the gas velocity in reactor *3*. A baffle *5* may be provided to aid in the separation. . . . The separated powder collecting on the inclined floor *6* of chamber *4* flows into vertical return pipe *7*, the lower end of which is connected to inclined leg *2*. Pipe *7* functions as a cooling chamber . . . pipe *7* down to valve *8* is ordinarily filled with a relatively solid mass of powder having a bulk density of 65–80 pounds per cubic foot and control over the made by rearranging the gaseous molecules of less valuable hydrocarbons. The process was known as vapor-phase catalytic *cracking*; the catalyst was a highly porous material based on Al_2O_3 and SiO_2; the rearrangement occurred in a matter of seconds while solid catalyst and gas were in contact at temperatures of about 500°C. During contact the activity of the catalyst was reduced because of carbon deposition on its surface. The activity of the spent catalyst could be usefully increased by burning off the carbon with the liberation of heat; that is, by *regenerating* the catalyst. The petroleum industry had learned that fluidization was advantageous during both cracking and regeneration.

The direct process is chemically more complex than the regeneration of a cracking catalyst. The two resemble each other, however, in this regard: both require close temperature control so that a reaction between a gas and a fine powder—a heterogeneous reaction that liberates heat— may proceed at reasonably uniform rate throughout the reaction zone. If fluid dynamic methods are best for cracking-catalyst regeneration, perhaps a *Fluidyne* (from *fluid-dyn*amic) would do for the direct process. So thought Reed and Coe. Needless to say, success in the simpler regeneration did not guarantee success in the more complex direct process.

On May 22, 1943, 4 months before the filing of the patent application [3] and 6 days before the "jiggling apparatus" officially became "the Fluidyne," Coe described it in part as follows:

This apparatus was projected from results obtained with a smaller reactor of the same type . . . . It was evident from the earlier experiments that a longer time of contact was necessary, if a decent yield on the methyl chloride was to be

---

downward flow of this powder is exercised by means of valve *8*. . . . The fluidized mixture of powder and gas ascending reaction tube *3* has a bulk density varying from 3–40 pounds per cubic foot. . . . Since the effective pressure produced by the column of powder above valve *8* exceeds the pressure at the bottom of *3*, relatively high bulk density powder flows down tube *7* and powder dispersed in gas to a lower bulk density flows up tube *3*. The result is a continuous circulation of silicon powder through a closed cycle comprising a . . . [reactor], a separator chamber and a cooling chamber. Practically the entire quantity of methyl chloride admitted through *1* flows up tube *3*. . . . Reaction products carrying small quantities of unseparated silicon powder flow from the reaction head through conduit *11* and enter cyclone separator *12* where a major portion of any silicon powder entrained is dropped out into receiver *13*. From cyclone separator *12* the reaction products flow through transfer line *14* into filter *15*, where residual fine silicon dust is removed by filter cloth *16*. The reaction products finally pass to condenser *17* provided with cooling coils *18* where the products are wholly or partially condensed. The condensate is collected in receiving drum *19*. Any noncondensible vapors are vented from the system through exhaust line *20*."

obtained. The yield on the silicon was raised to better than 60% by increasing the reaction pressure. . . .

The new jiggling apparatus, therefore, has been constructed exactly like the first, except that it is much larger. The reactor tube is about 31 feet long, and is made of $1\frac{1}{2}''$ pipe. The powder return tube is $2''$ pipe. The reactor section of $1\frac{1}{2}''$ pipe is replaceable with $2''$ pipe if it is desired to increase the capacity . . . the $2''$ sections have been made up.

...

Assemblage: The whole reactor, which weighs about 700 pounds with all its insulation, is hung from the roof of the [second] penthouse. It is fastened with wires and springs in several places to keep it vertical and prevent it from swinging. The bottom of the apparatus expands downward $2\frac{1}{4}$ [in.] when heated up to 300°C.

How the Fluidyne was designed to function is explained in Figure 6-2, which is based on the Reed and Coe patent [3]. The reader is asked to regard the Fluidyne as a loop traveled by the fluidized bed and consisting of a vertical *reactor* (left) opening into a *separator chamber* (top) joined to a vertical *cooling chamber* (right) and closed by an *inclined pipe* (near the bottom). Continuous operation of the Fluidyne could be envisaged; CH_3Cl and silicon would have to be supplied as needed, and unreacted CH_3Cl and gaseous reaction products would have to be removed as they left the cooling chamber.

In accord with what has been said earlier, the emphasis in the Fluidyne was on good temperature control and temperature uniformity. Both were achieved by having present in the reactor (*3*) a large enough *excess* of Si-Cu to take up the heat liberated in (6-1), without unduly increasing the temperature of reaction. This heat was subsequently removed through the walls of the cooling chamber by a liquid (not shown) before the Si-Cu re-entered the reactor. Note that uniform temperature (no "hot spots"!) assures constant rate of reaction, other things being the same. Among these things the reactivity of silicon is important. This reactivity may well be different initially for different kinds of silicon. But the extreme turbulence of the fluidized bed probably enhances the reactivity of sluggish silicon by the grinding action of the abrasive silicon particles on each other, and collisions of silicon and copper particles may produce some of the "wiping" of copper on silicon that Patnode envisioned [Chapter 4]. In a fluidized bed the direct process is sure to proceed differently than in either a fixed or a stirred bed.

The Fluidyne did in fact give results that were highly promising by the standards of 1943. The main trouble was the plugging at various points in the loop beyond the reactor. This plugging was aggravated by various materials such as reaction products and chlorides of copper and iron associated with the Si-Cu. On April 30, 1945, Henry A. Fremont, who had taken over the Fluidyne when Coe left for service in the U.S. Navy, reported in a letter to Marshall:

Fluidyne Run 9W was shut down after 95 hours [of] running time on Sunday, April 29. The cause of the shutdown was a powder plug in the powder return tube. Similar troubles have been encountered in the past in runs which were made after the reactor had been shut down for an appreciable period of time. The reactor will be cleaned and another run initiated as soon as possible.

Soon thereafter, Fremont began to investigate a stirred-bed reactor called the Rotodyne. Neither Fluidyne nor Rotodyne was ever installed at Waterford.

A note for the future: The Fluidyne experience did not mean the end of the fluidized bed. The plugging that plagued a small reactor might not occur in one much larger. Unfortunately, money, manpower, time, and facilities for work on large Fluidynes were not available to the Chemistry Section.

STIRRED-BED REACTORS

The investigation of stirred-bed reactors in the Chemistry Section began long before Fremont [5, 6]. The kind of stirred-bed reactor that put Waterford on stream in 1947 was developed in Building 77, mainly by Jesse E. Sellers and John L. Davis [7] of what was then the Glyptal and Varnish Products Engineering Department. Abrasion by silicon powder is a threat to all reactors with moving parts.

The Sellers and Davis stirred-bed reactor (Figure 6-3) resembles a fixed-bed reactor of the kind in the Chemistry Section's pilot plant, to which has been added an ingenious "stirrer" that doubles as a *helical ribbon conveyor*, carries the Si-Cu upward through the (outside) annular column adjacent to the reactor walls, and dumps it centrally downward through the (inside) annular column surrounding the shaft. Effective heat removal through the reactor wall is possible because most of the reaction (hence liberation of heat) occurs while the Si-Cu is being lifted.

The Sellers-Davis reactor served Waterford well until 1962, when it was retired because the present fluidized-bed reactor had by then been developed to the point where it could produce all the methylchlorosilanes needed.

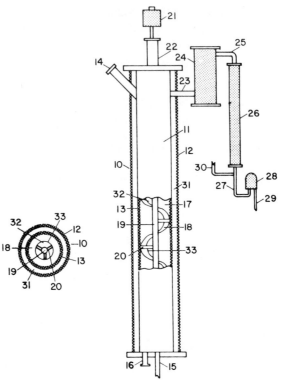

Figure 6-3 Stirred-bed reactor of Sellers and Davis [7]. See text. In the cross-sectional view the "outside annular column" is identified as *18*, and the "inside annular column" as *32*. The part of the system denoted by *23–30* serves the same functions as the corresponding part of Figure 6-2. See patent for complete description.

WAR AND THE SILICONE PROJECT

By December 7, 1941, there was no doubt that the silicone project would enlarge silicone chemistry, and there was promise that it would lead to a new industry. Under such conditions research workers worth their salt need no further motivation. World War II, which brought a 6-day week, nevertheless provided more. It created a sense of urgency nourished by expectations that silicones would improve the performance of needed military equipment.

On February 17, 1942, Dr. Coolidge telephoned Rochow about an upcoming meeting on high-temperature insulation with Commander (now Admiral) Rickover, which was held on March 10, 1942, in Newark, Ohio, with the Owens-Corning Fiberglas Corporation as host. At the meeting Marshall demonstrated that a coil with silicone insulation performed much better at 300°C than a coil built with Formex® and Glyptal®.

This silicone-insulated coil is a landmark in the history of electrical insulation. It was 2.25 in. in diameter and consisted of 675 turns of copper wire (diameter, 0.025 in.) covered with fiberglass coated with fiberglass impregnated with methyl silicone resin, brushed with a methylphenyl silicone varnish, and suitably baked.

Along with Owens-Corning and General Electric, the Bureau of Ships, Westinghouse, Mellon Institute, and Corning were represented at the Ohio meeting, which was followed on March 24, 1942, by another in Schenectady, at which Rochow described the direct process, and Marsden discussed silicone resins.

Admiral Rickover has kindly summarized his early sponsorship of silicones as follows [8]:

. . . Since records are no longer available to me, what I am writing represents the events to the best of my memory.

Early in World War II it became apparent that it was important to develop electrical rotating machinery with higher temperature ratings. Use of silicone varnish was proposed as the most logical approach to obtain these improvements.

I asked Dow Chemical Company in Midland, Michigan and General Electrical Company, Schenectady to undertake this development on a large scale. I offered to use my influence with the War Production Board to obtain rapid depreciation of the facilities involved by each company if they undertook the development.

Dow Chemical Company agreed and erected the necessary building at Midland, Michigan. General Electric decided not to engage in this venture at that time. Using the Dow silicone varnish, Westinghouse built several high temperature motors. One was, I believe, a 10 HP motor rated at 500°F.

By March 1942 the Chemistry Section was convinced that the Rochow route to silicones would eventually supersede the Kipping route (then being used by Dow), the advantages and disadvantages of which were well known. But it was much too early for the designing and building of a plant based on the direct process, which was not yet ready "for development on a large scale." After all, Scheiber did not start operating his pilot plant until July of that year, and the distillation epic was in the future.

Admiral Rickover's early sponsorship of silicones was crucial for the industry, and is memorable even in the face of his connection with nuclear energy. No direct funding of the silicone project was sought or accepted by the Company. But there were then matters more important. The close and continuing relationship of the silicone project with the armed forces, in which Admiral Rickover's influence was instrumental, led to the granting of favorable priorities for needed materials and services, resulted in draft deferment of men who could ill be spared from the project, and, above all, acquainted the Chemistry Section with military needs urgent enough so that the cost of materials to fill them was almost irrelevant. These needs afforded priceless opportunities for silicones to prove their worth. The opportunities, as we have seen in the cases of silicone-rubber gaskets and of water-proofed steatite insulators, arose because silicones had unusual and useful properties.

During World War II Admiral Rickover visited Schenectady more than once in connection with the silicone project [9]; one visit, perhaps more memorable than the others, is described by Flynn [9]:

The second visit was a short-notice visit [during which Rickover] wanted us to show him components and test data on components suitable for submarine propulsion units. By putting in several 24-hour days, we were able to do so. He expressed himself as pleased with [the] results, and then (I think) put pressure on us—i.e., GE—to produce a unit for test. This we did with apparently good results—good that is until we tested the [totally enclosed direct-current] motor. Then we ran into the "celebrated brush-wear" problem* . . . [a disastrous climax to heroic effort].

On April 3, 1942, the Department of Commerce sent notification that a silicone patent application had been withdrawn from issue for the duration of the war. Secrecy orders and classifications of silicone work soon followed. These necessary measures, however, did not dampen enthusiasm for silicone research. Any such effect they might have had was outweighed by the feeling of urgency generated by the war, by the enthusiasm attending the influx of new research personnel, and by the easier availability of materials and equipment, which sweetened the imposed restrictions.

---

* In direct-current motors, carbon "brushes" serve as current collectors. At the elevated test temperature, silicones low in molecular weight were volatilized from the motor insulation. In these *totally enclosed motors,* volatile silicones eventually appeared as silica at the interface between the carbon brush and the (rapidly rotating) copper commutator, with the result that the brush was quickly ground away. Not only was the brush wear prohibitive, but also the resulting carbon powder settled on the insulation and reduced its resistance below safe levels. This experience was shattering and unforeseen. I do not believe that silicone resins are suitable even today as insulation in motors *of this kind.*

ENTRY INTO THE SILICONE BUSINESS

In today's Company entry into the silicone business would be simpler and speedier than it was in 1943. Components that make chemicals occupy important positions in the organizational structure of today's decentralized Company. The success of the silicone project helped put them there.*

By July 1943 [2] three important and independent silicone efforts were under way in the Company: the silicone project in the Chemistry Section, begun after the 1938 visit to the Corning Glass Works; the work in Building 77 (Resin and Insulating Materials [RIM] Division); and that in Pittsfield, Massachusetts (Plastic Division). Formal top-level Company approval had nowhere been needed. Such approval was mandatory for the building of a silicone plant, however, because no Company component was authorized to make the large capital investment required.

At that time the Engineering Council of the Company had the responsibility of advising the President about the undertaking and financing of large new ventures. Mr. Roy C. Muir, Vice President, Engineering, headed the Council, and Mr. Harry A. Winne (who eventually succeeded Mr. Muir and became well known in national nuclear energy circles) was an important member. Chemistry and chemical engineering were not represented strongly on the Council, if indeed they were represented at all.

On December 21, 1943, the silicone project was presented to the Council by Marshall (introduction), Rochow (direct process), Patnode

---

* The crucial role of the silicone project in shaping the future of chemical operations in the Company was mentioned in Chapter 1. The nature, the size, and the timing of the project all contributed to this result. Chapters 4 and 5 deal with nature and size. As regards timing, the project of course benefited from World War II, but there was an added, indirect effect also. Because of the explosive growth of the Company during that war, Mr. R. J. Cordiner "was asked to study the new problems of organizing and managing a rapidly growing enterprise" [10, p. 45]. In the decentralized organization that resulted, chemical operations were virtually assured of an organizational status that called for a vice president as general manager—always provided that the silicone project *did not fail*.

The naming of Company components changed confusingly over the years. Pre-Cordiner, *departments* included *divisions,* and *works* (known by their locations) were more important than either. Today single *operating* components are divisions, departments, and sections in the direction of lower positions on organizational totem poles; today *services* components may not use these names. In addition, the catch-all term *operation* describes components that, for one reason or another, are not given the more permanent names. Juliet's "That which we call a rose . . ." does not apply here; names do matter, and the reader who wants to understand modern industry will have to struggle with them. Even Humpty Dumpty might feel discouraged.

(silicone surfaces and silicone oils), and Marsden (silicone rubber) [11]. E. L. Feininger and C. S. Ferguson spoke for the RIM Division. Within a week or so, informal word came back that the Company had committed itself to establishing a silicone business, that RIM would have the manufacturing and marketing responsibility (which it had expected to assume), and that the needed funds would be forthcoming provided that the Board of Directors approved.

Although any attempt to reconstruct the deliberations of the Council would be pure speculation, it is possible to make one's own summary of how the silicone project stood when the decision was made to expand it into a business. Here is mine:

1. Chlorosilanes could be produced in fixed-bed reactors, but at costs prohibitive for large-scale commercial applications. Prospects were good that stirred-bed reactors could be developed and that silicone precursors could be made in this way at lower cost.

2. The Fluidyne appeared to be the most promising of all reactors, but it was beset with plugging troubles that interfered with continuous operation. The funds, manpower, and facilities needed for a definitive test of the Fluidyne were greater than the Chemistry Section could command.

3. Distillation with packed columns on a large scale looked troublesome.

4. Silicone resins (varnishes), which had promised to permit operation of electrical equipment at higher temperatures and consequently had supplied the incentive for starting the project, had not lived up to early expectations. (More below.)

5. The generation of water-repellent silicone surfaces on steatite insulators was a successful application being carried on by Rasmussen (Schenectady Works Laboratory) with methylchlorosilanes produced in the Chemistry Section. (This arrangement continued until the application disappeared at the end of the war.) Undesirable production of HCl during the generation of these surfaces threatened to limit the application.

6. Silicone oils had proved disappointing as general-purpose lubricants. They promised to be useful in special cases, however, and unexplored possibilities existed of using additives or of modifying structure to make these oils better lubricants. The oils were being tested as hydraulic and damping fluids, but no business based on oils was in sight.

7. Silicone rubber, though costly, weak, and unsatisfactory in other ways, had been used in successful gaskets for two military nonelectrical applications that no other material could fill as well. It was being made

by Flynn (Schenectady Works Laboratory) from gum supplied by the Chemistry Section, a stop-gap arrangement that could continue until a silicone plant was ready to take over.

8. The silicone project had been unusually successful in uncovering new materials with interesting properties useful both at low and at high temperatures. Adequate scientific information about these materials was being obtained. Prospects for extensive but unforeseeable applications, electrical and nonelectrical, seemed good provided that the materials would not be more costly than their useful properties warranted.

9. The silicone project had served the government well during World War II.

10. No doubt the question of internal versus external silicone business was thoroughly discussed. Anyone responsible for the manufacture of equipment such as motors and generators might welcome an internal silicone business as leading to improved Company products, but oppose it because selling silicones externally might lead the chemical industry to purchase electrical equipment from competitors of the Company.

In 1944 Marshall concluded an address to the rarely assembled Chemistry Section as follows:

What of the future! At the present time consideration is being given to the idea of forming a Chemical Department within the Company with its own operating Vice President. Such a department at the start would involve the RIM Division and the Plastics Division. It is assumed that the Research Laboratory would do the Research and Development for such an organization.* The executive officers of the Company are enthusiastic about the future possibilities of such an organization and would expect it to expand rapidly. When such an integration takes place it should create many new opportunities for all of us.

You are all aware by now that the Board of Directors has authorized an appropriation of $1,500,000 to build a combined production and pilot plant unit for silicones which we expect will ultimately produce 2,000,000 pounds of crude chlorosilanes per year. The enthusiasm with which this project was undertaken and the size of the initial appropriation are the best measure of the confidence our management has in our ability.

Marshall almost certainly knew more than he told the Section. "At the present time consideration is being given . . ." in all likelihood had reference to a comprehensive study being conducted for the Company's

---

* The future decentralization of the Company forced a modification of this assumption.

top management by Dr. Zay Jeffries* to determine whether the Company ought to bring all its chemical activities into one organization, and whether it ought to expand its role as a manufacturer of materials. The silicone project thus became an important—perhaps a crucial—part of the broader picture that Marshall and Patnode had envisioned for years. The report submitted by Dr. Jeffries to Mr. Charles E. Wilson, President of the Company, was favorable, and the Chemical Department (later Division) came into existence on January 1, 1945, with him as Vice President and General Manager.

SILICONE RESINS. A BUSINESS? HOW SOON?

The silicone project had been launched in 1938 with the expectation that it would soon lead to a commercial *resin* business along two paths: that of chemical products (here loosely but conveniently called the "Glyptal path" for historical reasons), and that of improved performance in electrical machinery (similarly, the "Formex path"), made possible chiefly by insulation satisfactory for higher temperatures.

The "Glyptal path" to a silicone resin was logically the responsibility of the RIM Division (Building 77). By 1945 early methyl silicone resins, which crazed and cracked badly, had been superseded by the "precondensed" methyl resins, transferred to that Division from the Chemistry Section. To these the Division had added several silicone resins of its own, a methylphenyl resin included. On November 4, 1944, Schubert recorded that 47 people in toto were assigned to silicones in Building 77. Many of these were concerned with resins, which were the only silicones ever made there in any amount. As the Division was also developing the stirred-bed reactor, as the planning of the Waterford plant was in the offing, and as the silicone resins of that time were costly and in some respects unsatisfactory, it is not surprising that the "Glyptal path" had not by 1947 met the expectations of 1938.

Fortunately for the silicone project, E. J. Flynn and his associates in what was then the Schenectady Works Laboratory were exploring the "Formex path." The reader has had an inkling of why these men could

---

* Dr. Jeffries had behind him a distinguished career in metallurgy at the Case School of Applied Science, where Reed did his undergraduate work, and in the Company's Lamp Division, both in Cleveland, Ohio. For more about Jeffries see W. D. Mogerman, *Zay Jeffries,* American Society for Metals, Metals Park, Ohio, 1973; and H. A. Liebhafsky, *William David Coolidge: A Centenarian and His Work,* John Wiley & Sons, New York, 1974.

be expected to contribute, and a more explicit statement follows. It was only a step from the Works Laboratory to the Chemistry Section, and doors were no barrier. Because of past achievements the Works Laboratory was highly regarded by most of the many operating components in the Schenectady Works; contacts with those responsible for the design and manufacture of motors and generators were particularly close and cordial. The Laboratory had the equipment needed to make the parts of which electrical machinery is built.

The motor in which silicone insulation led to catastrophic brush wear resulted from only a small part of the effort along the "Formex path." The following story about induction motors* will take the reader a step beyond Table 1-1 toward understanding insulation problems, and will indicate why founding a business is seldom easy. Each test of a part must be designed with the property profiles of the other components in mind: there is no point in having one silicone-insulated part that lasts forever if another part of the motor fails within a week.

Before parts can be treated, they must be made. Among those made and tested, which carried silicone resins, were wire, tapes, coated fabrics, and laminates. Making these meant discovering a new art, for silicone resins were new materials for the processing of which there were no guides. In the beginning the "resins never seemed to respond twice alike"; often they were available only in pint lots, "sometimes [in] quarts, and on a big day a gallon or so" [9].

A silicone wire enamel—a "high-temperature Formex"—was of course diligently sought, but the poor physical properties of silicone resins made it necessary to abandon the search.† Instead, wires coated with fiberglass were brushed with silicone varnishes to provide adequate insulation, thus taking us back to the silicone project in 1938 and dimming hopes for a Formex-like silicone.

Of course, the experience acquired in making the silicone compo-

---

* An induction motor is only one kind of such machine. I hope the reader will acquaint himself with the others.

† Silicone resins adhered poorly to various substrates. Thus the first glass cloth used had to be burned to rid it of its hydrocarbon sizing, a costly operation, before silicone resins could be satisfactorily applied. No useful wire enamel resulted from putting resin directly on the usual copper, on oxygen-free copper, or on oxidized or nickel-plated copper. An adherence problem of a different kind gave silicones a bad name among factory people. A small amount of silicones sufficed to contaminate ovens, presses, and other equipment, producing silicone surfaces that would not wet properly in subsequent use. All these phenomena are traceable to the nature of silicone surfaces, which—turned to good account in many cases, as in the waterproofing of steatite insulators—proved to be a serious drawback in others. These troubles, as silicone films on glass had shown themselves to be, were forerunners of good and unforeseen things to come.

nents was not lost, and components that failed in one application could often be used in another.

To produce an induction motor, components or parts serve as coils, phase insulation, ground and slot insulation, wedges, leads, and binding cords; there must be impregnation with varnish as needed. All the problems that presented themselves here were eventually solved; when methylphenyl silicone resins became available, the motors were much improved. "The motors gave a good account of themselves in heat runs and moisture tests, and a considerable number of Navy motors were made later" [9]. Other motors and generators were made also. Silicone resins, varnishes, and greases were used successfully on other electrical military equipment. The many attempts made along the "Formex path" by the Schenectady Works Laboratory came to this: silicone resins served successfully in a number of military applications *where cost did not need to be counted,* but none of these successes was as spectacular as those scored by silicone-rubber gaskets and by the waterproofing of steatite insulators.

A successful demonstration of a significant advance (some prefer "breakthrough")—and the heating of silicone-insulated coils to temperatures out of reach for ordinary insulation was such a demonstration—generates a feeling of euphoria in all concerned. But harsh questions must find favorable answers before such an advance can lead to a successful *civilian* business. *What is the advance really worth in the actual marketplace?* Now, there is no doubt that insulation capable of successful operation at high temperatures is desirable if it does not cost too much, if it does not have other weaknesses (again, see Table 1-1), if more attractive or less costly materials do not come along, and if no satisfactory way of lowering the operating temperature of the equipment can be found. Mr. Flynn's experience [9] during and after World War II proved that for the silicone resins of that time the answers were often not favorable enough: the "Formex path" was rocky then and would so continue.

COMMENTS IN CONCLUSION

• As to implementation. The silicone project gave evidence that augmentation, until restrained by management, tends not to die—or even to fade away. Implementation on a large scale has to be a different story. It must be more tightly managed at every stage, for it is costly, the costs are visible, the stakes are high, and management is more clearly at risk.

Had there been no war, the silicone project would have included little or no implementation. Pilot-plant operation would probably have been kept at a level adequate to fill the needs of research within the Section, and of product development within the Company. All of this activity would have been properly classifiable as augmentation, and as chemical engineering research on the direct process. But the demands of the government had to be satisfied by making chlorosilanes and silicone rubber for sale. Consequently, the Section had to undertake implementation even though the sales were not expected to bring a profit.

Implementation on a scale great enough for a large peacetime silicone business did, as we have seen, require top-level action within the Company. Transition of silicone activity to operating components, mainly to Building 77, began before the Chemical Division was founded on January 1, 1945. Nevertheless, that day was a turning point in the history of the Company, and it became so because implementation of silicone research was eventually successful.

During implementation principal emphasis had to move from research to engineering, especially to chemical engineering.* Marshall's decision to reduce silicone work in the Chemistry Section was therefore logical. We shall say that the silicone project, begun in 1938, ended in 1950, at the minimum in Figure 5-1. Twelve years had elapsed, and a good many hard dollars had been spent.

The decision to implement may delay the augmentation and subsequent implementation of a major discovery. Although Reed and Coe applied for their Fluidyne patent on September 27, 1943, it was 1962 before a fluidized-bed reactor took over commercial methylchlorosilane production. Reed and Coe had certainly made a major discovery, and it was clear to them that, *in principle,* the direct process had best be carried out commercially in a fluidized bed. Chapters 6 and 7 explain

---

* Although competent chemists and chemical engineers both appreciate the important features of the problems with which they deal, the differences between their disciplines make chemical engineers better suited to implementation. To illustrate: from the beginning Patnode was concerned both about the mechanism of the direct process and about the kinds and amounts of products it could yield under practical conditions. Reed from the first let mechanism guide his thinking but introduced space-time yield in the analysis of the direct process [2]. Space-time yield in this case was the chemical engineer's way of describing how much of a product (or products) formed per hour per cubic foot of reaction volume when so-and-so many pounds of CH_3Cl were fed per hour per unit volume into a reactor containing Si-Cu under known conditions. (The use of empty-reactor volumes facilitates comparisons.) As implementation progresses, mechanisms as models tend to fade into the background. "What is the role of the free radical CH_3?" becomes less important than "Under what conditions do I get the best yield of $(CH_3)_2SiCl_2$?" Usually, final answers to the second question are obtained during implementation in runs planned and carried out without much attention to the first.

why almost 20 years elapsed before this technique came fully into its own. The matter is mentioned here because an incomplete account could make it seem that the Company had deliberately withheld an important invention from public use. If this ever happened, I do not know the case.

As this chapter has shown, the Company took serious though undeterminable risks when it decided on large-scale implementation of discoveries made in the Chemistry Section. In addition, formidable competition from the chemical industry was to be expected. The Dow Corning Corporation had been founded in 1943, and commercial silicone successes were sure to intrigue other chemical manufacturers. In short, faith was needed that the Company's silicone dreams would come true. As a consequence of the "four wise men's" confidence in new materials [Chapter 1], and of ample justification by the Research Laboratory of that confidence, the faith was there.

- As to research tradition. What we have called the "Whitney atmosphere," as well as the influence of World War II, made it possible to carry out silicone research, work in chemical engineering included, on a scale that was unusually broad and diversified. Increased opportunities for implementation were one result.

- As to our look at industrial research. In Chapter 5 we took a four-ring circus as a model of industrial research and the activities consequent upon it. The four rings represented discovery, augmentation, implementation, and marketing. The first three have been described. The Chemistry Section's marketing activity* was too embryonic to mention.

Why was "research and development" ("R & D" for short) not used? The answer is that it is too imprecise, and that it does not include all the activities we must consider. "Research" is itself broad and vague even though our definition ("scientific inquiry") is concise; we have divided "research" into "discovery" and "scientific augmentation of discovery." "Development" seems hopeless. One might possibly get agreement that

$$\text{Development} = \text{R \& D} - \text{Research (!)}$$

but consensus for a concise definition of "development" is not likely.

In juxtaposing "major discovery" and "scientific augmentation,"

---

* During World War II the silicone project's market was an intermittent *seller's market* with the government as sole buyer. The demands were so urgent as to make prices a minor consideration. We have seen that prices were set below costs.

we have done something like separating Christopher Columbus from the captains of ships that followed his to bring settlers to the New World; some of these captains made major discoveries in their turn, just as such discoveries sometimes occur during scientific augmentation. In making "scientific augmentation" a component of "augmentation," we call attention to the difference between them, which consists mainly of services performed, for the most part, with the objective of establishing whether implementation ought to be undertaken.

Whether this four-ring circus* is a better metaphor than R & D, the reader must judge for him- or herself. Certainly, any activity for which the annual expenditures total many billions of dollars needs to be better understood than R & D now is. Perhaps this look at industrial research will help!

---

* Just as the four rings are an incomplete model of the entire circus organization, so the four activities named above do not begin to describe everything that goes on in a large corporation or even in its corporate research laboratory. To spare the reader, our look at industrial research moves from laboratory bench to market without taking in accompanying activities. For a research laboratory, public information, liaison, accounting, personnel functions, purchasing, plant maintenance, and security—all these are among the activities not described.

Chapter Seven The Silicone Business Grows

The growth of the silicone industry has been dependent upon . . . the development of hundreds of product grades . . . [and of the requisite] commercial processes.

C. E. REED, 1956 Edgar Marburg Lecture [1]

In his travels Gulliver found Brobdignag different from Lilliput. We shall find Waterford* different from Schenectady, 20 miles away. In Schenectady, industrial silicone research dominated silicone business; in Waterford, the reverse is true. In the Research Laboratory, publication after filing patent applications was traditional, as was the opportunity to do research that enlarged and refined prior knowledge. This liberal policy was generally supported by those doing the research as furthering their scientific careers. It is not a suitable policy, however, for an operating component in a highly competitive field. A business in such a field must— and ought—to protect itself as appropriate by a wall of confidentiality: it will often have trade secrets. Industrial research in a corporate laboratory can differ considerably from industrial research closer to operations.

---

* When permissible, "Waterford" will be used to designate either the Silicone Products Department or the silicone plant located there.
Note. Brackets are used to enclose reference numbers and inserted material. For bracketed references, see References and Notes at the end of the book. Numbers in parentheses on the right of pages are often used in the text to identify reactions and the like: thus (7-1) means Reaction 7-1. Reference to laboratory notebooks is by name and date. CRDA stands for "Chemistry Research Department Archives."

Fortunately, we have not needed to present any proprietary material in this book. We are interested in Waterford, its processes, and its products primarily as they relate to earlier work done in Schenectady, our principal purpose being to promote the understanding of industrial research. This purpose can be served by relying mainly upon published material and reasonable inferences therefrom. Selected patents, especially General Electric patents that relate Schenectady to Waterford, will be our most important source of information about processes for making commercial silicones. They will give the reader a general insight into the silicone industry.

THE CURTAIN RISES AT WATERFORD

Toward the close of 1947, the General Electric *Monogram* [2] proudly described the new silicone plant in a lead article that began: "Just outside the town of Waterford in upstate New York, a stern mass of pipes and tanks and clean brick walls overlooks the Hudson River." The prospects of the plant were sterner than this view from the Hudson.

Prior chapters have implied that prospects for early success in the silicone business were not good. Military applications—and there had been no others—substantially disappeared when World War II ended. The serious failings of rubbers and resins tended to offset their undoubted advantages. All silicones, oils included, were then too costly for most commercial applications. The processes essential to making silicones were beset by difficulties unusual in diversity and number even for a chemical plant. But that was not all: Waterford turned out to be a multimillion dollar investment that considerably exceeded the initial appropriation. These dollars were harder dollars, and dollars harder to come by, than those of today. It was to be a long time before Waterford could return interest and depreciation charges, not to mention show a profit or liquidate research costs incurred before it was built.

A little sunshine brightened the early gloom. Bouncing putty (also called "Silly Putty"), described as Gilbert's Little Accident in Chapter 4, soon came into vogue as a product that, properly used, could amuse the young in years and the young at heart.* An inadequate foundation

---

* Norton Mockridge in a syndicated column that was a memorial to Peter C. L. Hodgson, a friend of his, related that Hodgson, "a not very successful advertising and marketing executive, got fascinated by a blob of useless stuff, and ran it into a $7-million-a-year national craze." He learned in 1949 that "General Electric scientists, searching for a synthetic rubber, had developed this blobby gloop," borrowed $147, and soon started to sell the gloop as Silly Putty, a name that has now ornamented our language for some 26 years. Gilbert Wright would have endorsed the name!

Mockridge's column appeared in the *Schenectady Gazette* of September 2, 1976.

Aerial view of Waterford in its early days.

for a business, the putty did produce some income and did bring silicones to public notice; even today, it is the first silicone many children get to know. There were sounder early applications as well. For example, silicone rubber had been used successfully to make grommets (small eyelets—diminutive "O-rings") for the sealing of capacitors in which the dielectric was liquid. Fortunately, a little of the costly rubber was enough for a grommet.

At the time the marketing situation was this: once silicones had been improved and could be made at lower cost, successful applications seemed certain; but to speed that happy day, applications yet undiscovered were needed. "If we had ham, we could have ham and eggs—if we had some eggs!"

In the silicone markets that had to be created if the silicone business was to survive, formidable competition was to be expected from highly regarded companies experienced in the chemical industry. The earliest competitor, the Dow Corning Corporation, had already done the country a great wartime service by developing silicone greases for use as ignition sealing compounds in military aircraft.* Dow Corning was fortunate in having the Dow Chemical Company, a respected member of the chemical industry as one parent, and the Corning Glass Works, an important manufacturer of glass products, as the other; and in having both parents experienced in bringing the fruits of chemical research to the marketplace, ethyl cellulose, a successful Dow product that found many applications in the 1930s, being a notable example.

Let us look at what this might mean as regards the marketing of silicones. Silicone products were to become large in number and bewildering in diversity—often to be bought in small amounts. Prospective customers were to be found in many industries. In such a situation marketing experience and contact in fields related to silicones were priceless assets. Waterford's dilemma was this: most of the new silicone markets lay outside the electrical industry, and most of Waterford's new customers had never before bought chemical products from the General Electric Company.

* A strong spark is needed for satisfactory ignition in an aircraft piston engine. Such a spark cannot be generated if there is appreciable external electrical leakage from the high-voltage terminals of the spark plugs. Heavy rains can cause such leakage and starting difficulties will then occur.

The remedy is to use suitable external insulation. Greases are suitable, but ordinary greases cannot withstand the temperature (400°F) near the external terminals. Silicone greases can and did. Dow Corning greases for such applications amounted to perhaps three-fourths of all wartime silicone production. The silicones for these Dow Corning greases were all made via the Kipping route. The spark-plug application was the first of many in which silicone greases now serve electrical equipment in similar ways.

Of course, the General Electric Company had had wide experience in marketing electrical products and some experience in marketing chemicals, notably alkyd resins. Because most of the chemical products were sold to original equipment manufacturers ("OEM accounts"), few salesmen were needed. Because of lack of applicable experience, it is not surprising that the original silicone marketing effort, as the accompanying plate suggests, was small and grew slowly.

Waterford was designed by General Electric (Reed and his associates in Building 77) and built by Blaw-Knox. It went on stream in 1947, almost 7 years exactly after Rochow discovered the direct process. In the functional Company organization then existing,* Waterford was managed from Building 77 by H. K. Collins, who was also in charge of chemical manufacturing activities centered in Schenectady. Engineering, which then included research and development in the operating components of the Company, was managed for Collins by C. Stewart Ferguson. Robert A. Nisbet was superintendent of the Waterford Works, as the plant was then called.

The early years at Waterford were hard. The business needed to become profitable, but annual sales did not reach $1 million until 1950. The emphasis before that time had to be upon implementation of research, not upon discovery or augmentation. Also, Waterford lacked individuality—in the Company organization then existing, silicones were not recognized as a separate business.

Such recognition, probably spurred by adversity, began to emerge with the arrival in 1949 at Waterford of James W. Raynolds, who came there from the Sun Chemical Company and had had long and extensive experience in the chemical industry. I have been told that Mr. Raynolds began his opening address to the assembled employees with, "This business is flat on its back. There is no way to go but up." He had caught Waterford at its lowest point, and he made himself a sound prophet.

In 1947, so the *Monogram* tells us, about 125 people were employed at Waterford, and its projected capacity was in the range from 1 to 4 million lb of silicones per year [2]. Many of the employees had transferred from Building 77; many others were new to silicones. By 1950 seven Waterford staff members had transferred from the Chemistry Section.

Meanwhile Reed was reporting to the general manager of the Resin and Insulating Materials Division in Building 77 and was responsible both for the design of Waterford and for seeing that the plant got built. Soon he had an office and laboratory space there also, and he was in charge of engineering work on processes that Waterford would need. He never

---

* Names of the old organizational components will not be given.

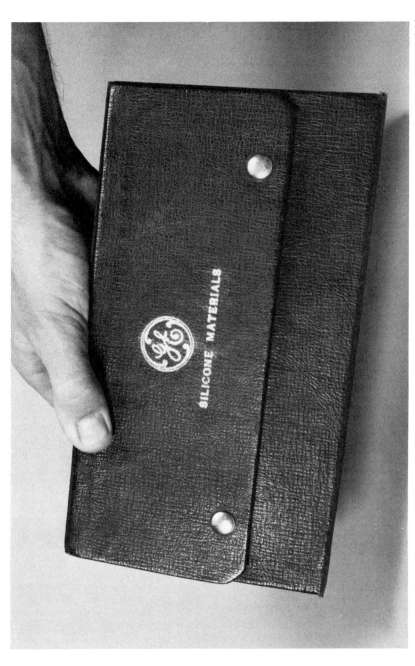

A measure of the first effort to market General Electric silicones. This "salesman's sample kit," of which a few were made in the pre-Waterford era, was large enough to hold all the samples then available for showing to prospective customers with the object of establishing silicone markets. Kindly lent to H. A. L. by Dr. Patnode, who used it for storing his photographic lenses.

lost faith in the fluidized bed as the best way to make silicone precursors on a large scale.

With the Cordiner reorganization,* the silicone business completed its evolution into a recognized Company component. Raynolds was promoted to a divisional position at Pittsfield, and Reed became General Manager, Silicone Products Department, effective March 1, 1952. The following reported to him as managers of the four sections in the functionally organized department: Engineering, Robert O. Sauer; Manufacturing, James R. Donnalley; Marketing, Mark R. Howlett; Finance, Paul D. Williams.

PROCESS DEVELOPMENT

A typical modern oil refinery looks pretty much like an enlarged silicone plant. Both kinds of plants show that chemical engineering is indispensable for the implementation of chemical discoveries. The similarity, however, goes deeper. As Chapter 6 has shown, the fluidized-bed reactor for the direct process is related to the catalytic cracker of the petroleum industry. Stills dominate plants of both kinds. With hydrolysis-condensation, a third process essential to silicone manufacture, the resemblance ceases.

As the chapter unfolds, the reader will see that chemical engineering research and product development were needed to make industrially useful the silicone products and processes that left the Chemistry Section.

The Direct Process. On June 30, 1948, Gilliam wrote Marshall to announce the termination of experimental work on what had become a joint program with Waterford to study the direct process to obtain information that Waterford would need. Among the results were a reduction in the proportion of silver needed for the acceptable production of phenylchlorosilanes, convincing evidence that Si-Cu need not be ''hydrogen fired'' to react satisfactorily with methyl chloride, and proof that phenylchlorosilanes could be made in a fluidized bed.

---

* Waterford, with its many complexities and uncertainties, was a prime example of the need for a decentralized organization in which on-the-spot decisions were possible and mandatory. ''As I See It,'' a revealing interview with R. J. Cordiner, *Forbes,* October 15, 1967, pp. 30–37, describes the difficulties in changing from the old functional organization of the Company to the new decentralized organization characterized by operating departments, each a complete business where possible, and each with a general manager who, broadly speaking, was told what he must not do, and otherwise left free to do what he would.

The reader will remember the long earlier struggles with Si-Cu; Table 4-1 lists three major discoveries involving the direct process for methylchlorosilanes. To ensure adequate reactivity, plans had been made to heat in a large furnace with a hydrogen atmosphere all the Si-Cu used at Waterford. The discovery by Gilliam and Meals that this costly step could be eliminated was consequently welcome and important. It was confirmed by the results of interesting work done in Building 77. In their patent Ferguson and Sellers report, "It was quite surprising and unexpected to find that the use of cupreous [*sic*] powder . . . resulted in better [and more reproducible] yields [of $(CH_3)_2SiCl_2$] than could be obtained with . . . finely divided metallic copper or with copper oxide. . . ." [3]. This "cupreous [not to be confused with cuprous] powder" is essentially finely divided copper—"friable metallic copper core particles"—carrying a thin, protective film of cuprous oxide that inhibits further oxidation of the metal underneath. "Cupreous powder" could be mass produced in an ingenious way patented by D. S. Hubbell [4], who was led to this powder via work on cement and on antifouling paints—work that may well have brought the material to the attention of Building 77. In any case the favorable experience of Ferguson and Sellers with Si-Cu containing cupreous powder finally buried "hydrogen firing" of Si-Cu as a preliminary to the direct process. Cleaning up the silicon surface by reaction with hydrogen conferred no enduring benefit.

So much for the route to unfired powders: now should be mentioned a stepping stone to the modern fluidized-bed reactor, in which these powders are widely used to make methylchlorosilanes *continuously* and *cheaply enough for the market to bear.* No integrated silicone manufacturer can survive in a competitive economy if he must rely upon an inferior reactor for his methylchlorosilanes.

In making phenylchlorosilanes, Gilliam and his associates used a reactor that may be regarded as descended from both the pilot-plant static-bed reactor and the Fluidyne. Please turn to Figure 6-2, eliminate from the Fluidyne the two sections in which the arrow points downward, and add a heat-transfer jacket taken (conceptually) from the pilot-plant reactor. Gilliam's stepping stone results. Its reaction chamber was a 2-in.-diameter steel pipe, perhaps 6 ft high, in which C_6H_5Cl traveled upward through the Si-Cu bed at a pressure of 100 lb/in.^{2}, fluidizing the bed and reacting at 575°C with the silicon. There was no plugging. Of course, the conditions were quite different from those under which plugging plagued Reed, Coe, and Fremont in their work with the Fluidyne (Chapter 6). A great difficulty with the direct process is the translation of results from one set of conditions to another.

Again conceptually, the shift from the Fluidyne in Gilliam's reactor includes a change in the way that the heat of reaction is removed. Reed and Coe, in their great concern for a controlled and constant reaction temperature, used excess Si-Cu to transport this heat out of the reactor into the "down leg" of their Fluidyne. Gilliam showed that effective heat removal could be accomplished under his conditions directly from the reactor itself—but only from a reactor so narrow as to be useless for making the methylchlorosilanes needed in a large plant. There was much left for chemical engineering to do.

It is worth repeating that Reed never faltered as champion of the fluidized-bed reactor, some form of which he felt sure would prove superior to the stirred-bed reactor patented by Sellers and Davis [5]. He believed that the troubles which bedeviled the Fluidyne in the Chemistry Section would disappear in reactors large enough to meet Waterford's needs. Faith was needed and courage also. Plugging and thermal runaway in laboratory reactors were unwelcome setbacks that generated new experiments; on an industrial scale they could become catastrophes.

As we saw in Chapter 6, the principal assignment for an industrial methylchlorosilane reactor is to bring about the change

$$2 \ CH_3Cl + Si = (CH_3)_2SiCl_2 + heat \qquad (6\text{-}1)$$

quickly, reliably, and completely, with minimum consumption of CH_3Cl. Hard-won experience described in earlier chapters, and general considerations governing reactions between gases and solids, both suggest that this assignment will be most satisfactorily discharged when silicon and copper are in *intimate* contact with CH_3Cl in a reaction bed as nearly *uniform* as possible. These two requirements will be met to an increasing extent the more nearly the reaction mixture *approaches a true fluid* at all times. Clearly, finely divided solids uniformly dispersed in CH_3Cl are needed. In such a model of a reaction bed, "grinding" action due to frequent collisions of solid particles should help to keep them highly reactive, and the escape of reaction products from their surfaces should be free and easy, while the heat generated by reaction should be simple to remove.

The Process Development Unit at Waterford must have had some such model of a reaction bed in mind during the work that eventually made it possible to install the fluidized bed as the sole source of supply of crude methylchlorosilanes. I believe Waterford was the first silicone plant in the world to do this.

Details of commercial chlorosilane production by the direct process are naturally considered confidential by the industry. We are fortunate in having the patent [6] for Dotson's *fluid energy mill* to rely upon. The patent not only describes the mill, but also, in demonstrating the mill's effectiveness, gives the results of significant direct-process experiments. The mill as part of a fluidized-bed reactor system is shown in Figure 7-1.

The fluid energy mill is added to the reactor system in order to make the reaction bed approach more closely the model described above. In the mill a nonoxidizing gas at velocities near that of sound issues from peripheral jets into an annular chamber and reduces the size of ("comminutes") the solid particles introduced into the chamber. We quote from the patent [6], omitting details less pertinent here:

[The system] was operated employing methyl chloride gas as a feed for the reaction between methyl chloride and silicon to produce a mixture of methylchlorosilanes including methyltrichlorosilane and dimethyldichlorosilane. The reactor had a diameter of one foot and was 15 feet high. In a series of runs the cylindrical fluidized bed reactor was charged with 400 pounds of a powdered mixture of silicon having an average particles size of about 90 microns with about 60 percent by weight of the particles having actual diameters of from 20 to 200 microns, 30 pounds by weight of the copper oxide coated copper catalyst de-

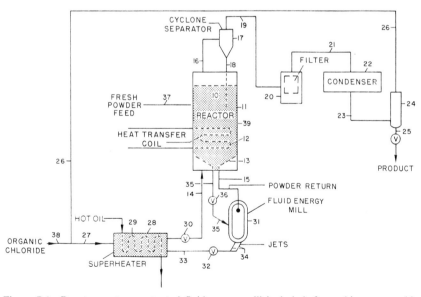

Figure 7-1 Reactor system, patented fluid energy mill included, for making organochlorosilanes continuously by the direct process [6]. When CH_3Cl is fed in at *38*, the product at *25* is a methylchlorosilane crude. Valve at *30* permits admitting a desired fraction of the CH_3Cl directly to the reactor, the rest passing first through the fluid energy mill *31*. Other numbers are explained in patent.

scribed in the aforementioned patent [3], and one pound of finely divided zinc. Methyl chloride, which had been superheated to a temperature of 270°C., was passed into the base of the fluidized bed reactor at a rate sufficient to provide an average methyl chloride velocity through the reactor of 0.5 feet per second at a pressure of about 50 p.s.i.g. at the top of the reactor. This methyl chloride flow rate was sufficient to cause elutriation of about 2 pounds per hour of the powdered mixture in the reactor. In each case the reaction was initiated by heating the fluidized bed reactor to a temperature of about 340°C. After initiation of the reaction, the temperature was allowed to drop to the steady state temperature of about 280 to 290°C. During the reaction powdered make-up silicon, make-up copper catalyst and make-up zinc were continuously added to the fluidized bed reactor . . . [so as] to maintain 30 pounds of copper . . . and one pound of zinc in the fluidized bed. This procedure resulted in a production of about 100 pounds per hour of mixed methylchlorosilanes. In Examples 1, 2, and 3, continuous comminution of a portion of the solids in the fluidized bed was conducted. This was accomplished by withdrawing at a uniform rate about 100 pounds per hour of the solids from the fluidized bed and passing these solids into a fluid energy mill having an annular chamber 2 inches in diameter. The fluid energy mill was activated by passing about 80 percent by weight of the methyl chloride feed through the fluid energy mill. This methyl chloride used in the fluid energy mill was then passed into the reactor. Example 4 differed from the first three examples in that no comminution was employed and the reaction was merely continued to the point described below. The mixture of methylchlorosilanes and unreacted methyl chloride and elutriated solids [was] passed out of the top of the fluidized bed reactor and the particles having an average diameter greater than about 5 to 20 microns were separated and returned to the fluidized bed reactor. The finer particles were then filtered from the unreacted methyl chloride and mixture of methylchlorosilanes, and the unreacted methyl chloride was then separated from the mixture of chlorosilanes and returned to a storage tank, from whence it was again fed into the reactor.

In order to compare the results of Examples 1, 2, and 3, which are within the scope of the present invention, with Example 4, which is typical of the prior art, the ratio of methyltrichlorosilane to dimethyldichlorosilane was continuously

Example	Comminution	Run Time (hours)	Methyl-chlorosilane Production (pounds)	Avg. T/D Ratio	Silicon Utilization (percent)
1	Yes	225	23,500	0.212	88.9
2	Yes	234	23,700	0.192	87.3
3	Yes	269	27,000	0.196	85.2
4	No	65	6,500	0.228	80.9

$[T = CH_3SiCl_3; D = (CH_3)_2SiCl_2]$

determined in each run. Each run was continued until the ratio of methyltrichlorosilane to dimethyldichlorosilane (the T/D ratio) reached a value of 0.35. In the table below [are] listed the total time for each run, . . . whether grinding was employed, the total production of mixed methylchlorosilanes, the average T/D ratio during the course of the reaction, and the silicon utilization, which is the percentage of the silicon fed to the reactor which was converted to methylchlorosilanes.

Before we comment on the fluid energy mill, let us look at the broader significance of the first three examples listed above. By 1960, when the patent application was made, Waterford had brought the fluidized bed to the point where a reactor 1 ft in diameter and 15 ft high could make 100 lb of crude chlorosilanes an hour from *commercial* silicon for 10 days running with no indication that continuous operation might not be greatly extended. In importance to the industry, this achievement rivals the discovery of the direct process.

The patent also says, "The specific size and shape of a fluidized bed reactor which can be used in the process of the present invention can vary within wide limits and the choice of a particular fluidized bed reactor is within the knowledge of those skilled in the art." Here "wide limits" is a further key to the Waterford accomplishment. It suggests that reactor diameter could be increased to the point where a single reactor produced the many millions of pounds of crude methylchlorosilanes needed annually by a large silicone plant: after all, if the production is proportional to the square of the reactor diameter, increasing the diameter of Dotson's reactor to 5 ft would increase its output to some 22 million lb annually. We have, of course, not allowed for "downtime" or "outage," which might well decrease as the reactor is widened; on the other hand, Dotson's reactor output may be low by today's standards, and a reactor with an internal diameter exceeding 5 ft might operate satisfactorily.

Figure 7-1 has more to tell us. With central injection of the 80% CH_3Cl that goes through the fluid energy mill, reaction will be fastest near the center of the reaction bed and the spent powder will tend to move toward the walls, where it is replenished. The height and the position of the heat transfer coils indicate that most of the reaction is over before the powder enters the upper regions of the reactor. Violent motion within the fluidized bed helps heat transfer and hinders plugging. The cyclone separators and the filter trap the particles carried out by the moving gas; these of course tend to be "fines" (Dotson's "elutriation losses"). Such losses could be reduced by doubling or tripling the reactor height.

According to Lichtenwalner and Sprung [7, p. 520]:

The composition of the crude mixture [from a fluidized-bed reactor] varies, depending on operating conditions, but generally falls within the ranges given below:

dimethyldichlorosilane	>50%
methyltrichlorosilane	10–30%
methyldichlorosilane	<10%
trimethylchlorosilane	<5%
other monosilanes	about 5%
higher-boiling residue	up to 10%

Separation of the mixture is accomplished by fractional distillation.

The datum ">50%" is particularly significant, for it measures the extent to which (6-1) occurs. If Dotson computed the average T/D values in his first three experiments by taking the mean of an observed minimum in this ratio and 0.35 (see above), his minimum must have been near 0.05, which would indicate a yield of $(CH_3)_2SiCl_2$ gratifyingly above 50%.

The ideal reactor would produce a crude of composition controllable by changing the operating conditions. Dotson proved that the T/D ratios increased with time in his experiments, suggesting that further experiments like his might point the way toward a controllable, improved crude. His work shows that particle size and surface condition are variables of first importance in studying the reaction mechanism for a fluidized bed: the postulation of free radicals is not enough. If $(CH_3)_2SiCl_2$, which (6-1) suggests is easy to form, could be induced not to linger on the solid surface, its yield might increase.

On May 10, 1940, Rochow made "at most 5 cc" of methylchlorosilane crude on a working day, lunch hour included [Chapter 3]. By 1960 Dotson could probably have made 1000 lb of crude in the same time. Why was so much work needed to get from Lilliput to Brobdignag, and why did it take so long? To answer would be to retell our story. The reader will have seen by now why the Introduction contains a mild demurrer to Lord Snow's statement that "technology is rather easy."

Distillation. The unhappy distillation epic of Chapter 6 was a consequence of Marshall's decision to test Stedman packing, then new and promising, on methylchlorosilane crudes. Marshall took a justified risk. At the time, distillation was far behind where it is today. The silicone

Gilliam with his first vacuum system (Room 605, Building 37, January 8, 1940) for the isolation of pure methylchlorosilanes, a process that is essential to the modern silicone industry and gradually became routine on a large scale after knowledge and experience had been painfully acquired through work on necessarily small amounts of materials. His still is on the left.

project demanded *batch* as opposed to *continuous* distillation: chlorosilanes were not yet available in tank-car lòts, and stills that could operate on smaller volumes (small "hold-up"), as packed columns can, were needed. Packed columns, especially analytical columns, had served the project well. Satisfactory Stedman columns were expected to require little headroom, an important consideration in Building 37.

At the time in Chapter 6, Stedman columns were new, but distillation itself was old. Early stills used *trays*, one variety of which is sketched in Figure 7-2. *Bubble-cap* trays (not shown) date from 1818, and *sieve* trays from 1832 [8]. Largely because of the growth of the petroleum and chemical industries during the last half-century, stills with improved sieve trays can now be routinely built to order. Such stills are capable of handling continuously and satisfactorily the large volumes of chlorosilane

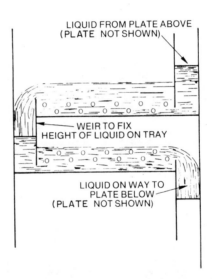

Figure 7-2   Schematic diagram of two trays in a distillation column. To assist boiling, the trays are often perforated (perforations not shown); such trays are called "sieve trays." Ideally, each "tray" is a theoretical plate, which means that each tray is supposed to deliver liquid and vapor at the compositions prescribed by the vapor-liquid diagram (see Figure 6-1) for the temperature of the tray. Toward this end, each tray has openings that permit vapor to pass upward through a layer of liquid supported by the tray. The turbulent passage of vapor through liquid provides the contact needed for the approach to vapor-liquid equilibrium. The vertical spacing between trays must be great enough to prevent the underside of a tray from being contaminated by liquid from below. Operation is considered satisfactory when three trays make about two theoretical plates.

crudes produced by industrial reactors, even though, as the Chemistry Section had discovered, the distillation task is very difficult.

Fortunately the silicone industry did not need to devote nearly as much chemical engineering effort to distillation as to the direct process. The importance of high-quality distillation to the industry is shown by the following example.

Distillation has to yield $(CH_3)_2SiCl_2$ pure enough for making satisfactory intermediate materials, such as D_4, for subsequent conversion to silicones such as oils and rubbers. Of course, the purer, the better, but up to 0.5% CH_3SiCl_3 by weight could be tolerated; today's stills *routinely* deliver $(CH_3)_2SiCl_2$ *nearly 99.9% pure.*[*] The distillation has to be economical, a term here synonymous with producing $(CH_3)_2SiCl_2$ of acceptable purity quickly and at lowest overall cost; *continuous* distillation is preferable to *batch*. The distillation can be done outdoors so that fractional-distillation ("rectifying") columns can be built to reach the sky; penthouses, as in Building 37, are not needed.

Schubert and Reed, with the costly and hard-won distillation experience of the Chemistry Section behind them, made patent application in 1948 (patent granted in 1951 [10]) for a sophisticated distillation system to serve a large silicone plant. Their patent teaches all we need to know; see Figure 7-3 and the notes thereto.

The silicone industry has found that chlorosilane crudes present no special distillation difficulties: provided that wetting in the still is satisfactory and no chemical reactions (especially hydrolyses) are allowed to occur, distillation proceeds according to the vapor-pressure relationships among the components (azeotropes included); see Figure 6-1. The gloom generated by the distillation epic in the Chemistry Section gave way to sunshine at Waterford.

Continuous Hydrolysis. Because commercial silicones are of many kinds, and because complex hydrolysis reactions are needed in making them, the hydrolysis processes are not usually continuous.[†] One fortunate exception is the reaction of $(CH_3)_2SiCl_2$ with water to make (among other things) D_4 for subsequent conversion to oils and rubber. The following description of a continuous process for this reaction is modeled upon Lichtenwalner and Sprung [7].

---

[*] Voorhoeve [9] gives an excellent modern synopsis of industrial methylchlorosilane distillations on p. 332, and stills are prominent in several of the photographs of silicone plants that follow his p. 333. Note how much these plants look like petroleum refineries.
[†] During the pre-Waterford period, Building 77 played a major role in carrying hydrolysis-condensation processes beyond the scale attained in the Chemistry Section.

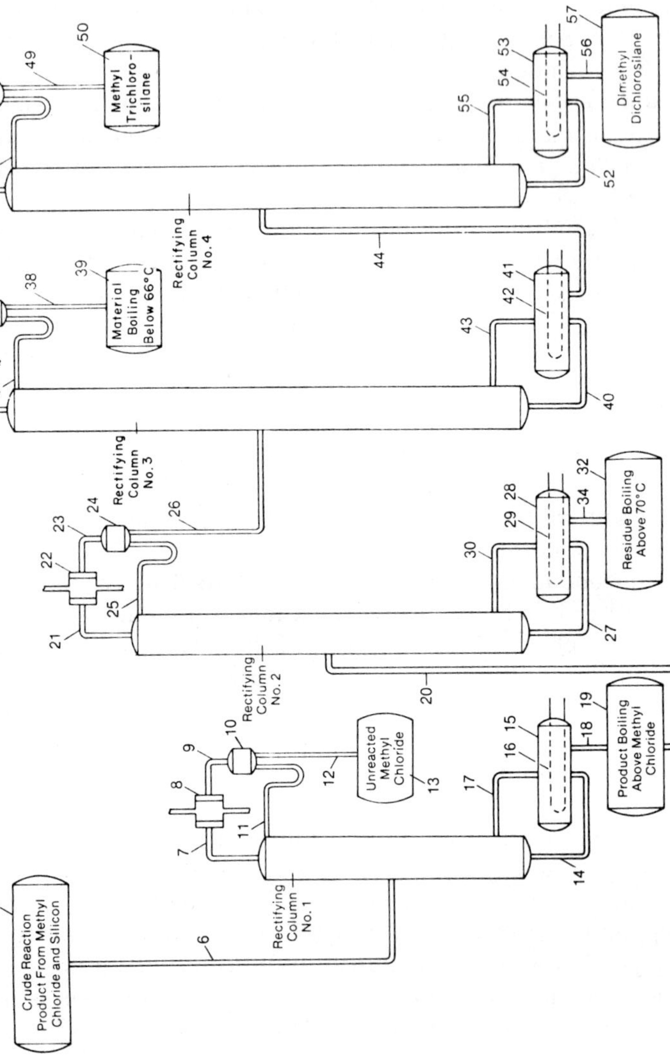

Figure 7-3

252

The reaction may be written as

$$(m + n)[(CH_3)_2SiCl_2] + [2(m + n) + 1] H_2O$$

$$= [(CH_3)_2SiO]_n + HO[(CH_3)_2SiO]_mH + [2(m + n)] HCl \quad (7\text{-}1)$$

Were $n = 4$ and $m = 0$, D_4 would be the only silicone formed; it never is: m and n are never integers, as they usually are in simple chemical reactions. The complexity, thermal and chemical, of reactions such as (7-1) was touched upon in Chapter 2 with $SiCl_4$ serving as the simplest example. An additional point here is that aqueous hydrochloric acid is the source of water for the continuous hydrolysis.

Aqueous hydrochloric acid is an azeotrope (constant-boiling mixture) when the HCl content is nearly 21% by weight. At 37% HCl the acid is

---

Figure 7-3   System for the continuous distillation of methylchlorosilane crudes on a large scale [10].

Notes Based on the Patent

1.   Each of the rectifying (= fractional distillation) columns has one "recycle loop" overhead and another at the bottom of the still. These loops return once-distilled materials to the column ("If at first you don't suceed . . .") and supplement the recycling (= refluxing) that occurs within the column. Taller columns are a less desirable alternative. Here one is trading off increased energy consumption (recycle loops) against increased capital investment (taller columns) and other factors.

2.   Each *top* recycle loop needs a condenser (*8* and analogues) and an accumulator (*10* and analogues) from which part of the condensate is returned to the column and the remainder is forwarded. The top recycle loop for Column 1 has provisions (not shown) for venting hydrogen, methane, and other "noncondensables."

3.   Each *bottom* recycle loop needs a heater (*16* and analogues) within a "reboiler" (*15* and analogues); the reboiler also serves as accumulator.

4.   Liquid products are taken out of the system at six points as shown. The three most important normal boiling points are as follows: CH_3Cl, $-23.7°C$; CH_3SiCl_3, $66.1°C$; $(CH_3)_2SiCl_2$, $70.0°C$; the azeotrope boils at $65.3°C$. No product is removed at the top of Column 2 because further distillation of the condensate is needed to accomplish the desired separation. It would be pointless to remove product from the bottom of Column 3.

5.   The difficulty of separating $(CH_3)SiCl_3$ from $(CH_3)_2SiCl_2$ by distillation is corroborated by the following quotation from the patent (which see for the meanings of the numbers shown in the figure): "Under practical operating conditions with a practical number of theoretical plates installed in the last rectifying column, it was found possible to make simultaneous products analyzing about 90 per cent methyltrichlorosilane in the overhead product and 99.5 per cent dimethyldichlorosilane in the bottoms product."

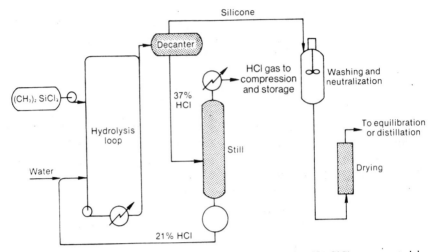

Figure 7-4 Flow diagram for the continuous hydrolysis of $(CH_3)_2SiCl_2$ as reported by Lichtenwalner and Sprung [12].

Notes

1. Two condensers identified by jagged arrows are shown.

2. Reaction occurs in the hydrolysis loop on the left of the figure; products are removed from the right.

3. The concentration of HCl in the loop can be reduced below 21% by adding water.

4. Corrosion-resistant materials of construction must be used as needed to eliminate corrosion by HCl, which can be a serious hazard. Dry HCl is not corrosive.

5. More sophisticated methods than the action of gravity may be used to separate the silicones from the acid.

6. The silicones produced will contain from 20 to 80% of cyclamers with D_4 predominating.

of a density great enough so that it can be conveniently separated from the lighter, oily silicone layer by decantation. Finally, upon distillation 37% HCl yields the azeotrope and (almost) dry HCl gas.

Knowing these facts, William A. Schwenker, who had worked extensively on the direct process with Gilliam, patented the continuous hydrolysis of $(CH_3)_2SiCl_2$ [11]. We follow here the account published by Lichtenwalner and Sprung [12]; see Figure 7-4 and the notes.

It would be nice if silicones could be manufactured profitably without making HCl, which can become a burden to the manufacturer and a threat to the environment. At Waterford this disposal problem was eliminated by reconverting the HCl to CH_3Cl according to Du Pont patents, the last of which expired in 1964. The reaction is

$$CH_3OH + HCl = CH_3Cl + H_2O \qquad (7\text{-}2)$$

In this way the silicone industry can "close the chlorine cycle" and largely replace CH_3Cl with CH_3OH as a raw material.

The reader seeking more complete information about the processes that must be carried out in a diversified silicone plant is advised to examine the "simplified flow sheet" presented by Meals [13] in his Figure 1(a) and (b).

SILICONE RUBBER AS AN EXAMPLE OF PRODUCT DEVELOPMENT

When Waterford went on stream, silicone resins needed further improvement before they could be expected to support a silicone business; extensive past experience indicated that such improvement would be slow and hard-won. Oils, on the other hand, were much further advanced as products; here, markets and applications at the time logically had to take precedence over possible improvements. Applications that were founded on the nature of silicone surfaces (e.g., their water repellency) seemed unlikely to provide large markets. That left silicone rubber, and Waterford decided in its early years to make this product its principal concern. The decision was well taken. Silicone rubber turned out to be not only the most interesting of the silicone products but also the one susceptible to greatest improvement. Imagine, then, the sharp jolt Waterford took when Dow Corning began to market a stronger silicone rubber. The jolt was felt elsewhere in the Company, notably in Schenectady, and at Pittsfield in the laboratory of the (then) Chemical Division. Fierce competition in the marketplace naturally led to fierce competition in the laboratory.

As the reader will remember from Chapters 4 and 5, the road to silicone rubber had been long and tortuous, and the quality of the rubber suffered because ferric chloride was not a good rearrangement catalyst. Glennard R. Lucas was first in the General Electric Company to use D_4 and KOH to make a silicone gum from which a rubber comparable with that introduced by Dow Corning could be made [14].

Dow Corning's contribution was of major importance. Hyde's patent, discussed in Chapter 5, teaches the use of solid KOH *in trace amounts* as rearrangement catalyst: not only were these amounts insignificant beside those of ferric chloride, but also the KOH could be neutralized, or washed out of the finished gum, which meant increased stability for the silicone rubber. Nor was the use of D_4 as starting material* less important. The high purity of D_4, especially the absence of T units, meant that cross-linking, the importance of which is shown by Figure 4-3, could be controlled better than it was with ferric chloride.

Dow Corning also took the lead in a further improvement of silicone rubber, which resulted from the use of *reinforcing fillers*, the best of which were finely divided silicas—costly materials once, but no longer. Two patents granted to Earl L. Warrick [15, 16], who was at the time Senior Fellow in the organosilicon research being supported at the Mellon Institute by the Dow Corning Corporation, tell us all we need to know. Warrick uses a *stress-strain efficiency*, which is "the tensile strength of the rubber in pounds per square inch times the per cent elongation at break divided by 1000"; that is, an efficient rubber will both stretch a great deal and support a high load. Incorporating a *reinforcing filler* such as carbon black into natural rubber was known to increase its stress-strain efficiency; but silicone rubber, for which such blacks and earlier silica fillers both gave efficiencies below 75, was a different story. Warrick discovered that, with traces of KOH in a solvent as rearrangement catalyst and by use of finely divided silica as filler, D_4 could be made into a rubber of efficiency "at least 150" (tensile strength, 500 lb/in.^{2}; elongation at break, 300%). In the second patent [16] Warrick recognizes two classes of fillers, *nonreinforcing*† (such as metallic oxides, diatomaceous earths, and clays) and *reinforcing* (silica aerogels and "fumed" silica, both characterized by high surface areas). The silica produced in the flammability experiments of Chapter 5 would have made a reinforcing filler for the silicone rubber being worked on next door in Building 37,

---

* The history of D_4 as starting material to make other silicones is an interesting chapter in the story of silicone rubber. J. F. Hyde, U.S. Patent 2,438,478: applied for, February 26, 1942; granted, March 23, 1948, claims D_4 as a novel composition of matter on the basis of silicon content and molecular weight. Patnode concluded his more complete characterization late in 1941 [Chapter 5], and used D_4 as a starting material for oils in 1942 [Chapter 4]. Neither knew what the other was doing.

Newer rearrangement catalysts, more economical of H_2SO_4 than Patnode's [Chapter 4], are those of E. C. Britton, H. C. White, and C. L. Moyle, U.S. Patent 2,460,805: applied for, December 6, 1945; granted, February 8, 1949, and assigned to the Dow Chemical Company. These catalysts are "bleaching earths," hydrous aluminum silicates such as bentonite, impregnated with H_2SO_4 (of which these earths absorb comparable weights), dried, and comminuted to fine powders, which are easy to handle and can be stored without deteriorating. These powdered catalysts are mixed with the siloxane to be polymerized, and filtered out when no longer needed. The earth holds the acid so strongly that an uncontaminated product is achievable. As the product becomes thicker, an organic solvent may be added to expedite filtering; as thickness (viscosity) increases, it eventually becomes advisable to shift to KOH as rearrangement catalyst.

† On occasion, "nonreinforcing" fillers give some reinforcement, but their main function is to act as *extenders* of the rubber.

but in 1943 no one thought to try. Warrick [16] gives initial stress-strain efficiencies of 75 and 300 for the two kinds of fillers; clearly, improved rubbers were being made.

The reinforcing action of these finely divided fillers is further evidence of the special relationship that siloxanes bear to silica, one aspect of which is diagrammed in Figure 4-1. But there can be too much of a good thing, as all those concerned with making silicone rubber soon discovered. Presumably because of hydrogen bonding (see the right-hand side of Figure 4-1), the mixture of gum and filler becomes "structured," that is, difficult or impossible to process in the normal way, as on rolls. A remedy must be found if the mixture is to yield a satisfactory rubber when it is eventually cured [1]. Of various patented remedies we mention only that of Lucas [17], which relies on our old friend D_4 and is noteworthy for its language. According to Lucas:

the compounded material [gum and filler] becomes tough and nervy . . . [as] recognized by the presence of an undesirable snap. [On storage] . . . this toughness and nerve increased to such a point that excessive milling times are required to form a plastic continuous film . . . [needed] prior to further processing . . . such as for purposes of incorporating other fillers and additives . . . or for "freshening" the filled compound . . . [which] will not knit readily . . . , in some instances will not knit at all.

Well, the various terms beginning with "structured" (architecture) and on through "tough and nervy" (cops and robbers), "snap" (garters), "freshening" (dairy cow), and "knit" (stockings) ought to convince the reader that making silicone rubber had its moments. Treating the filler with D_4 vapor tames it well enough—probably because of its reaction with acidic sites on the filler—so that a satisfactory product can be made in a more comfortable and less exciting way.

Unfortunately, there is not room here for a complete history of the development of even the one product, silicone rubber, we have chosen as example. We must omit the history of other improvements, such as that in compression set,* and conclude with a quick, grossly oversimplified sketch of *room-temperature-vulcanizable* (RTV) silicone rubber.

Thanks to product development, we now speak of silicone *rubbers,* not rubber. This bounty has come about because competing silicone manufacturers have anticipated requirements and have moved quickly to fill needs called to their attention by customers. The markets thus created were essential to survival. Let us relapse and again speak momentarily of a single silicone rubber in a general way. We have here a material, not

* See Figure 4-3, Note 6.

superior in mechanical properties to the usual (organic) rubbers at room temperature, but far superior in such properties at low temperatures and at high; far better as regards resistance to ozone, ultraviolet light, and air at all temperatures; and possessed of other virtues as well. Here, then, is a material about which extravagant-sounding claims are justified. Customers will soon want to use it in ways that rule out curing (vulcanizing or cross-linking) by the use of heat. These customers will want to apply it by brushing, pouring, spraying, dipping, or caulking. They will want a liquid or paste that turns itself into a rubber at room temperature after it has been put in place. In short, they will demand RTV.

To give them what they want, the silicone manufacturer must find a system that can be induced to form cross-links of the right kind and at the right rate at room temperature. ("Right rate" means slowly enough to permit easy application, but quickly enough to forestall long waiting for the rubber to form.) Silicone chemistry suggests schemes such as that in Figure 7-5 [18].

Figure 7-5 implies that three species (the **D** backbones terminating in OH groups, the cross-linking agent, and the catalyst) participate in the condensation reaction. In two-part RTV systems the components are supplied so that the parts must be mixed before rubber results.* In one-

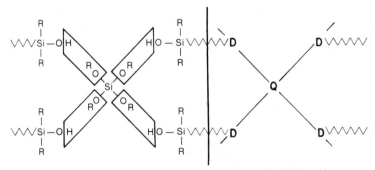

Figure 7-5   Prototype of condensation reaction suitable for RTV rubber.

Notes

1.   A common cross-linking agent for the **D** backbones is $(C_2H_5O)_4Si$; others can be used, some of which have less than four RO groups attached to the silicon.

2.   The **D** backbones, represented by /\/\/\/\/\ , ideally have an OH group at each end.

3.   When complete, the (catalyzed) condensation reaction would result in a product cross-linked as shown in shorthand on the right side of the figure.

* The reader may have used epoxy cements at home. These become cements only after the contents of two tubes have been squeezed out in the fashion of toothpaste and properly mixed. One tube contains the reaction catalyst. The two-part RTV system is analogous.

part RTV systems these components are mixed at the factory, but the rubber will not begin to form until the mixture is penetrated by water vapor from the surrounding air. Always, the customer who insists on RTV rubber must be willing to sacrifice somewhat, principally as regards physical strength, the desirable properties of heat-cured silicone rubbers.

Satisfactory RTV systems did not come overnight. Hyde [19] was among the first to disclose such a system: he produced by use of either H_2SO_4 or P_2O_5 (reminiscent of Wright and Patnode; see Chapter 5) a gum which he then cross-linked after the manner of Figure 7-5. Today, the two acid catalysts are gone, but the method of cross-linking remains. Later patents reveal improved materials, some of which originated at Waterford [20] and attest to the importance being attached there to RTV development. Today RTV bids fair to become the finest jewel in Waterford's silicone crown.

FURTHER RESEARCH

General. The Cordiner decentralization of the Company no doubt helped invalidate Marshall's assumption, reported near the close of Chapter 6, that the "Research Laboratory would do the Research and Development for [the future Chemical Division]." Decentralization required the Division to do much of its own research. Also, Marshall must have thought that silicones had been the major project of the Chemistry Section long enough, and that Waterford could prosper with less help. No one could foresee at that time how difficult it would be to improve silicone rubber, or how it was to be done. All these matters contributed to a sharp decline in silicone activity within the Chemistry Section. At Waterford's request the decline was abruptly reversed in 1950, as Figure 5-1 shows.

Rearrangement of Chlorosilanes. We have seen in Chapters 4 and 5 that siloxane molecules can be rearranged to yield desired products. Silicone precursors, the *chlorosilanes,* offered a similar opportunity, of which Sauer and Hadsell [21] took advantage in 1948 before Sauer transferred to Waterford. Their work made it possible to change the products of the direct process into silicone precursors more urgently needed or more profitable.

This rearrangement of organic groups attached to metal or metalloid atoms has interested chemists for a long time. The history of such reactions contains the names of old friends (Ladenburg, Friedel, Crafts, Stock, Somieski) from Chapter 2 and References and Notes. While the silicone project was under way, a group headed by Calingaert was active in the field, their interest having been kindled by rearrangements possible with tetraethyllead [Pb(C_2H_5)_4] and other antiknock compounds. In the seventh paper of their series [22], they reported that $Si(C_2H_5)_4$ and $Si(C_3H_7)_4$, when heated for 5 hours near 175° with $AlCl_3$ as catalyst, gave all possible compounds of the type $Si(C_2H_5)_n(C_3H_7)_{4-n}$, and in amounts called for by the laws of chance, which govern the rolling of dice. In the early days of the silicone project, these discoveries caused concern that thermal rearrangements might limit the usefulness of silicones at high temperatures. Fortunately, this has not happened.

Sauer and Hadsell [21] discovered that rearrangement of CH_3 groups and chlorine atoms in methylchlorosilanes could be accomplished at temperatures of 300°C and higher with $AlCl_3$ as catalyst. The amounts of the products were not governed by the laws of chance. Although CH_3 groups could be exchanged for chlorine atoms in phenylchlorosilanes, the C_6H_5 groups themselves appeared not to migrate.

The Sauer patent [23] on rearrangement was closely followed by one assigned to Dow Corning [24], and some years later by another granted to Harry R. McEntee of Waterford [25]. McEntee found that by introducing a little $SiHCl_3$, another old friend, into Sauer's reaction system, useful rearrangements became possible at temperatures much below 300°C.

Among the possible chlorosilane rearrangements, the reaction

$$(CH_3)_3SiCl + CH_3SiCl_3 = 2\ (CH_3)_2SiCl_2 \qquad (7\text{-}3)$$

is, as the reader may have guessed, of particular industrial usefulness.

Discoveries of Marginal Commercial Value. Unfortunately, there is no correlation between the scientific merit of a discovery and its commercial value; the latter is determined by competition in the market place and may be different tomorrow from what it is today. Furthermore, a successfully implemented discovery is difficult to displace. As a consequence, discoveries—even when augmented by well-planned applied research—are sometimes not implemented because they are judged to be of marginal commercial value. Several such examples will be briefly described to emphasize the chancy nature of research and the need for good judgment in its implementation.*

The commercial value of silver as catalyst in the direct process for phenylchlorosilanes will be reduced as the price of silver increases rela-

---

* In selecting these examples, I have drawn upon silicone information that I absorbed years before this book was thought of. The earlier disclaimer relating to Waterford is particularly apropos here.

tive to that of copper, although the reduction can be mitigated by recovering silver from spent residues. An offsetting advantage of silver could be an increased proportion of $(C_6H_5)_2SiCl_2$ in the crude reaction product. If a satisfactory crude can be made acceptably by using copper, the increasing price of silver would make it (and I think has made it) a catalyst of marginal commercial value.

The number of possible silicones with groups other than CH_3 or C_6H_5 attached to silicon is almost endless. Such silicones could contain hydrocarbon groups other than the two named, or they could have other elements replacing hydrogen. Even in the early days of the silicone project, it was clear that much research to find new and different silicones could be done, and it was expected that many useful substances would be found among them. I think that the number of these has actually been disappointingly small.

Polar silicones are a case in point. The "weak intermolecular forces" in silicones have some undesirable consequences, one of which is the marked swelling of the rubber caused by various solvents such as gasoline, toluene, and carbon tetrachloride. Appropriate introduction of a polar (electrically) charged group into organic groups attached to silicon ought to increase the (electrical) attractive forces between the silicone backbones, tighten up the rubber structure, and decrease swelling. The strongly polar CN group was successfully introduced into the rubber, and reduced swelling did result. But, as so often happens, the product had other weaknesses, and it could not in its existing state compete with a fluorine-containing silicone rubber then being introduced by Dow Corning. A great deal of fine research in organic chemistry thus failed to attain its primary commercial objective, although it proved useful in other ways.

The irradiation of silicone gum or rubber, as by electron beams of high energy, produces cross-links without leaving a residue that might cause later deterioration of the rubber. In this respect irradiation is preferable to using a catalyst such as benzoyl peroxide. Also, irradiated rubbers with satisfactory mechanical properties can be made. Nevertheless, such rubbers have not replaced conventional silicone rubbers in ordinary applications.

Replacement of KOH as rearrangement catalyst by a volatile ("transient") base ought also to improve siloxane polymerization by yielding a cleaner, more stable product, and it should open the door to continuous polymerization processes. If NH_4OH were a stronger base, it might make a good transient catalyst. Bases that have four organic groups (instead of four hydrogen atoms) bonded directly to the nitrogen are known to be much stronger than NH_4OH. Such bases were prepared and tried as rearrangement catalysts for D_4 as early as 1948 [26]. By 1959 a successful continuous process [27] was available in the laboratory for changing D_4 into gum: volatile organic bases with phosphorus instead of nitrogen served as catalysts. The continuous process, though scientifically satisfying, has found a stiff competitor in the batch process that uses KOH.

The examples cited above have much in common. Each was an outgrowth of sound reasoning based upon reliable scientific knowledge. Each was an opportunity for promising applied research. In each the scientific objective was achieved. Each resulted in creditable additions to scientific knowledge. But so far none has been a complete success.

Further Augmentation. Chapter 5 established by describing selected examples that the early silicone discoveries offered unusual opportunities for scientific augmentation of various kinds. If anything, these opportunities had multiplied by the time Waterford went on stream.

The following sampling of topics investigated and papers published will give the reader an idea of the kind and extent of augmentation carried out in the Chemistry Research Department, immediate successor of the Chemistry Section, during the years after 1950. In scanning the topics, the reader may replace "polydimethylsiloxane" with D_n, for a gum or a rubber is the usual subject.

1. Hydrolysis of unusual organosilanes [28].
2. Rate of silanol condensation [29].
3. Kinetics of the polymerization of D_4 [30].
4. Mechanism of the base-catalyzed rearrangement of organopolysiloxanes [31].
5. Mechanism of the equilibration of siloxanes [32].
6. Chemical stress relaxation as a clue to the mechanism of rearrangement in polydisiloxane elastomers [33].
7. Transient (volatile) catalysts for the polymerization of organosiloxanes [34].
8. Photochemical chlorination of methyl silicones [35].
9. Ultimate properties of cross-linked polydimethylsiloxanes [36].
10. Radiation chemistry of polydimethylsiloxanes [37].
11. Swelling and elasticity of irradiated polydimethylsiloxanes [38].
12. Vibrational modes in polysiloxane chains [39].
13. Molecular motion in polydimethylsiloxanes as revealed by nuclear magnetic resonance [40].
14. Effects of radiation-induced cross-linking in polydimethylsiloxanes as revealed by nuclear magnetic resonance [41].

How nearly did this scientific augmentation, samples of which have just been cited, fulfill the hopes that led it to be done? Well, it went a long way toward repaying the Company's debt to the earlier scientific literature; it deepened our understanding of silicones, and thus contributed indirectly to the silicone business; but it did not yield the hoped-for discoveries of major benefit to Waterford: once again, "gold is where you find it."

WATERFORD TODAY

[Annual expenditures] *in product and application research and development* . . . [are large enough to suggest] *that classification of the silicone industry as an engineering service would be almost as appropriate as its inclusion under the usual category of chemical manufacturing.*

C. E. Reed, 1956 Edgar Marburg Lecture [1]

Over almost three decades, time, new buildings, and landscaping have combined to soften the sternness of Waterford's "pipes, and tanks, and clean brick walls." The farm buildings, old in 1947, are gone; and the Van Schoonhoven burying ground, which the General Electric Company has agreed to keep forever neat and clean, remains as perhaps the only Waterford feature unchanged since then.* The silicone business continues to grow and to prosper; its managers have come and gone, but the bones of Jacobus Van Schoonhoven, founder of Waterford village and patriot of the American Revolution, lie undisturbed by the activities in the modern chemical plant that would now cost well in excess of $100 million to build.

Today the Silicone Products Department is one of more than 40 "strategic business units" within the Company. (This designation is reserved for unique components that formulate their own business plans as separate units.) In accord with Figure 7-6, the functions now carried out at Waterford are manufacturing, sales, research, product development, market development, strategic planning, international development, legal

---

* The environmental control facility at Waterford, one of the newer installations, will even help to preserve the gravestones, although its main function is of course to assure compliance at Waterford with New York State and federal regulations pertaining to air and to waste materials, liquid and solid. The conversion of HCl to CH_3Cl, mentioned earlier in this chapter, was an early step in a large and growing investment in environmental control.

Aerial view of Waterford in 1977. Please use the water tower (near center of plate) as landmark for comparison with plate on p. 238.

This final plate shows a General Electric employee in the act of fulfilling a corporate responsibility—the care in perpetuity of the Van Schoonhoven burying ground, which appears as a small rectangle in the lower right of plate on p. 264. The banks of the Hudson River (lower right) are not far away.

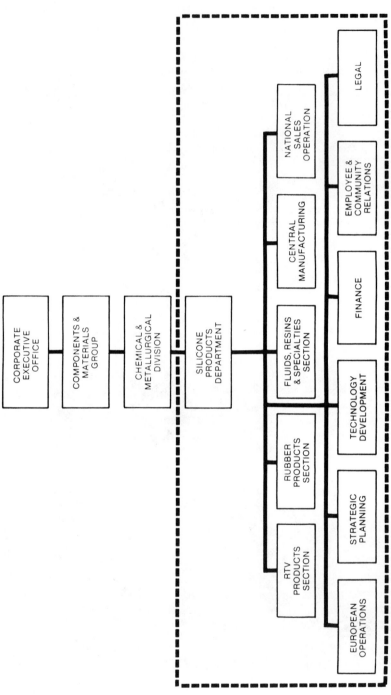

Figure 7-6 The chart shows the organization of the Silicone Products Department as of January 1977, and the position of the Department in the Company. I am grateful to Mr. T. H. Fitzgerald, General Manager, and to Mr. Donald V. Brown, Manager, Business Development, for the chart.

CORPORATE EXECUTIVE OFFICE

COMPONENTS & MATERIALS GROUP

CHEMICAL & METALLURGICAL DIVISION

SILICONE PRODUCTS DEPARTMENT

RTV PRODUCTS SECTION

RUBBER PRODUCTS SECTION

FLUIDS, RESINS & SPECIALTIES SECTION

CENTRAL MANUFACTURING

NATIONAL SALES OPERATION

EUROPEAN OPERATIONS

STRATEGIC PLANNING

TECHNOLOGY DEVELOPMENT

FINANCE

EMPLOYEE & COMMUNITY RELATIONS

LEGAL

matters, employee relations, and finance. The Department depends for its existence upon three major Product Sections, namely, RTV; Rubber (conventional, heat-cured); and Fluids,* Resins, and Specialties. Each is headed by a product general manager responsible directly to the Department General Manager, Mr. Thomas H. Fitzgerald.

The Department currently employs well over 1000 people in the United States. Not all of these are at Waterford, for the Department maintains district sales offices to cover the country. Its products "are marketed in Europe through a subsidiary company based in the Netherlands. In Japan, G.E. shares a joint-venture company, Toshiba Silicones, Ltd. In Canada, Canadian General Electric is in the silicone business, and in the third world, national distributors market silicones under the Monogram in virtually all developing countries of any size."

The General Electric Company now occupies a strong number-2 position worldwide in the highly competitive silicone industry. The Dow Corning Corporation is first; its products are marketed in many foreign countries where silicones are not produced, and it has sizable silicone businesses in Europe, Japan, Latin America, and Australia. At present, the world's annual silicone business has a value of well over $0.5 billion. The diversity of silicone products and the complexity of the industry make it impossible to give the *amount* of silicones corresponding to this sum. Waterford's annual silicone production is several tens of millions of pounds, and its chlorosilane production correspondingly greater.

The preceding quotation and the one to follow were written by Mr. Donald V. Brown,† whom I thank for most of the information on the Waterford of today.

A trip through the Waterford finished-goods warehouses would impress any visitor, not only with the quantities of materials involved, but with the great variety of products and packages, from tank wagons and 55-gallon drums to squeeze tubes and aerosol-spray cans.

Silicones have found . . . [uses as shown] . . . in virtually every major industry including:

---

* For the sake of simplicity, we have spoken of silicone *oils. Fluids,* after all, may be gases, and the word could also be used to describe a fluidized bed. Nevertheless, good arguments can be made for using *fluid: oil* is associated with *petroleum.* One compromise is to speak of silicone *oil* when the product is used as an intermediate in the making of other products such as automobile polishes, and silicone *fluid* when it is sold for use as a liquid. Page Humpty Dumpty once more!

† Mr. Brown began his career in the Company in 1947 by working on silicone resins in Building 77, Schenectady. In 1949 he transferred to Waterford, where he has been successful in a number of different assignments.

Automotive	Electrical/Electronic
Wire insulation	Motor and transformer insulation
Transmission seals	Wire and cable insulation
Spark-plug boots	Circuit board laminates
Special lubricants	Telephone wire connectors
Hydraulic bumpers	Transistor encapsulants
Truck hose	Circuit encapsulants
	Television insulation
	Rubber tapes (adhesive)

Military/Aerospace	Paper
Aircraft seals	Antistick surfaces
Firewall insulation	Process defoamers
Special lubricants	
Heat shields	

Textiles	Rubber
Water repellents	Tire release coatings
Fabric softeners	
Dyeing-process defoamers	

Food	Construction
Coffee defoamers	Window and building sealants
Bread pan coatings	Roof coatings
Milk-carton release coatings	Masonry water repellents
Cooking-process defoamers	Weather-durable paints
	Heat-resistant paints

Plastic Tooling	Consumer Products
Furniture molding	RTV sealants
Vinyl shoe molding	Tile grout
Jewelry molding	Shoe water repellents
	Eye-glass tissues
	Lubricant sprays

Chemical Specialties and Cosmetics
Auto and furniture polish
Antiperspirants
Hair sprays
Hand creams
Bath oils
Foaming agents

Note. This list was not intended to be complete. Even in 1954 Rob Roy Mc-Gregor in Chapter 4 of his *Silicones and Their Uses* (McGraw-Hill, New York) showed that silicones were being used by 20 industries in a total of 425 ways!

Table 7-1 Some 1975 Waterford Patents

U.S. Patent No.	Date of Issue	Inventor(s)	Title
3,859,321	1-7-75	F. J. Traver	Polysiloxane Composition Useful as a Brake Fluid
3,861,939	1-21-75	D. F. Merrill and P. J. Lavan	Process for Coating Substrates Using Fast-Curing, Silanol-Containing Organopolysiloxane Resins
3,865,759	2-11-75	R. A. Smith	Room-Temperature-Vulcanizing Silicone Compositions
3,865,766	2-11-75	D. F. Merrill	Process for Body Organopolysiloxanes
3,865,778	2-11-75	G. Christie	Heat-Vulcanizable Silicone Rubber Compositions Resistant to Oil Degradation
3,872,054	3-18-75	H. R. Shaw	Flame-Retardant Room-Temperature-Vulcanizing Organopolysiloxane Composition
3,881,290	5-6-75	G. J. Bouchey	Glazed Impervious Sheet Assembly and Method of Glazing
3,882,033	5-6-75	J. H. Wright	Silicone Grease for Semiconductors
3,882,083	5-6-75	A. Berger and B. B. Hardman	Latent Addition Curable Organopolysiloxane Compositions
3,884,866	5-20-75	E. M. Jeram and R. A. Striker	High-Strength Organopolysiloxane Compositions
3,887,514	6-3-75	D. F. Merrill	Bodied Organopolysiloxanes and Process Therefor
3,888,815	6-19-75	S. J. Bessmer and W. R. Lampe	Self-Bonding, Two-Package, Room-Temperature-Vulcanizable Silicone Rubber Compositions
3,896,123	7-22-75	G. P. De Zuba, A. Berger, and T. G. Selin	Silyl Cyanurates
3,897,376	7-29-75	W. R. Lampe	Room-Temperature-Vulcanizable Silicone Rubber Composition

Despite earlier hints in this book that the silicones are versatile and find diversified uses, Mr. Brown's list, though incomplete, will surprise the general reader. Moreover, it continues to grow.

Research at Waterford. The list given above suggests that applied research for new products must be the principal research activity at Waterford. To show that the suggestion is valid, there are listed in Table 7-1 some of the silicone patents issued to Waterford in the first 7 months of 1975.

Table 7-1 invites comparison with Table 4-1. In the early days the emphasis was on the processes that built the industry; today, it is on the products that have made the processes useful. As might be expected, the products and applications are now more numerous and more sophisticated than they were before 1950.

A LITTLE DIVAGATION

Since 1938, when the General Electric silicone project began, the Company and its Research Laboratory have undergone changes that are strong evidence of Adamsian multiplicity and complexity at work.

Today the Company has 228 manufacturing locations in 38 states and Puerto Rico, and 109 in 23 countries other than the United States.* In 1938 the total was approximately 35. From 1938 to 1976 Company growth was predominantly internal. An unusual and noteworthy exception occurred in December 1976, when the Company added important mineral resources to its other assets by merging with Utah International, Inc., a large mining concern. Among many other matters we mention only that the General Electric Credit Corporation has become a large and successful financial operation, and that the Company has a dominant position in the manufacture of engineering polymers—a position it might never have reached had its silicone business been unsuccessful.

The Cordiner decentralization of the Company enabled it to cope with multiplicity (growth). Increased complexity, attributable mainly to the Company's great diversity, and intense competition in its many markets eventually brought additional management innovations. One of these, greater emphasis on strategic planning, was particularly significant for research.

---

* In 1938 a subsidiary, the International General Electric Company, handled the Company's foreign business. The situation today is more complex. Also, it is no longer simple to categorize the Company: "international" seems to have fallen from grace; "multinational," "world," and "global" are likely replacements. Humpty Dumpty must act!

Naturally, the Research Laboratory also changed. Shortly after Dr. C. Guy Suits succeeded Dr. Coolidge as its director in January 1945, the planning of new quarters to house it was begun. These new quarters were built on a bluff ("The Knolls") above the Mohawk River, some 5 miles northeast of Schenectady. The move there from Buildings 5 and 37 was completed in 1950. When Dr. Suits retired in 1965, the erstwhile General Engineering Laboratory, which in the meantime had expanded from Building 5 into Building 37, was merged with the Research Laboratory to form today's Corporate Research and Development, of which Dr. Arthur M. Bueche, Vice President, Research and Development, took charge.

Soon after the move to the Knolls, the Chemistry Section was merged with the Mechanical Investigations Section to form the Chemistry Research Department with Dr. Marshall as Manager, Chemical Research, a position to which Dr. Bueche succeeded when Dr. Marshall retired in 1961. That Department has since been disbanded, and its work is now being done by several newly created components.

Today, Corporate Research and Development has 1900 employees, of whom 650 are professional and 360 have the PhD degree.

Nothing in this little sketch would have surprised Henry Adams, particularly not the accelerating rate of change since World War II.

COMMENTS IN CONCLUSION

The interaction of the Company and Waterford is an interesting chapter in the history of the life and growth of American corporations. A complete study of this interaction would have to include a necessarily speculative estimate of what Waterford might have grown into as a subsidiary of, but separate from, the Company. The comments that follow are no more than guidelines to such a study.

• As to the Company and Waterford. Waterford came to be founded owing to these circumstances internal to the Company: (1) early recognition, discussed in Chapter 1, of the urgent need for new materials*

---

* A comparison with the automotive industry at this point will assure the reader that different policies as regards research and diversification have been successful in different industries. In its early days the automotive industry does not seem to have emphasized the need for new materials; machines were its dominant concern, and older readers will remember the famous dictum by Henry Ford that a customer could have a car of any color "so long as it was black." Note, however, that new materials soon came into their own, and that two General Motors discoveries ["antiknock fluid," $Pb(C_2H_5)_4$; and Freon 12 or F-12 refrigerant, Cl_2CF_2] are among the most notable contributions to industry of applied research in chemistry.

to ensure the progress of the electrical industry, (2) the consequent founding of the Research Laboratory, (3) the liberal research policies established by Whitney and continued by his successors, (4) in accord with these policies, the laying, by the Chemistry Section, of a broad scientific and technical foundation for a silicone business, (5) the broad first objective of the Company, quoted at the head of Chapter 6,* under which the silicones easily qualified as "related materials," (6) the consequent willingness of top management to risk the founding of Waterford and to support Waterford through difficult times, and (7) adequate financial resources.

Any risky venture, such as Waterford was, needs stout-hearted champions. As the silicone business grew, championing by managers in the Research Laboratory (Marshall, Coolidge, Suits) gradually yielded to championing by managers (Feininger, Raynolds, Reed, Jeffries, and their successors) in operating components of the Company, notably reinforced by Cordiner in the Executive Office. Faith in silicones probably faltered, but it did not fail: the flame flickered, but it did not go out.

Championing is easier when a seller's market exists or is in prospect. Normally, new products with valuable properties ought to be blessed with such a market. In Chapter 6 we saw that there was in fact a seller's market for silicones during World War II. Peace brought a sharp change: a buyer's market sprang up with sellers strongly competing. This situation has continued to the public's benefit. Probable reasons are the high

---

* In the quoted objective of the General Electric Company (Chapter 6), "diversified" is the first word that describes its business. Let Alfred P. Sloan, Jr., speak for General Motors:

". . . Our nonautomotive business accounts for roughly 10 per cent of our civilian sales. And yet there have always been limits to our product diversification. We have never made anything except 'durable products,' and they have always, with minor exceptions, been connected with motors." A. P. Sloan, Jr., *My Years with General Motors,* Doubleday, Garden City, New York, p. 340.

In 1924 the Ethyl Gasoline Corporation was formed by General Motors and Standard Oil of New Jersey to market $Pb(C_2H_5)_4$ for antiknock purposes, the material itself being made initially by Du Pont. In 1962 each partner sold his interest in the Ethyl Corporation (see Mr. Sloan's book) to a third company.

In 1931 Du Pont and General Motors formed Kinetic Chemicals, Inc., for the manufacture of refrigerant chemicals, Du Pont having 51% ownership. In 1949 General Motors sold its interest to Du Pont. (See *Moody's Industrial Manual,* Vol. 1, 1975).

In both cases the final action by General Motors is in accord with the quotation from Mr. Sloan's book.

An article subtitled "Regulations, Costs, and Consumer Demands Are Changing GM's Research Approach" in *Business Week,* June 28, 1976, will bring the interested reader up to date.

manufacturing cost of silicones; the capacity of the industry, especially in its early days, to produce in excess of demand; and the determination of all producers to compete worldwide.

The Cordiner decentralization of the Company came at the right time for Waterford; that is, it came in the early 1950s, not long before the silicone business "took off." Departmental status for Waterford meant that Reed and the general managers to follow would have the authority, responsibility, and accountability needed to help Waterford succeed in the new, rapidly growing, highly competitive silicone markets that were as yet uncertain.

• As to Waterford and the Company. Waterford has returned more than money to the Company. By surviving and succeeding, it smoothed the way for subsequent ventures in the manufacture of materials. Today the Company has two divisions (the Plastics Business Division, and the Chemical and Metallurgical Division) that do nothing else, and they do not make all the materials that the Company manufactures. The value of such materials made annually is measured in hundreds of millions of dollars.

The Company was a successful manufacturer of various materials when Waterford was founded. But that does not diminish the critical importance of Waterford in extending the scope, character, and size of this activity. The manufacture of silicones was, by a wide margin, the most costly, the most extensive, the most diversified, and the most risky venture the Company had yet undertaken. Had it failed, the Company might not today have the strong position, mentioned earlier in this chapter, as a manufacturer of engineering polymers.

The experience gained in chemical engineering, product development, and marketing at Waterford has been invaluable to the Company. It is no accident that before 1977 seven Waterford employees had become vice presidents of the Company, with two among them subsequently advancing to senior vice president.

• As to Waterford's beginning. To repeat, the early years were hard. It was not easy to design and successfully operate a sophisticated chemical plant, the first such in an electrical company. It was not easy to sell costly products, initially few in number, for which markets were virtually nonexistent. "Ten days from oblivion" was often heard at Waterford during those years: the phrase joins the time required to fill the orders usually on hand to a grim prediction that never came to pass.

• As to chemical engineering. Waterford is a chemical engineering triumph because chemical engineering made the direct process and distillation, both unsatisfactory when they left the Chemistry Section, suitable for continuous commercial operation.

• As to discovery and scientific augmentation. Chapters 3, 4, and 5 reveal this pattern in silicone research: an early period rich in important discoveries, followed by a longer period during which extensive scientific augmentation *enlarged* and *refined* knowledge, but the frequency of important discoveries decreased. The silicone research done in the Chemistry Research Department after 1950 continued the earlier pattern. We shall return to this matter in Chapter 9.

• As to the public interest. The wide and diversified usefulness of silicones, perhaps unique among modern specialty products, shows that it was in the public interest to get a competitive silicone industry on its feet as quickly as possible. That this could have occurred in the early 1950s—or, the difficulties considered, after a very short time—is a tribute to American industry.

Chapter 3 revealed that industrial silicone research at its beginning took two separate paths. These paths were separate and different because the Corning Glass Works, the General Electric Company, and the Dow Chemical Company had different backgrounds, histories, and research traditions. The knowledge and the know-how essential to a successful silicone industry could not have been generated so quickly—if at all—by a single company.

But such generation was not enough. The knowledge and know-how had to become accessible to expected competitors if silicones were to be quickly available at a fair price. Fortunately, the patent system provides patent licensing as a mechanism to make this possible. In this way Dow Corning could make early use of the direct process, and Waterford could soon improve its silicone rubber. In these rapidly changing times of Adamsian multiplicity and complexity, patent licensing is of increasing importance. Trade secrets have their place, but there is no quicker way to build a new industry than by the licensing of needed patents, if such exist. There must, of course, also be adequate resources and the willingness to risk them. The silicone industry is an outstanding case in which patent licensing served the public interest by hastening and broadening effective competition and placed a high premium, especially at the outset, on effective marketing.

• As to industrial research and the silicone story. The principal part of *Silicones under the Monogram* has now been told. What it teaches about industrial research may be summarized as follows:

1. Industrial research benefits from prior knowledge, which has usually accumulated because of pure research done outside industry.

2. Industrial research changes character as it moves from test tube through pilot plant to commercial plant.

3. The silicone story told here is a sobering corrective to "Gee Whiz!" research publicity. By no means do all research acorns turn into great oaks. Silly Putty would not have been enough.

Silicones grew into a large industry only because the complete spectrum of needed and costly activities—discovery, augmentation, implementation, product development, market development, and marketing—was successful. Without exceptional success in the last four, the industry would have foundered.

A further sobering thought is this: although the work in the Chemistry Section was unusually successful in that it gave Waterford the types of silicones (methyl and phenyl; "first was best") that are even now its mainstays, although it showed silicones to be promising as resins, oils, and rubbers, although it yielded the direct process—it failed to deliver products of high quality and in the great diversity needed to sustain today's industry; it failed to deliver an acceptable reactor for the direct process; and it failed to deliver the best type of still.

4. Industrial research on silicones followed a chronological pattern showing a sharply decreasing frequency of major discoveries despite greatly increased scientific augmentation.

5. The growth of the silicone industry to the public's benefit was expedited by the interaction of competing companies.

A noteworthy addendum: On January 28, 1977, the *Schenectady Gazette* reported two actions by Waterford's General Manager, Mr. Thomas H. Fitzgerald, which suggest a continuing bright future for Waterford's activities abroad and for the silicone industry in general. Dr. Hart K. Lichtenwalner was appointed Managing Director for General Electric Silicones, Europe, and Dr. Joseph Wirth, Manager of Technology Development at Waterford, was given operating responsibility for Toshiba Silicones, the Japanese joint-venture company mentioned earlier in this chapter.

Chapter Eight

A Silicone Panorama for the Bicentennial Year

The utility of the silicone polymers will depend somewhat on our skill in manipulating them. It is no longer sufficient to expect that simple hydrolysis of alkyl chlorosilanes will give the desired properties. . . .

W. I. PATNODE, July 30, 1943

SILICONE RESEARCH AND INNOVATIONS

1561. *A desire to innovate all things . . . moveth troublesome men.*

The Oxford English Dictionary

Patnode proved a better prophet than he knew. He could have replaced "somewhat" by "enormously" and not overstated his case. When he took his 1943 look ahead, he could not have foreseen that silicone research and innovation would follow the Adamsian law of acceleration and result in an almost overwhelming multiplicity.

Note. Brackets are used to enclose references and inserted material. For bracketed references, see References and Notes at the end of the book. Reference to laboratory notebooks is by name and date. CRDA stands for "Chemistry Research Department Archives."

276

Consider research. Rochow, in his summary of the silicone project before August 1, 1945, wrote, "As one product of our research [1938–1945], we have accumulated (at the time of this writing) a group of about 75 reports on various phases of work connected with the preparation of organosilicon compounds, the formation of silicone polymers, and the evaluation of such polymers for specific uses." At that time the important silicone discoveries in Table 4-1 *had all been made.* Many of the 75 reports Rochow mentions dealt with scientific augmentation of these silicone discoveries: the ratio of reports to discoveries was still small. Noll's *Chemistry and Technology of Silicones* [1], which was published in 1968 and will serve as a mainstay for this chapter, contains 2875 citations, some of which inevitably are duplications and peripheral references. But the disparity between 2875 and 75* does testify to an Adamsian acceleration—an acceleration that has spawned an increase in complexity so great that Noll, himself a ranking silicone authority, enlisted several other experts to write sections that our panorama must include.

Every 3 years an international symposium on organosilicon chemistry is held. The fourth of these was opened in Moscow on July 1, 1975, by K. A. Adrianov,† Chairman of the Organizing Committee, who was followed by E. G. Rochow speaking on "The Polymer Side of Organosilicon Chemistry." Thirteen countries and West Berlin were represented.

At the Fourth Symposium the Institute of Chemical Process Fundamentals, Prague, Czechoslovakia, made known the following facts about its documentation of organosilicon chemistry: "So far over 20000 publications with the content of more than 40000 compounds (the earliest from 1830) have been taken in and processed. *The annual increment amounts to over 1200 works, over 600 patents and around 5000 new compounds*" [italics mine].

As to innovation, we must place this alongside discovery, augmentation, implementation, and marketing. The reader deserves to know why.

The provocative citation dated 1561, quoted above from *The Oxford*

---

* Numbers games must not be allowed to get out of hand. The ratio of reports to major discoveries in the Chemistry Section seems to be much smaller than a similar ratio for Noll's book.

† In January 1940 Rochow's notebook shows that Earl H. Winslow, an analytical chemist who had learned Russian, translated silicone articles by Adrianov. This would have been a better book could I have done justice to the work on silicones outside the monogram, as in Russia and elsewhere abroad.

I owe the information about the Fourth Symposium, and much else, to Dr. W. F. Gilliam, Waterford, who attended.

Precursors	Stocks	Modified forms
	Oils	Emulsions Antifoaming agents Greases Agents to control foamed structures
	Gums	Rubber compounds Rubber dispersions RTV rubber constituents
Organochlorosilanes		
	Resins	Resin solutions Silicone-organic copolymers Resin emulsions and foams[a] Resin powders Resin composites
	Siliconates	Solid impregnating agents Impregnating solutions

[a] Not widely used.

Figure 8-1 An incomplete overview of silicone manufacturing. The figure is intended to complement the incomplete list in Chapter 7 of silicone products used by various industries. Throughout the figure, which is based on information by Noll [1, p. 387], "silicone" is taken as understood.

Explanatory Notes

1. The complexity of the silicone business makes inconsistencies unavoidable. For example, oils listed above as *stocks* may be sold unchanged as *fluids*; then they are *products*. The *general* intention is to present *modified forms* of silicones as sold to industrial customers or used to make silicone products.

2. *Emulsions* are made by blending two liquids with additives as needed to give stable suspensions of one liquid in the other.

3. *Greases* are made by dispersing fatty materials such as soaps in silicone oils. Such greases usually serve as lubricants. There are two inconsistencies: the Dow-Corning "grease," filled with silica, is really a "paste"; a "paste" made from a silicone oil and MoS_2 is a lubricant.

4. *Pastes* are made by dispersing solids in silicone oils. The distinction between greases and pastes, not yet well established, will not be made in the text.

5. Rubber *compounds*, sometimes called "pastes," are not chemical compounds. They are gums with the additives needed to make silicone rubber by curing the compound.

6. *Resin solutions* (some of which could be used as *varnishes*) are dissolved resins, a volatile hydrocarbon such as toluene often being the solvent.

7. *Resin powders* (also called "molding powders") are resins mixed with fillers.

8. *Resin composites* (e.g., laminates or varnished "cloth") are made by compressing resins with glass or asbestos cloth, the resin being the meat in the sandwich, for which the

English Dictionary, is one of many which indicate that "innovation" and its derivatives have long been used to describe, favorably and unfavorably, changes in religion, politics, law, and language. In retrospect it seems strange that "innovation" was not used earlier to describe changes brought on by the scientific revolution. This use, now becoming widespread, seems here to stay.

Utterback [2] defines "innovation" as "*technology* actually being *used* or *applied* for the *first time*" [italics mine]. In the spirit of his definition, we shall say that an innovation* is a product, process, or application introduced as new *outside the laboratory*. The innovation may or may not have originated in the laboratory; Aries [3] has emphasized that many do not.

Perhaps "technological innovation" would have been desirable as being more specific and restrictive. Attempts at precise definition seem foredoomed. Two cautions are in order. "Innovation" is *not* synonymous with "discovery" or "invention" [Chapter 5], for most discoveries and inventions are never implemented. Nor is innovation linked exclusively to marketing; for example, process innovations in manufacturing are frequently made.

Before 1950, when commercial silicones were few and came largely from the laboratory, innovation as a concept was not needed. Today is another story. Consider these examples of actual and conceivable silicone innovations. Someone outside the silicone industry and skilled in the art of making polishes decides to add silicone oils to an already successful product. If the addition is worthwhile, he or she will have leap-frogged much research that someone else paid for, and made an innovation valuable to the silicone business, in this way helping those who paid for the research. Or the person might have made an innovation even more easily by finding a new use for the oil itself, or by asking the manufacturer to tailor an oil to fill a need he or she had in mind. In today's civilization

---

* "Innovation" also means the *act* or *process* of innovating. We shall avoid this meaning here.

---

bread may be a board or a cloth; also made by varnishing the cloth and heat-treating it if necessary. The historical importance of the last composite is clear from Chapter 3.

9. *Siliconates* are salts formed by the action of strong bases on chlorosilanes and by subsequent condensation reactions. Upon acidification, siliconate solutions form gels when, as is usual, T units are present.

10. The definitions given above are imprecise. (Page Humpty Dumpty.) Usage is loose. The definitions are only guides to the text, in which further information is usually given.

any family of new materials, not prohibitive in cost, that has a variety of useful and unusual properties is sure to lead to numerous and continuing innovations. The multiplicity of silicone innovations, foreshadowed in Chapter 7, is implicit in Figure 8-1 and is confirmed by the sketch of the Dow Corning Corporation that will appear later.

Silicone research has grown, and silicone innovations have multiplied, far beyond what seemed likely at the end of World War II. Silicone research, academic and industrial, no doubt grew in part because most research boomed after the war's end. But there were special factors also. When wartime secrecy was lifted, the world became aware of new silicone chemistry, of new and promising materials, and of new ways in which they could be made. By contrast the early increase in innovations had a simpler cause: the silicone industry had to innovate or die.

Ultimately, as always, nature made possible the growth in silicone research and in silicone innovations. Silicone chemistry cried out for further scientific augmentation. Even though the old belief of a close analogy between silicone and carbon compounds had been corrected, the sheer number of organic compounds must have continued to spur research aimed at preparing and studying silicone analogues. The uniqueness of silicones among polymers also inspired further research. A large number of innovations became possible because even methyl and phenyl silicones could be made in a variety of forms with unusual properties to an extent not easily matched, even by versatile compounds such as the fluorocarbons. It seemed that silicones could be usefully modified almost *ad infinitum,* and that important undiscovered compounds remained in the wings. Only a few modifications, notably the RTV's Dow Corning fluorinated silicone rubber, have been mentioned. Waterford sells methylalkyl silicones, chlorophenyl silicones, and carboxyester silicone fluids. Ethyl silicones are said to be favored in Russia. Other examples are easy to find.

I do not know how to relate silicone innovations and silicone research. Many papers presented at the Fourth International Symposium would not have made strong ripples at the first one. Scientific augmentation of course dominated major discovery: the trend observed in the silicone project, notably in Chapters 4 and 5, seemed universal. Perhaps research on silicone drugs, to be described later, will bring welcome surprises. Meanwhile, innovations are keeping the silicone industry healthy by compensating for the slowdown in major discoveries.

A final word: the Adamsian multiplicity and complexity of the silicone industry continues to rest upon the original, slender but strong foundation of CH_3, C_6H_5, and vinyl groups; Si—O—Si bridges; and copper as catalyst for direct processes. In these vital respects first has proved best.

WHY SILICONES?

The unusual properties of silicones were mentioned as they were discovered. The time has now come for a summary that shows why the world buys more than $0.5 billion worth of silicones annually, and will be willing to buy successively more each year in the foreseeable future. Silicones as "miracle polymers" have long since been displaced by subsequent wonders, but silicone applications continue to grow.

Of course, as a consequence of Adamsian acceleration and multiplicity, we can make no adequate summary. The best we can do is to speak of silicones generally, becoming specific in only a few instances. Fortunately, the reader seeking precision and detail can turn to Noll and his collaborators [1, particularly pp. 437–527], to Lichtenwalner and Sprung [4], to Meals and Lewis [5], or to McGregor [6]. Our references will be mainly to Noll.

In our summary we shall rely upon **D** backbones as the principal silicone prototype. This restriction is less serious than might be imagined. After all, these backbones exist over an enormous range of molecular weights, which begins with the single **D** in **MDM**; extends through silicone oils, some of which have molecular weights up to 500,000, where polymers other than silicones are solids; and terminates in silicone gums, in which the number of **D** units in the backbones is so large* as to permit the end groups to be neglected. More serious than this disregard of end groups in our summary will be the scant attention paid to cross-linking and to the presence of groups other than **D** units in silicones not further described.

Chemical Stability

1. Silicones properly manufactured for use at high temperatures will not change significantly over a long time in vacuum below, say,

---

* On May 5, 1954, W. T. Grubb of the Chemistry Research Department recorded the results of viscosity measurements on a "super high M.W. [molecular weight] silicone" gum he had prepared by rearrangement of D_4 with KOH as catalyst. The molecular weight he calculated from these measurements was $6.9(10^6)$, surely among the highest ever recorded. As a single **D** unit has a molecular weight of 74, Grubb's backbones contained nearly 93,000 **D** units—a highly approximate number of course. The gum was slightly cross-linked; that is, it was on the way to becoming a rubber.

250°C. Here "properly manufactured" means that materials of low molecular weight ("volatiles") and all catalyst residues have been removed.

2. As the Chemistry Section soon discovered, silicones are attacked by steam. Their sensitivity to acids and bases, a blessing when they are being tailored by rearrangement polymerization during manufacture, becomes a curse when residual catalysts cause rearrangement (= decomposition) during service. The ease with which cyclamers in a continuous series (D_3 . . . D_n) form and disappear during rearrangement is a distinctive feature of silicone chemistry.

Incorporation of C_6H_5 groups can increase chemical stability.

3. Silicones are greatly superior to alternative materials in resistance to oxidation. The Si-C bond in silicones is relatively invulnerable below about 200°C.

Incorporation of C_6H_5 groups is helpful.

4. Silicone rubbers are greatly superior to alternative materials in resistance to attack by ozone, as well as by corona, the electrical discharges (which can produce ozone) sometimes generated in air under service conditions.

5. Silicones resist weathering admirably anywhere on earth. Their water repellency helps them because rain does not wet them so long as they are clean.

Physical Stability

6. Silicone oils properly tailored can be poured at temperatures of −40°C and below. The significance of these low pour points was clear to Patnode, who established them when he began his work on methyl silicone oils as lubricants. As one might expect from the behavior of oils, the stiffening and the crystallization temperatures of silicone gums and rubbers are also low.

If −40°C is not low enough, one can lower the pour point by introducing C_6H_5 groups or by changing the backbone from linear to branched. Cross-linking raises the pour point.

7. Methyl silicones have lower boiling points than carbon analogues of comparable molecular weight; **MDM** is an example. The statement is true also for organosilicon compounds *not* containing a siloxane bridge.

Physical and Physicochemical Features.   Once Patnode had begun his systematic work on silicone oils and on the water repellency of silicone surfaces, it became increasingly clear that silicones differ from other substances in physical and physicochemical ways, and differ to an extent that had been wholly unanticipated. These valuable features helped the silicone business survive the dark days that followed World War II, and they are contributing to its growth today. The most significant features follow.

8. Important properties of methyl silicones, such as viscosity, and certain electrical and mechanical properties change very little with changing temperature. Consequently, a silicone, though only comparable to an alternative material in respect to a property at room temperature, may be greatly superior at 200°C.

9. Methyl silicone oils spread quickly on liquid and solid surfaces to form films that need be only one molecule thick. In these films the CH_3 groups are exposed, generating a hydrocarbon-like surface; the siloxane bridges are anchors. The water repellency, release ("nonstick") action, and polishing effectiveness of silicones and of preparations containing them all derive from this property. Qualitatively, this idealized model of a silicone film helps one to understand many properties of silicone surfaces.

The behavior of methyl silicone oils on surfaces is in accord with the low surface tensions (hence low surface energies) of these oils. In liquids of low surface tension the molecules separate easily so that the liquids spread quickly. Somewhat surprisingly, the silicone oils show high compressibilities at higher pressures, which means that the molecules are easily brought close together under these conditions [1, pp. 324–327].

10. Methyl silicones, alone or made into copolymers with organic materials, show high *surface activity;* that is, they concentrate at the surfaces of liquids to which they are added in small amount. They are *surfactants.* This property added to Item 9 makes such silicones capable of *producing, preserving,* and *destroying* foams; of acting as flotation agents for minerals; and of improving paints.

Electrical Properties

11. Silicones are satisfactory with respect to dielectric constant, loss factor, specific resistance, and electric strength for most applications over a wide range of temperatures. Many are satisfactory *at some temperature*, but *not* over a wide range. ·

Physiological Properties

12. With isolated exceptions, usually compounds of low molecular weight, silicones are outstanding as regards compatibility with living systems. More than chemical inertness and lack of toxicity are needed:

for example, the body must not reject silicone implants, and the implants must do no harm over the long term. In this respect silicones have an excellent record.

SILICONE PROPERTIES EXPLAINED (?)

Nature and Nature's laws lay hid in night: God said, "Let Newton be," and all was light.

ALEXANDER POPE, Epitaph Intended for Sir Isaac Newton

Pope shows no hint of Holmesian skepticism and had no forewarning of Adamsian multiplicity. The flash of enlightenment in his couplet resembles those our television commercials try to give us today; this modern art form would have suited Pope admirably. Unfortunately, the public, when it listens, is likely to get the impression that with each sunrise modern Newtons make "all was light" out of something new.

Not in silicones! To make this chapter's panorama realistic, we shall have to see how far theory and experiment have succeeded in explaining the unusual properties listed as numbered items in the preceding section.

Let us begin with Items 1 through 5. Descriptive chemistry suggests that the organic part of a silicone will be hydrocarbon-like, and that the siloxane backbones will resemble silica and the silicates [Chapter 2, and Reference 1, pp. 287–317]. Saturated hydrocarbons (e.g., paraffin) attract most substances (e.g., water) weakly or not at all, and paraffins are chemically inert (Item 9). Methyl groups, though inert, cannot, so descriptive chemistry teaches, be expected to make methyl silicones as stable toward oxidation as are those containing C_6H_5 groups, which exhibit "resonance" and have other special properties [1, pp. 440–442]. This expectation is realized. Silica is impervious to all ordinary electrical attack, and virtually impervious to chemical attack *except as mentioned below*. Siloxane backbones, with their partially ionic bonds, are electrically unbalanced and can therefore anchor the hydrocarbon-like silicone surfaces to many liquids and solids.

Descriptive chemistry long known also guides us to important chemical instabilities shown by silicones. As silica is immune to oxidation, and saturated hydrocarbons resist it, why should not the Si-C bond in many silicones be the one most vulnerable to oxidative attack? Silica is attacked in varying degrees by alkalis [Chapters 2 and 4], why not the Si—O—Si bridge in silicones? (Prior descriptive chemistry could scarcely have been expected to prepare one for Hyde's important discovery [Chapter 5] that 0.01% solid KOH is a good silicone rearrangement catalyst, indicating that silicone backbones are unusually mobile and flexible.) Also, the ease with which HF attacks silica makes predictable its destruction of Si—O—Si backbones in siloxanes. However, the action of acids such as H_2SO_4 on siloxanes [Chapter 4] was not predictable from their relative inertness toward SiO_2, an inertness perhaps attributable to the failure of these acids to form products that readily leave the SiO_2 surface.

So far we have *explained* nothing. We have merely shown that certain properties of silicones are in accord with the descriptive chemistry of related compounds, much of which was prior knowledge when the silicone project began in the Chemistry Section. Items 1 through 5 were gratifying discoveries, but they did not come as major surprises.

Items 6, 7, 8, and 9 present a greater challenge. Patnode's work on methyl silicone oils [Chapter 4] has been extensively augmented in the hope of explaining the remarkable properties he observed—properties with which these items are concerned. We shall follow this augmentation along two lines, marked by years as way stations, but we shall not make every stop possible on the journey. Line 1 will be called the "low-intermolecular-force line," and Line 2 the "cyclamer-equilibrium line" (see Item 2), which began with Patnode's discovery of rearrangement polymerization and the characterization of D_3, D_4, etc. The same intermolecular forces act throughout.

Line 1: 1948. Roth and Harker [7] concluded that a weakness of intermolecular attractive forces in silicones results from free rotation of the "$Si(CH_3)_2$ assembly" within the silicone molecule, the silicone in O—Si—O being "free to move as in a ball-and-socket joint"; that is to say, the Si-O bond is a "soft" bond in that it permits easy rotation. This rotation was supposed to keep silicone backbones abnormally far apart, with lowered attractive forces between them as the result. These conclusions were based on X-ray diffraction work in which silicon and·oxygen (*but not* hydrogen) atoms were "located." **1955.** Rochow and Le Clair (Harvard University) [8] used nuclear (here *proton*) magnetic resonance absorption, which can "detect" hydrogen (*but not* silicon or oxygen) atoms, to show that the CH_3 groups rotate about the Si-C bond "with unusual freedom," and that other motions are likely. **1960.** Huggins, St. Pierre, and Bueche [9, 10], also using the proton magnetic resonance method, reported results "compatible with a rigid lattice and free rotation of the methyl groups." Evidence of chain translation was also found. **1961.** Scott and associates at the Bureau of Mines, Bartlesville, Oklahoma

[11], after an exhaustive thermodynamic investigation of **MDM**, found (among other things) that the intermolecular forces in its vapor* were normal, and concluded that the "low-intermolecular-force" explanation must be rejected, and that the "explanation of the unusual physical properties of methyl silicones as arising from relatively free internal rotation about Si-O bonds is confirmed." **1965.** McCall (Bell Telephone Laboratories) and Huggins (CRD) [12] reported after studying self-diffusion in linear dimethylsiloxanes that there is very little difference between silicone oils and liquid n-paraffins in the way that the self-diffusion coefficient changes with the chain length—surprising in light of all the "backbone differences" we have been stressing. **1970.** Rochow [13] summarized his work along Line 1 as follows: "The *type* of motion that differentiates silicones from organic polymers, refinements of technique now indicate . . . [to be] a combination of C_3 rotation,† double rotation, and chain translation." So much for Line 1.

Line 2.   1946. Scott and Wilcock, as we saw in Chapter 5, began Line 2 with a mathematical treatment that described the equilibria between cyclamers and backbones (of course with end groups attached) in **M-D** systems. Great progress has since been made along Line 2, which is summarized by Flory (Nobel Prize, 1974) in an important book [14] that gives references (not all repeated here) to the pertinent original work. Line 2 becomes "macrocyclization equilibrium" when the cyclamers are large. **1950.** Jacobson and Stockmayer gave a theoretical treatment of the **M-D** system in which it was assumed that the angles at the bond junctions in the Si-O-Si backbones could assume all values with equal probability. **1965.** Brown and Slusarczuk greatly extended the molecular-weight range over which the cyclamer equilibrium data were known by using gas chromatography [Chapter 5]. **1966.** Flory and Semlyen improved the theoretical treatment of Jacobson and Stockmayer by introducing *real* backbones to replace their "freely jointed" model.‡ The remarkable success of the 1966 treatment is evident from Figure 8-2.

---

* Experiments on neither a vapor nor a crystal [7] need give results directly applicable to a polymer backbone.

† C_3 is the descriptive designation assigned to the important intramolecular rotation that involves neighboring CH_3 groups. Ethane, $H_3C—CH_3$, is the simplest example. In such molecules rotation of the CH_3 groups encounters three energy barriers—hence the designation C_3.

‡ In both theoretical treatments an important quantity is the "characteristic ratio," which is the quotient of r^2, a mean-square end-to-end length of a backbone (chain), divided by the product nl^2, in which n is the number of bonds in the backbone, and l is the bond length. Characteristic ratios can be obtained by measuring the viscosities of polymer solutions or the light scattered by them.

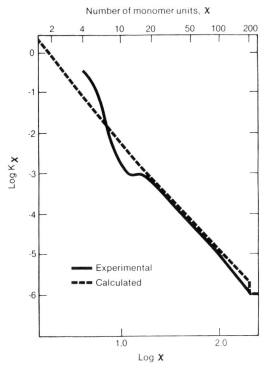

Figure 8-2 Exquilibrium constants K_x for cyclamers in the **M-D** system as a function of x, the number of **D** units in the cyclamer. The curve is modeled after Fig. 1, Appendix D, Reference 14. The experimental results are those of J. F. Brown and G. M. J. Slusarczuk, Chemistry Research Department, with those below $x = 4$ and above $x = 200$ omitted, as the investigators considered them ''less reliable than the remainder.'' The experimental curve dips abruptly downward below $x = 4$. See *J. Amer. Chem. Soc.*, **87**, 931 (1965), Fig. 1.

Figure 8-2 and the theoretical work related to it will stand as a permanent landmark in polymer history. To see why this is true, let us sketch what was done. *Real* backbones were used as models: this means that measured values of bond angles* ($\langle$O—Si—O $= 110°$, very near the tetrahedral angle; $\langle$Si-O-Si $= 143°$) and measured values of bond distances were used; that rotations† centered about silicon and oxygen encounter rotational energy barriers assumed to be low in accord with independent

---

* The situation as regards $\langle$Si—O—Si is complex, but it is noteworthy that values centering around 143° are found in silica and silicates as well. See Noll [1, pp. 300, 301].
† These rotations are different from those (e.g., the C_3 rotation) used in the interpretation of X-ray and nuclear magnetic resonance results.

experimental evidence; and that the forces existing between different parts of the molecule during rotation were considered. The calculations for Figure 8-2 took into account the fact that a cyclamer cannot form unless the bonds at the ends of a **D** backbone are suitably oriented.

The significance of the large difference ($143° - 110° = 33°$) between successive bond angles in the **D** backbones was first appreciated by Birschtein, Ptitsyn, and Sokolova [14, p. 175] and made full use of by Flory and his colleagues. In polymer backbones (e.g., C—C—C) where no comparable difference exists, linear backbones that lie in a plane have their units disposed about a rectilinear axis, which keeps their ends far apart—in sharp contrast to the **D** backbones, where the ends often approach each other closely enough to make cyclamer formation possible: the **D** backbones are *highly flexible* and the intramolecular rotational energy barriers are *small*. Flory [14, pp. 46, 178] points out that this difference of 33° in the alternating bond angles leads to a positive temperature coefficient (the largest he lists) for r^2 [footnote, p. 286], which means that the ends of a **D** backbone will on the average be further apart at higher temperatures. Line 2 has led to a successful and *direct* theoretical treatment of *real* **D** backbones in accord with diverse experimental evidence.

After all this, where are we? What has scientific augmentation wrought? Noll [1, p. 324] says, "The intermolecular forces in the methylsilicones are unusually weak. This property is· most important for an understanding of the technological behavior of these materials." The statement must not be construed as implying that a knowledge of the forces led to a prediction of the properties—that understanding preceded discovery.

The understanding resulting from scientific augmentation is valuable. Lines 1 and 2 both relate the remarkable properties of silicones to the remarkable siloxane backbone—a good step forward. Confusion then begins, as is to be expected from the complexity of the problem. So much may be said: the periodic table and descriptive chemistry tell us that silicon and oxygen are different in a way that gives pronounced ionic character to the Si-O bond and leads to pronounced average differences in successive bond angles along the siloxane backbone. Next, we have a paradoxical conclusion. The (attractive) intermolecular-force components attributable to the ionic-bond character of the siloxane backbones are abnormally low for siloxane backbones when normally separated. (The other intermolecular-force components, such as those between neighboring CH_3 groups, are normal.) Now, different free and easy motions have been experimentally demonstrated for siloxane chains. Among these motions certain rotations are particularly effective in keeping neighboring backbones abnormally far apart. As electrical attraction decreases with distance (witness Coulomb's law), this abnormality leads to weak resultant intermolecular forces in the silicones. Most of the evidence cited above does not conflict with this interpretation, which had its genesis in the work of Roth and Harker [7]; details, results, language, and models differ from case to case in the subsequent work that has supplemented theirs. Noll's statement, though oversimplified, is a fair summary of the present, incomplete knowledge.

Dr. Schultz explained to me that the viscosity (and other properties; see Item 8) of methyl silicone oils ought to change little with changing temperature. The increase with temperature in r^2 for such oils (see the earlier account of Flory's work along Line 2) means that the ends of their molecules will be further apart at higher temperatures—this expansion should *increase* viscosity. Such an increase tends to compensate for the *decrease* in viscosity to be expected at increasing temperatures for all liquids—a general decrease that results because molecules at higher temperatures move more rapidly and are on the average more widely separated. Such compensation tends to make viscosity independent of temperature* and has a bearing on the rest of Item 8. There is more to come.

Bridgman [15; Nobel Prize, 1946] measured, up to high pressures and at room temperature, the viscosities and compressibilities of various liquids, several methyl and one methylphenyl silicone included, with the following results.

1. The silicones are unique in the marked increase in relative viscosity they begin to show at pressures above 2000 kg/cm². Very roughly, the silicones at 10,000 kg/cm² have the same relative viscosity as other liquids at three times that pressure.† To quote: "at atmospheric pressure . . . [MDM] is *five* times more viscous than *i*-pentane [i-C_5H_{12}], whereas at 10,000 kg/cm² it has become *5000* times more viscous" [15, p. 127; italics mine].

2. To his surprise Bridgman found that he could not freeze by increasing the pressure any silicone of molecular weight exceeding that of MM. He suspected that this stubborn and unusual tendency to stay liquid resulted because high viscosity inhibited crystal nucleation and growth. He later proved himself correct [16]. Failure to freeze therefore tells us nothing about the normal intermolecular forces in the silicones.

* Linear saturated hydrocarbons enjoy no such compensation because the "expanded" molecules (r^2 large) are lower in energy than the others and hence become rarer at high temperatures.
† Results recorded by Noll [1, pp. 465, 655] were obtained at pressures too low to show the viscosity increase found by Bridgman.

3. For the compounds **MM** through **MD₆M**, Bridgman found compressibilities that "fall within the range ['of those] of the other liquids measured," the silicones being "among the more compressible"; and he concludes that "the mechanism responsible for viscosity must be entirely different from that responsible for compression" [15, pp. 141, 145].

Bridgman attributes the viscosity increase to an "interlocking" of his silicone molecules at high pressures. According to the model presented above for the siloxane backbone, such interlocking could plausibly be brought about by a "crushing" of the backbones at his high pressures. To be sure, **MM** seems a little small for interlocking, and the suggestion just made might require higher compressibilities than those Bridgman measured.*

One thing must be borne in mind for all comparisons of physical properties between silicones (e.g., Items 6 through 8) and other substances: descriptive chemistry teaches that reference substances for such comparisons must be chosen with care; if, for example, surface tension is being considered, the comparison should be made with materials lower, as well as higher, in surface tension than the silicones.

If this panorama is to be realistic, we must examine at least a few organosilicon compounds that are not silicones, but lend themselves to building families of *silicone analogues*. These compounds are esters, simple and complex, of orthosilicic acid, $(HO)_4Si$.† The complex esters may be regarded as condensation polymers, and they contain Si—O—Si (siloxane) bridges but lack Si-C bonds: they are not silicones. See Noll [1, pp. 639–662]. We shall look only at $(n\text{-}C_4H_9O)_4Si$ (here called "butyl ester") and its polymeric derivatives (here called "butyl polyesters"). Noll points out the following: (1) the polyesters have uncommonly low pour points (near $-100°C$); (2) both ester and polyesters show the low dependence of viscosity on temperature that we have come to regard as a hallmark of methyl silicones; and (3) the polyesters are *better lubricants* than the methyl silicone oils. Fortunately for silicone sales the ester derivatives of silicic acid are all sensitive to hydrolysis and subject to objectionable decomposition at high temperatures. But they are useful in certain spots; they certainly bear watching; and they are worth scientific augmentation along Line 1 to provide results for comparison with those already available for silicones.

---

* I gladly thank A. R. Shultz and W. L. Roth, Corporate Research and Development Center, for helpful discussions of all the preceding material related to siloxane backbones.
† Patnode had tried unsuccessfully to make useful polymers (not silicones!) out of organosilicon compounds, notably out of $(C_2H_5O)_4Si$ in Pittsfield before he joined the Chemistry Section in 1933. See Chapter 3.

SILICONES AND THE ELECTRICAL INDUSTRY

The reader will remember from Chapter 3 how applied research in silicones started. According to Dr. Suits, who succeeded Dr. Coolidge, "it was not the sale of silicone materials as such which provided the incentive for this work. It was rather the prospect of a greatly improved performance of electrical machinery . . ." [17]. We will here give the electrical industry short shrift for three reasons. First, it has had considerable space already. Second, Noll [1, pp. 532–559] has done what we shall not. Third, most silicones are sold outside the electrical industry.

The following are informed estimates* of the percentages of various kinds of silicones used *outside* the electrical industry: oils, 90%; resins, 75%; heat-vulcanized rubbers, 50%; RTV rubbers, 90%. Dow Corning [18, p. 10] estimates that only one-fourth of its total sales were made to the electrical industry over the period 1970–1974. Cost considered, silicones have served the industry well. Cost neglected (if only we could!), silicones probably enjoy the distinction of being the family of polymers that could find the widest spectrum of uses in the industry because they meet satisfactorily so many of the criteria set by Table 1-1. This versatility of course explains the attractiveness of silicones to industries outside the electrical.

In today's world silicon is more important as semiconductors than as silicones. Semiconductors are made by adding minute amounts of certain materials to the purest obtainable silicon—"hyperpure silicon," so called—in order to obtain the desired electrical characteristics. The relationship between the silicone and semiconductor industries is much closer than one might think. Dow Corning [18, p. 10] makes hyperpure silicon mainly from $SiHCl_3$, presumably from $SiHCl_3$ formed in the direct process and purified to a large extent (perhaps completely?) by distillation. The reader will remember from Chapter 4 that near the close of the last century Combes made impure $SiHCl_3$ by passing HCl over "copper silicide." He would be surprised to learn that his work can justifiably be linked not only to silicones, but to transistors, rectifiers, integrated circuits, microprocessors, and computers as well.

The Chemistry Section's fondest silicone dream, an *unsupported* wire enamel, has not come to pass. Silicone resins simply do not have the strength, toughness, and continuing tight adhesion to copper that made Formex® wire enamel a spectacular success in its day. (Ironically,

* For these estimates and for much other help (primarily directed toward, but not confined to, Chapters 7 and 8—as a lawyer might say), I thank Dr. Hart K. Lichtenwalner, Manager, Fluids, Resins, and Chemical Products Section, Waterford, in 1976.

copper seems to be the *worst* among common metals as regards silicone adhesion.) Now that organic successors to Formex® can be used quite well at higher temperatures than the original product, the prospects for a successful unsupported silicone wire enamel are poorer than they seemed in 1938. But *glass-served* silicone insulation, though costly and bulky, is indispensable for electrical equipment that must be operated at the highest possible temperatures. This composite insulation is of course a close relative of the silicone-impregnated glass tape that interested Corning Glass in Kipping's silicones before 1938.

Another unrealized dream is a practical silicone fiber, which could have found wide use not only in the electrical industry; it might even have made a little more likely the miracle tires "foreseen" at the first General Electric press conference on silicones.

The high intrinsic electric strength of silicone oils, which was discovered quite early, of course suggested their use as insulating liquids in transformers and capacitors. They do not have properties comparable to hydrocarbons, but if used in properly designed equipment, electrical breakdown of silicone oils becomes very unlikely. The flammability of the oils, though not zero, is small enough to be acceptable; furthermore, flames in these oils do not tend to propagate, largely because of the formation of a silica film on the surface. Oils with suitable viscosities and heat-transfer coefficients can be made. The high cost of silicone oils has made it difficult for them to displace the "PCBs" (polychlorinated biphenyls) as insulating liquids. However, PCBs bioaccumulate and tend to work up the food chain when they are present as contaminants in waterways. Therefore their use in industry and commerce is being eliminated.

SILICONES AS LUBRICANTS

After Patnode thought of silicone oils as lubricants in 1941, the belief quickly grew in the Chemistry Section that silicones would become general-purpose lubricants useful over the range from very low to very high temperatures. Even a cursory glance at silicone properties shows that this optimism was justified.

Lubrication problems are so complex as to dampen any hope for a satisfactory general lubricant. Bearings are made of metals and alloys, vary in shape, undergo different motions under different loads, operate at different temperatures—these are only some of the obstacles a satisfactory lubricant must surmount. Silicone oils, although they will never replace hydrocarbons as lubricants, are doing about all that could reasonably be expected of them: they are in fact acquitting themselves well under conditions (mainly at high and low temperatures) *where nothing else will do.*

Methyl silicone oils as lubricants for various kinds of service can be improved by introducing phenyl or long-chain alkyl groups, by resorting to halogenation, and by using additives. In choosing an additive, a soap, for example, it must be remembered that a chain is thermally no more stable than its weakest link.

Silicone greases are perhaps the most successful among silicone lubricants: much depends on the filler; proper combinations can be unsurpassed. Some of Dow Corning's Molykotes®, a family of outstanding lubricants for service at high pressure and extreme temperatures (both high and low), are blends of a silicone and MoS_2, itself a useful solid lubricant.

SILICONE TREATMENTS TO IMPROVE SURFACES

The commercial use of silicones to give improved surfaces has by now gone far beyond the first example, which was the successful treatment of steatite insulators by the Chemistry Section during World War II. Attempts to extend this treatment to materials such as paper, leather, and textiles miscarried because of the objectionable, necessarily concomitant, formation of HCl when chlorosilanes are the treating agents. Today no one needs to consider producing water repellency in this way if the HCl is a nuisance or worse; better ways are known. Silicone surfaces continue to be highly regarded.

The Chemistry Section [Chapter 5] attempted to generate such surfaces in other ways. Elliott and Krieble patented siliconates (see Figure 8-1) for this purpose; they found the strong alkali a drawback. Norton's 1942 patent anticipated modern practice, which is to use silicones that contain the reactive Si-H bond and are already condensed: these were regarded by Norton as silicone oils containing Si-H bonds, which, via oxidation, hydrolysis, or hydrogen bonding, would promote anchoring of the silicone surfaces to the substrates.

With the exception of silicone-containing paints, which were a special concern of Building 77, I do not believe that the Chemistry Section went beyond the achievements mentioned in the preceding paragraph in envisioning the important ways in which silicones are now being used to improve surfaces. One consequence of the many innovations of this kind made over the years has been a strong market for silicone oils, which are often applied as solutions in organic solvents or as aqueous emulsions.

(In some cases, silicone resins are preferred to silicone oils.) At this point the reader may ask, "If oil and water do not mix, how can silicone oils, noted for water repellency, ever be mixed with water to give emulsions that are stable during storage and shipment, and that can be diluted with water before they are used?" The answer is, "Gentle persuasion by artful means outside our scope." The long question and its pat answer are intended to inform the reader that Noll [1, pp. 566–619], in discussing the innovations involving silicone surfaces, cites other examples equally surprising. Emulsions and foams are related phenomena; we shall expatiate a little on foams.

The ways in which silicones improve surfaces are so diversified that many silicone properties come into play; the importance of each often changes with the application. The simple model of a silicone surface, namely, an outer hydrocarbon-like layer anchored by siloxane bridges, is often applicable; one may think of this outer layer as repelling substances other than water. The good spreading power of the silicones, which results from their low surface tensions (surface energies), is important, as are their solubilities and insolubilities in other substances.

When, for example, silicones act as *release agents,* often to release solid materials from a solid mold, the characteristics just named come into play, as do viscosity and vapor pressure. The releasing of automobile tires from their molds during manufacture is an illustrative example, historically important because it was an early innovation that helped the silicone business to survive. Most rubber articles are molded and removed from molds while they are hot enough to suffer damage if the mold surface is so sticky that undue force is required to separate the work from the mold. Silicones, sometimes used along with organic materials, have virtually eliminated these problems. In the simplest cases an aqueous emulsion, containing 30 to 40% of silicone oil as bought from the manufacturer, is diluted with water until the silicone content is in the range 0.1 to 1%, whereupon it is sprayed onto the hot mold as a fine mist. Many variations and many different applications exist.

Considered as a release problem, the baking of bread resembles the making of automobile tires. Here the mold is the bread pan, and silicone resins baked onto its inner surfaces have proved highly satisfactory release agents. Dow Corning deserves credit for the innovation. "The cost of coating a pan [in the early days] is about equal to that of grease and labor to process 100 pans by the conventional method" [6, p. 131]. Ordinarily, the number of bakings between silicone treatments comfortably exceeds 100, but this excess alone does not measure the advantage silicones give, which include cleaner air in the bakery, better-tasting bread, satisfactory release, and less damage to the pans. "A nonstick waffle iron is considered in some quarters one of the major blessings brought by modern technology" [6, p. 132].

The reader will have inferred that the use of silicones to achieve water repellency has flourished. A particularly interesting example is the protection of masonry against liquid water. These silicate building materials need to remain internally porous so that they can continue to be good thermal insulators, and to permit the exchange of gases (especially the *egress of water vapor* from the building) needed for healthful living. Filling of the pores by liquid water (rain) is harmful for many reasons [1, p. 605]. One of several successful treatments of masonry rests upon the formation of a water-insoluble resinlike silicone at room temperature during and after the evaporation of water from the outer *siliconate layer* formed by impregnating the masonry. The siliconates used are soluble sodium salts such as the following

$$
\underset{\text{Simple}}{\overset{\overset{\displaystyle CH_3}{|}}{\underset{\underset{\displaystyle ONa}{|}}{HO\!-\!Si\!-\!OH}}}
\qquad
\underset{\text{Slightly condensed}}{\overset{\overset{\displaystyle OH \quad ONa}{| \quad\;\; |}}{\underset{\underset{\displaystyle ONa \;\; OH}{| \quad\; |}}{CH_3\!-\!Si\!-\!O\!-\!Si\!-\!CH_3}}}
\qquad
\underset{\text{Highly condensed}}{HO\!\left[\overset{\overset{\displaystyle CH_3}{|}}{\underset{\underset{\displaystyle ONa}{|}}{-Si\!-\!O-}}\right]_n\!\!\overset{\overset{\displaystyle CH_3}{|}}{\underset{\underset{\displaystyle ONa}{|}}{Si\!-\!OH}}}
$$

siloxane bridges being formed by condensation in the alkaline solution to produce more complex species (right, above). Masonry is of course less vulnerable to the alkali in these siliconates than were the textiles that Elliott and Krieble tried to waterproof in this way.

We will digress to record what *now* appear as two missed opportunities for innovation. When equipment with silicone insulations was baked to "cure" the insulation during World War II, ovens in the Schenectady Works became objectionably coated with tenacious silicone films difficult to remove. A nuisance here became a blessing in the bread pan. During one night, near the same time, cooling water stopped flowing through a methylchlorosilane still on the sixth floor of Building 37. The escaping gaseous products reacted with the many windows* at the front of the Works and whitened the red brick of the outside wall nearest the still. The complaints from the other buildings were many, and the first (but

---

* A waterproofed window is a nuisance for the same reason that laboratory glassware (especially burettes) become such when silicone treated. A water-repellent surface never drains perfectly. A continuous film of liquid in such cases is to be preferred over isolated, adherent drops.

unintentionally) silicone-treated masonry on 37 remained white, and presumably well protected, for many years.

The textile and paper industries generally use silicones for imparting repellency to surfaces. The Xerox Corporation seems to have found silicone oil useful in a special way that makes use of other properties. Their copiers must usually be resupplied every so often with Xerox Duplicator Toner (the black "pigment") and with Xerox Fuser Oil (from a container marked "Contains silicone which can cause irritation upon contact with the eye"). I have been told that the oil is brushed upon the paper before the toner is electrostatically attracted to it. Heat then "fuses" the toner to make the admirable, sharp, permanent copy we have come to take for granted.

The ease with which silicone oils can be spread over a surface to give a thin film has made the cleaner and polish industry a ranking consumer of these oils. To paraphrase an old slogan, "Save the surface and save yourself." I have been told also that the first silicone-containing polish originated outside the silicone industry. There are so many such polishes and cleaners today that we shall dismiss the subject with two matters mentioned by Noll [1, pp. 604–605]. First, when wax is the principal polishing agent, silicone oil probably helps to form the final lustrous film by coating the wax crystals, which can then be easily smeared over a silicone-oil film that has already spread over the surface. Second, sometimes an old silicone film causes trouble when a surface is to be refinished. Fortunately, ways of mitigating this problem have been found.

SILICONES IN MEDICINE AND COSMETICS

Forever wilt thou love, and she be fair!

JOHN KEATS, *Ode on a Grecian Urn*

In retrospect once more, it seems that silicones with their unusual properties ought to help us as regards life, health, and the pursuit of happiness, always provided that they are not rejected by living tissue and produce no undesirable side effects. These provisos are so formidable as to threaten innovation in today's drug industry [19].

Silicones were lucky in being introduced early and successfully into medicine [Chapter 5]; only good seems to have resulted from this early work. As silicones do not occur naturally, there was no historical evidence that they were acceptable risks for medical use, but they have by now been proved so [1, pp. 516–527]. Organochlorosilanes, on the other hand, are of course clearly harmful. Other substances, of low molecular weight, either silicones or related thereto, have also been proved harmful, mainly to animals; but usually the harm has not been permanent. The silicones most useful in medicine and cosmetics are oils and rubbers; they are reported to have a clean bill of health except that silicone oils or creams containing them should not be put into the eye [1, p. 527].

Silicone oils in small proportion make ointments and the like water repellent and easy to spread; therefore the oils are being widely used in preparations designed to protect or beautify the skin [1, pp. 620–631, deal with this point and subsequent ones]. The RTV compounds are used in dentistry to make replicas of jaws, and in other ways as well. Silicones have become important auxiliaries in surgery, general medicine, and related fields: silicone rubber can be sterilized with impunity; it does not stick to other materials; and it contains no plasticizer. Silicone films still retard the clotting of blood, as they did in the Schenectady Works infirmary many years ago. Silicone oils in small amounts are useful in destroying objectionable foams arising during the digestion of food. The use of RTV as a liquid carrying iron particles makes it possible to seal aneurysms by "magnetic brain surgery" [20] because the substance can be guided to the ruptured artery, where sealing occurs as the RTV solidifies.

The *New York Magazine* of February 10, 1975, featured "Spare Parts for Humans. The Ultimate Cure for Bodily Wear and Tear." Silicone rubber ranks high among the materials of which such spare parts and implants can safely be made. Here we shall mention only two, one for each sex. Silicone mammary prostheses have been successful.* Less well known is a more recent innovation in which an implanted hydraulic system produces an erection. On demand, a silicone-rubber pump moves a hydraulic fluid, probably a silicone oil, to accomplish the desired result. For the clinical details, possible consequences of malfunction included, please see the article "Replaceable You," by Mr. Aaron Latham in the magazine cited. Mr. Latham concludes with this quotation from Kipling's "A Truthful Song":

Your wheel is new and your pumps are strange
But otherwise I perceive no change.

Silicones have made better prophets of Keats and Kipling both.

* In these prostheses silicone rubber is used. The injection of silicone oil to modify women's breasts has had undesirable consequences, is now condemned by medical authorities, and has been made illegal.

SILICONES AND FOAMS

When a fluid exists in the form of a thin film between other fluids, the great inequality of its extension in different directions will give rise to certain peculiar properties. . . .

J. WILLARD GIBBS, "The Equilibrium of Heterogeneous Substances" [21]

Silicones have helped make Gibbs's great statement into a profound understatement. When both fluids are liquids, we have emulsions. When a thin film of liquid encloses a gas, we have a bubble. When the bubbles form an aggregate, we have foams. No matter what we have, the complexities can be horrendous. Were not silicones among the most interesting of the materials that "give rise to [these] certain peculiar properties," and were not the consequences of great and growing importance, we should omit all discussion of silicones as *surfactants* (see below). As it is, we shall attempt no more on the theoretical side than a qualitative presentation of Gibbs's ideas as they apply to foams in local equilibrium.

Foams are often fragile and unstable, such as the foam on a glass of champagne (or ginger ale!). Foams can be strengthened and preserved, however, by adding proper amounts of surfactants, that is, materials that tend to concentrate at the surfaces of liquid films—witness the trouble phosphate detergents have caused on our lakes and rivers. Ultimate and permanent strengthening occurs when a surfactant is added to a liquid mixture that is foamed as the cell (bubble) walls are being *solidified by polymerization reactions*. In this way synthetic foam rubbers, on which we can sit or sleep, can be made, as well as rigid foams useful as thermal insulation in household refrigerators and in many other ways.

Foamed silicone resins are gradually finding favor in applications where their flame resistance and high-temperature stability adequately offset their brittleness and higher cost [1, p. 422]. These foams will not concern us further, as we shall restrict ourselves to silicone surfactants, which divide themselves into *foaming* and *antifoaming* agents. The former types of silicones preserve foams until the cell walls have solidified as was mentioned above; the latter destroy foams that are not wanted. The two types have different structures. For the former, see Figure 8-3; "Ann T. Foam," a polydimethyl siloxane, will appear later as an example of the latter.

The "peculiar properties" mentioned by Gibbs show themselves in foams about as follows. In a pure liquid the surface molecules are attracted inward by the molecules below the surface. This pull is felt only by the surface molecules because those in the interior, unlike those on the surface, are on the average uniformly surrounded by other molecules, so that no net pull exists. This inward pull places the surface *in tension,* and the liquid tries to relieve this tension by reducing surface area. This process tends to destroy a foam.

Pure liquids can scarcely be expected to foam. If a surfactant is added, foaming becomes possible because the surfactant lowers the surface tension by concentrating at the surface, especially if the surfactant is virtually insoluble in the liquid. The energy of the system decreases when the surface tension is lowered. So far, all surfactants would seem to behave alike, although we know that some can preserve films whereas others destroy them.

Gibbs's work suggests an explanation. Any liquid film forming the wall of a bubble is sensitive to physical disturbances that tend to increase or decrease its size. If the bubble is to survive, there must be compensating changes in surface tension. The surface tension always acts to decrease the area of the film, that is, to contract the bubble. If the disturbance tends to increase the area, the surface tension will also have to increase in order to block the disturbance; this will occur if the concentration of surfactant at the surface decreases. In the contrary case the concentration of the surfactant will have to increase so that a reduced surface tension can preserve the bubble from collapse. In making synthetic foam rubber, uniformity in the distribution of surfactant ought to help keep the bubbles uniform in size.

The first foam rubbers were natural rubbers, being made from latex into which air had been dispersed. Today, art has so improved on nature, with silicone surfactants to help, that the leading foam rubbers are the synthetic polyurethans.*

* Even Humpty Dumpty would be nonplussed by all the peculiarities of nomenclature that attach to "polyurethan." Here are a few. The polymer, sometimes called "urethan," is *not* made from urethan, which may be regarded as a compound of *general formula* NH_2COOR, one form of which is $NH_2COOC_2H_5$, the ethyl ester of a *nonexistent* acid. "Urethans" were (and are still by some) called "urethanes." The name had to be changed because it implies a dubious relation to ethane; the relation to urea is tenuous. Polyurethans are made by the *addition polymerization* of diisocyanates, R_2CNO, with a variety of organic compounds that have terminal OH groups. The nomenclature tangle tempts one to paraphrase Voltaire's famous "neither holy, nor Roman, nor an empire." Kipping's little slip in naming silicones [Chapter 2] is trifling by comparison.

The foregoing diminishes not a bit one of the great polymer successes of this century. Diisocyanates, the keys to this success, were made by Wurtz in 1849 but lay fallow until 1937, when Bayer began to study them for making addition polymers; in 1968 over 500 *million* lb of polyurethans were produced in the United States alone [22]. When industrial research succeeds on this scale, even Humpty Dumpty will forgive those who still prefer "polyurethanes" to "polyurethans." A little nonconformity brightens our look at industrial research.

A flexible polyurethan foam (mattresses, chairs) must be resilient, be of low density, and have the proper quota of *open* cells (formed by the rupture of bubbles, i.e., of cell walls). A rigid polyurethan foam for thermal insulation (refrigerators) must be of low density with predominantly closed cells. A foam for furniture construction will resemble the rigid foam just mentioned but needs to be denser.

Gibbs himself might have been intimidated by the difficulties of treating theoretically the formation of polyurethan foams. His treatment was an *equilibrium* one; he assumed *localized equilibrium* in thin films, not equilibrium overall. Actual foaming problems, particularly those associated with polyurethan foams, are often *rate problems* too complex for theoretical treatment [23].

We shall mention only the "one-shot" method of making polyurethan foams, which is an excellent illustration of the use of a silicone as an auxiliary in an important manufacturing process [22]. The "one-shot" operation must accomplish (1) the addition polymerization [Chapter 2] of a diisocyanate, and a polyether or polyester with terminal OH groups; (2) the generation of CO_2 to fill the foam cells, the CO_2 being produced by the reaction of water and diisocyanate; and (3) the production and preservation of foam cells of the proper kind. Catalysts are required for (1) and (2), and a silicone surfactant for (3). We thus have the problem of so mixing at least three reactants, two catalysts and a surfactant, all in proper proportions, as to keep polymerization and gas generation suitably "in phase" in order to make the uniform foam rubber desired. The method just sketched has many variations.

The manufacture of foamed polyurethans requires special surfactants used at proper concentrations. The introduction of suitable silicone surfactants was a significant innovation by the Union Carbide Corporation [24], an important silicone manufacturer not previously mentioned. Even though the proportion of surfactant is low, say 1% [4, p. 558], the huge amount of polyurethans manufactured annually makes these surfactants major products in the silicone business. Figure 8-3 gives more information about them [25].

Objectionable foams appear in stomachs, in steam boilers, and in many other places. Antifoaming agents can bring relief. They work in complex ways, but what follows is a starting point. We saw above that bubbles are subject to shocks and disturbances, which will destroy them unless surfactants are present under the special conditions (proper concentration, and proper change of surface tension resulting from quick change of concentration) needed for survival. During defoaming these special conditions must be avoided, and the antifoaming agent must get to the surface of the bubble walls to make the bubbles burst. In this

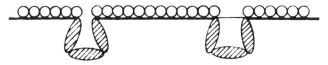

Figure 8-3 Structure of a silicone-surfactant film according to Gaines [25]. The surfactant is a *block copolymer* in which "blocks" of **D** units (open circles) are joined to "blocks" of an organic polymer (shaded lozenges).
Caution: The silicone surfactants actually used in making polyurethans, though block copolymers, are more complex [24]; they often contain silicone units more complex than **D** units. During the foaming the surfactant is obviously not on a water surface.

Nonetheless, Gaines's model is valuable because it suggests that the organic copolymers act as movable anchors during the foaming operation, it being possible that the siloxane bridges are not strong enough anchors themselves to keep bubbles at the proper size during the CO_2 evolution that causes foaming. Such a system would be far from the kind of equilibrium postulated by Gibbs. I wish to thank Dr. Gaines for help with this section of the book.

respect great spreading power, such as is shown by methyl silicone oils, is helpful. The concentrations of surfactant effective for foam destruction ought to be far less critical than those required for foam preservation.

Instead of continuing with prosaic scientific discussion, let us quote from a highly effective 1976 "Incredible Short, True Story" by Dow Corning* in which virtue triumphs as bubbles are [Bubbles Galore is (?)] smitten hip and thigh.

BUBBLES GALORE REVEALED

Ann T. Foam

We know all about your relationship with Bubbles Galore.

But you're not the only one who's involved. There are thousands of companies— like yours—where excess foam in processing operations is playing havoc with processing machinery, maintenance costs, and plant productivity.

There is hope. There is a way to make sure you won't see or be troubled by Bubbles Galore ever again. The solution is simple . . . and effective. Dow Corning® defoamers. Bubble for bubble, they usually last longer and give you faster, safer, and more efficient production than conventional defoamers.

. . .

As you know, Bubbles Galore can be pretty fickle. So Dow Corning offers a complete line of agents for industrial, chemical, and food-processing applications.

---

* I thank Dow Corning for permission to enliven these pages by quoting Ann T. Foam to introduce Bubbles Galore.

However and wherever Bubbles Galore causes trouble, Dow Corning has the defoamer to prevent, minimize, and destroy troublesome bubbles. In heat or cold. In aqueous or nonaqueous systems. In alkaline or acidic solutions.

Don't worry about your product's integrity. Dow Corning® defoamers are inert and substantially insoluble in aqueous media. They act physically, not chemically, to eliminate foam by stressing bubble walls to the bursting point. And they're safe. Food-grade defoamers are permitted for certain uses by the EPA and FDA.

...

What a happy ending to a frothy section!

SILICONE RESINS. A PARTING NOTE

Had it not been for Kipping's "sticky masses," precursors of our modern silicone resins, and the realization by Corning Glass Works that such materials might make a salable electrical insulation out of Corning glass tape, the beginning of the silicone industry would have been delayed— for how long, who knows? Silicone resins deserve a parting vote of thanks because they carried the silicone project at least until the "forks in the silicone high road" of Chapter 4 were discovered.

Even after that, the resins continued to command a great deal of work, the original impetus for which stemmed from Rochow's decision, made in September 1938, to concentrate on methyl silicones, and from Wright's subsequent demonstration that these resins could meet Rochow's expectations of good performance at high temperatures in composite insulation. Years followed years in which successful resin innovations were too few to justify the work that went into them. As we have seen, resins did *not* build the silicone industry, which could scarcely have survived without successful innovations that did not involve resins. No attempt will be made here to catalogue the early difficulties with these materials; most difficulties derive from the extensive cross-linking that resins must have. No cross-linking, no strength; ample cross-linking, cracking and crazing; added phenyl groups for product improvement, added cost—this is a short summary of the resin quandary. A further serious difficulty eventually appeared, namely, slowness to "cure" (complete the condensation reaction) in resin films. Extensive cross-linking does offer one interesting advantage: some resins make suitable adhesives, presumably because the numerous siloxane bridges in them anchor them well to various surfaces.

Figure 8-1 shows that silicone resins have (praise be!) arrived. The work done to mitigate their defects has been enormous; see Noll [1, pp. 404–426, 474–490, and elsewhere]. They became successful, not by virtue of theoretical guidance, but by long-continued experimentation based upon growing knowledge and upon an abiding faith in their virtues. Among silicones the resins are the Horatio Alger heroes—poor, deserving products that finally made good in the face of difficulties that ought to have finished them.

SILICONES AND ANALYTICAL CHEMISTRY

All chemists know that analytical chemistry has undergone a revolution since World War II, that the chemistry has been going out of it with a vengeance, that physical methods, instrumentation, automation, and the computer have been moving in, and that this proud science has resigned itself to becoming a service activity. The atomic bomb project alone would have made the change inevitable. Adamsian acceleration and multiplicity once again! But chemistry and the chemical industry would be far less efficient today had the change not occurred.

The change was visible early and clearly in the silicone project. As Chapter 5 shows, the classical analytical methods that revealed where the project was going in its early days gradually and irreversibly lost ground to physical methods. The trend continued at Waterford, where instrumentation was stressed. The comprehensive *Analysis of Silicones* [26] manifests the Adamsian multiplicity of all modern analytical chemistry. This book was written by A. Lee Smith, Manager, Analytical Department, Dow Corning Corporation, and 13 associates in his department.

SILICONES OUTSIDE THE MONOGRAM

The worldwide interest in silicones is evident from the representation of 13 countries at the Fourth International Silicone Symposium, which was mentioned above. In the United States the Dow Corning Corporation was founded in 1943 and for some years used the Kipping route to make silicones for greases. Since 1957 the Union Carbide Corporation has had a Silicone Division at Long Beach, West Virginia. In 1965 The Stauffer Chemical Company started up its silicone plant at Adrian, Michigan.

In view of the valuable properties of silicones, the reader may be surprised to learn that they have generally been sold in a *buyer's* market. Aren't silicones at least as attractive as the proverbial "better mousetrap''? Shouldn't customers have beaten paths to the manufacturer's door? Instead:

The sale of silicone . . . products historically has been *very competitive* both in the United States and abroad, and Dow Corning currently experiences *substantial competition*. . . . The competitive elements in the sale of Dow Corning's products are principally *price* and *product performance* and, in addition, *technical service* and *delivery''* [18, p. 12; italics mine].

One can only guess why these "miracle plastics''—a title that silicones held for a while—have not lured buyers strongly enough to ensure profits for the seller.* Logical guesses are (1) high cost, largely deriving from process complexity; (2) increasing competition from alternative materials that continue to grow in number; (3) a continuing need for silicone market development and innovation; and (4) reluctance of customers to maintain large inventories of costly materials.

An interesting article in *Chemical Week,* November 5, 1975, p. 42, carries the title "Versatility is Key to Success of Silicones. Steady Stream of New Uses Will Keep Demand Pushing Capacity.'' The title combines an acknowledged fact with what Hamlet would call a "consummation devoutly to be wished.''

The standard dimethyl silicone oils are among the few silicone products sold in carload lots. Their recent price history is shown in Figure 8-4. The sharp decline in selling price, which began before 1950, resulted from improved technology (probably mainly in that of the direct process), and the increased sales volume that followed, largely from the use of silicone oils in polishes and cleaners. The abrupt reversal of the price decline because of inflation is not surprising. After all, the cost of elementary silicon reflects not only the general price inflation but increases in the cost of energy as well: it takes a lot of energy to make a pound of silicon, and the costs of CH_3OH and CH_3Cl reflect the price of natural gas. No one concerned with silicones can take inflation or the energy

---

* *Moody's Industrial Manual,* Vol. 1, 1975, is the source of the information that follows. For the Dow Chemical Company the net income "before extraord. item'' as percent of net sales was 11.9 (1974), 8.8 (1973), and 7.9 (1972), according to data on p. 310. Comparable data for Dow Corning from p. 1328 are 10.9 (1974), 11.3 (1973), and 10.0 (1972). The year 1974 seems to have been an exceptionally good one for Dow Chemical. In that year the company began its attempt to correct for the effect of inventory profits resulting from inflation. The correction, which *lowered* net income, was made by shifting from the *fifo* (first in, first out) to the *lifo* (last in, first out) method of valuing domestic inventories.

It seems permissible to conclude from these figures that there has been no huge pot of gold at the end of the silicone rainbow. Dow Corning is a specialty manufacturer; Dow is an important factor in the overall chemical industry.

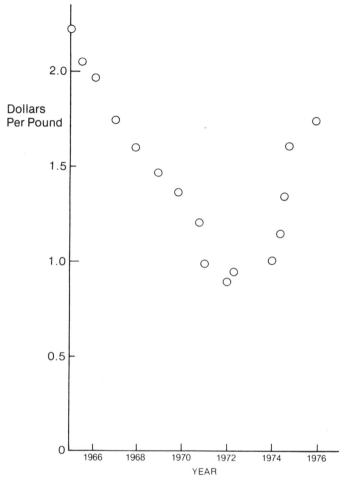

Figure 8-4 Recent January prices of standard dimethylsilicone oil in single drums (200 kg). For the earlier history, which begins with a price just under $6 per pound in 1945, see Reference 4, Fig. 1, p. 471.

crisis lightly. The early dream that silicones could become low-cost materials because they are based only on sand and coal was never justified and is now ended.

The prospectus [18]* relating to a $60,000,000 debenture issue marketed in 1975 completes our silicone panorama and enables us to salute

* Reference 18 is a prospectus giving information required by law when a corporation seeks public financing. I have not found comparable published information for other silicone manufacturers.

the largest silicone manufacturer, the Dow Corning Corporation, head-quartered in Midland, Michigan, whose 1974 sales were almost $278 million, about 40% made abroad. An increase of $53 million in 1973 sales over 1972 resulted mainly from an improvement in the economy and from increased sales at improved profit margins to the electrical industry. Price increases accounted for almost all the $36 million sales gain in 1974 over 1973 [18, p. 8].

Of the 1975 debenture issue, $13 million was earmarked for capital expenditures and most of the rest for the retirement of outstanding debt. A sign of the times was the "planned pollution control revenue bond financing . . . of approximately $7 million" [18, p. 4].

Dow Corning today sells more than 2500 (yes, 2500!) products: sili-cones, silicon, and related materials [18, p. 9]. In 1974, marketing and administrative expenses ($61 million) were more than one-third the man-ufacturing costs ($159 million), and very nearly double the net income. Dow Corning has about 200 full-time salesmen, about 1 employee in 25 being a salesman.

The foreign operations of Dow Corning involve sales of substantially all Dow Corning products, principally in Europe, Latin America, Canada, Asia and Australia. These products are manufactured either domestically or by one of the Company's foreign subsidiaries. Dow Corning owns less-than-majority interests in the Japanese companies [18, p. 11].

About four-fifths of Dow Corning's "aggregate investment in plant and equipment is represented by its United States facilities." Here are further details:

Dow Corning's largest manufacturing plant is located at Midland, Michigan. Other United States production plants are located at Carrollton and Elizabeth-town, Kentucky; Hemlock, Michigan; and in California, Connecticut, and North Carolina. Dow Corning's largest foreign manufacturing plant is located in Barry, South Wales, United Kingdom. . . . Other foreign manufacturing plants are lo-cated in Australia, Belgium, Brazil, Canada, Germany, Japan and Mexico. Dow Corning owns all of its manufacturing facilities. . . . [18, pp. 13, 14].

Dow Corning employs approximately 4,700 persons, of whom approximately 1,700 are production and maintenance employees and the remainder are technical, administrative, supervisory and sales personnel [18, p. 13].

In 1974, approximately 675 persons were employed in research, development and technical service activities. These activities are centered in laboratories located at or near Dow Corning's Midland, Michigan, plant and are also conducted in other locations throughout the United States and abroad.

Dow Corning's research and development expenditures totalled approximately $19.6 million in 1974 [18, p. 12].

Additional evidence that Dow Corning has faith in the future of silicones is this: the corporation added all of its substantial net income in 1973 and 1974 to earnings previously retained in the business.

Silicone innovations seem likely to continue at a satisfactory rate over the foreseeable future. For 10 years Dow Corning has had an extensive research program in which silicones of low molecular weight, tailored with organic molecules as models, are being studied for possible biological activity.* One of these new compounds, according to a joint announcement by Dow Corning and a Swedish pharmaceutical firm [27], has alleviated pain due to prostate cancer in 20 terminal patients. More recently, Dr. R. H. Rech, Michigan State University, has reported that $(C_6H_5)_2Si(OH)_2$ is useful in treating epilepsy [28]. These welcome results do not conflict with the position that *polymeric* silicones are physiologically inert.

The silicone panorama will continue to change over the years, but present indications are that it will become more impressive, both under the monogram and outside it.

COMMENTS IN CONCLUSION

• As to understanding and industrial research. Scientific augmentation, a component of industrial research, has often improved understanding of prior knowledge as an important aim. A few comments about understanding and industrial research are therefore in order.

The properties of liquid silicones and of silicate polyesters are poorly understood today despite investigations of different kinds that were crowned by the theoretical work of Flory and his colleagues, as a result of which we now have a quantitative explanation of the relations, experimentally observed, between **D** backbones and the cyclamers of general formula $\mathbf{D}_n$. This is a promising beginning, but we are still far away from explaining the physical properties of silicones, or from making reliable predictions based on such explanations. As a cynic once said, "Physical chemistry tells you where you've been."

Industrial research has often been successful when understanding

---

* This research program brings to mind the periodic-table relationship of carbon and silicon [Chapter 2], Combes's 1896 attempt to synthesize the silicon analogues of carbon dyes, and the early dreams of siliceous protoplasm.

was defective or lacking: witness Baekeland and his resins. Naturally, all research should strive for understanding, but industry cannot afford to wait upon it. Thus D_4 had to be used for making silicone rubber long before the Flory work was done; implementation went ahead even though, to paraphrase Pope, all was not light. Industry should welcome attempts to improve understanding, engage in them at least to the extent needed to keep its research staffs intellectually alive, and support pertinent academic work as partial repayment for the prior knowledge that industrial research finds useful. Even Mr. Justice Holmes, skeptic though he was, once wrote, "The men who teach us to disbelieve general propositions are only less valuable than those who encourage us to make them" [*Uncollected Letters*, p. 173]. And, with regard to the cynic's view of physical chemistry, let us remember Montaigne's observation that often the journey, not the arrival, matters. Although we may never arrive at a complete understanding of descriptive chemistry, we must continue the journey.

• As to the eventual displacement of silicones. The periodic table contains elements heavier than silicon that might be regarded as alternatives for the building of backbones and skeletons in materials that might displace silicones: germanium is one such element, on which Rochow tried his hand before he left the Company.

I have been told that a good many hundred man-years of effort have been directed to finding silicone analogues that might displace or supplement silicones. The work has led to no commercial products and brought no threat to the silicone industry.

The relevant descriptive chemistry, of which the periodic table is an incomplete model, must have been well known to many who tried to make these silicone analogues. But the knowledge and understanding were imperfect enough so that the extensive work was needed; in such cases the laboratory is still the court of last resort, which one flouts at his or her peril.

• As to the silicone industry. The 1976 silicone panorama shows that the silicone industry survived a sickly infancy to reach a healthy early maturity. Being concerned primarily with industrial research, we have emphasized this activity over others (e.g., implementation and marketing) that were also essential to this survival and growth.

The research pattern of the silicone project—decreasing frequency of major discoveries (Table 4-1) and increasing scientific augmentation—seems to have been the universal pattern since 1950, the last year of that project. The chances for future major silicone discoveries seem best in fields related to the life sciences, particularly the field of beneficial drugs.

Because of the attractive properties of silicones and the number of useful silicone products already available, silicone innovations seem likely to continue at a high rate—a rate sufficient to at least compensate for any declining frequency in major discoveries.

Inflation and increasing energy costs will bedevil the silicone industry by raising the prices of its already costly products.

The silicone industry has done the world much good and little or no harm. It will continue this course. Those who enabled it to survive and grow deserve the world's gratitude.

Chapter Nine The Silicone Story in Perspective

[A] *coat of many colors.*

Genesis 37:23

To be properly appreciated, the silicone story needs to be viewed against a background of industrial research. Some of this background has already been supplied. Completeness is clearly out of reach. As a necessary compromise, selected industrial research projects will be sketched briefly in this chapter. The reader who wishes to learn more about them can do so by consulting the references given.

The synthesis of ammonia was chosen as the first project because it belongs among the greatest accomplishments of industrial research. It has meant more to mankind, and probably to BASF, than the synthesis of indigo, crucial in its day [Chapter 1]. Next come three General Electric polymer projects, successors of silicones in the Research Laboratory, that might never have been born had silicone research been unsuccessful. A few projects from other companies round out the list.

Note. Brackets are used to enclose references and inserted material. For bracketed references, see References and Notes at the end of the book. Numbers in parentheses on the right of pages are often used in the text to identify equilibrium models and the like; thus (9-1) means Equilibrium Model (9-1). Reference to laboratory notebooks is by name and date. CRDA stands for "Chemistry Research Department Archives."

THE SYNTHESIS OF AMMONIA

The early ammonia story is well told in three of the most distinguished papers in the literature of chemical research [1, 2, 3]. The first two are Nobel Prize addresses (Fritz Haber, 1918; Carl Bosch, Joint Laureate with Friedrich Bergius, 1931). The third is the 1930 Bunsen Medal address by Alwin Mittasch. The following quotations therefrom, freely translated, summarize the story:

. . . a new richness of nourishment for mankind from the richer nitrogen-fertilization of the earth . . . Fritz Haber [1, p. 1049].

. . . the development . . . of the processes that serve for the undertaking of chemical reactions at elevated temperatures and pressures . . . of 100 atmospheres and higher, Carl Bosch [2, p. 127].

When the question of implementing Haber's discovery of the ammonia catalysis arose in 1909, Bosch gave his colleagues the assignment of finding more easily available catalysts or of improving to the point of practical usefulness the poor catalysts then known, Alwin Mittasch [3, p. 570].

The Haber-Bosch process changed the course of history because it enabled Germany to make nitrates for explosives during World War I. The consequent lengthening of the war now seems less important than the good being done by synthetic ammonia as the prime man-made source of the "fixed nitrogen" needed by plants for the synthesis of the proteins essential in food, and as the material from which many nitrogen compounds are prepared. The annual production of ammonia by this process is measured in *tens of millions of tons* (15.781 million tons in the United States during 1975 [4]), and liquid ammonia is today injected directly into the soil to help grow our food. The three papers record a need-discovery-augmentation-implementation sequence difficult to match—a sequence that invites comparison with the earlier chapters of this book. Comparisons with the direct process for methylchlorosilanes are particularly apt and will be given *within brackets*.

Haber. Discovery and Augmentation.   The world's growing need for fixed nitrogen was widely known when Haber began in 1904 to seek a chemical alternative to the mining, mainly in Chile, of nitrates that were being alarmingly depleted. Although Haber always held pure research in high regard, he says (freely translated) that he would scarcely have investigated the synthesis of ammonia from its elements so thoroughly had he

not been convinced of the "volkswirtschaftlichen Notwendigkeit" (pol-
itico-economic need) for chemical progress in the field of nitrogen fixation
[1, p. 1041].

As we shall see, it had long been known that iron would catalyze
the *decomposition* of NH_3 into N_2 and H_2 at high temperatures. By 1884
Ramsay (Sir William Ramsay, Nobel Prize, 1904) and Young were able
to show that this decomposition was *never complete* near 800°C, and this
argued for the existence of the equilibrium

$$N_2(gas) + 3\ H_2(gas) \underset{r}{\overset{f}{\rightleftharpoons}} 2\ NH_3(gas) \tag{9-1}$$

which we shall take as a model of the ammonia synthesis, with f and r
denoting the forward and reverse reactions. At equilibrium these two
reactions must proceed at equal rates: N_2 and H_2 *must* react to form
NH_3. Nevertheless, Haber says that, when he began work in 1904, this
reaction *was considered impossible* because no one had succeeded in
detecting NH_3 in synthesis experiments despite the application of heat
and pressure, and despite the use of spongy platinum as catalyst. Haber
undertook crude experiments to find whether (9-1) was a true model. He
discovered that it *was easy* to show at 1000°C with iron as catalyst that
(9-1) exists: equilibrium could be approached via the forward or the
reverse reaction, and iron was not the only possible catalyst. This great
discovery came so easily that one must wonder why his predecessors
failed. Haber attributes their failures to *Zufall* (chance, accident—our old
friends), and there we shall have to leave the matter. Nor have we room
here for describing the nitrogen-fixation story between 1904 and 1909—
a pity, because the period was noteworthy for progress in chemical
thermodynamics and for personal conflicts [1, pp. 1046, 1048]. Such
conflicts often enliven research and sometimes advance knowledge.

By 1909 Haber's work was being sponsored by BASF,* which had
serious misgivings about the commercial usefulness of processes based
on (9-1). What was then known about this model predicted that the partial
pressure of NH_3 at equilibrium would be greatest in the gas mixture that
contained three molecules of H_2 for each molecule of N_2, and that this
partial pressure could be *increased* by *increasing* total pressure or by

---

* BASF, where Bosch had been concerned with ammonia since 1899, was then and is
today an important German chemical manufacturer. In the interim it became a component
of IG Farben and remained such until after World War II.

Haber did his ammonia work at the Technische Hochschule Karlsruhe. It is doubtful
whether this close and effective academic-industrial cooperation could then have been
matched in the United States or in the British Isles.

lowering temperature. So much for *equilibrium*; *rate* was another matter. A satisfactory rate—one high enough to be commercially useful—was more likely at *high* pressure and *high* temperature (not low), but, above all, it was more likely if H_2 and N_2 could be put into contact with an *active* solid catalyst.

Also, by 1909, Haber and Robert le Rossignol [1, p. 1047] had found that at 700°C and 200 atms* H_2 and N_2 rapidly formed NH_3 in contact with either iron or manganese. This temperature was still too high. Haber used the periodic table as a guide to elements that might be better catalysts than iron, and soon found that NH_3 formed readily in contact with either osmium or uranium below 600°C at 200 atms. "Readily" means a *space-time yield* [Chapter 6] of several grams of NH_3 per cubic centimeter of reaction space per hour. To convince BASF, Haber built a pilot plant for *continuous* operation during which the gases were *recycled* and liquid NH_3 was condensed by use of liquid air.

Haber's "big day" was July 2, 1909 [5]. On that day two BASF chemists,† Bosch and his assistant Alwin Mittasch, came to Karlsruhe to see ammonia made. Things went badly at first, and Bosch left. Mittasch stayed and was rewarded late in the day by seeing the first 100 g of NH_3 ever made from its elements [5, 7]. Mittasch had found a career that led to the 1929 Bunsen Medal for his work, soon begun under Bosch's direction, on catalysts for the synthesis of ammonia [3].

[July 2, 1909, was as important for ammonia as May 10, 1940, was to be for silicones. Each process requires a metallic catalyst. Adsorption and desorption of gases are important in both. For both, an electric-discharge method was investigated as alternative. Both were plagued by unexplained early failures. Each displaced an earlier process.]

[There were also revealing differences. Although both processes were continuous, NH_3 was removed either by condensation or by absorption in water during *recycling* of the gases; in the direct process the products left the reactor as vapors. Heat liberated during reaction was a boon to Haber-Bosch, for it could be used to heat incoming gases and thus lower energy costs; in the direct process liberated heat was mainly a nuisance. In the direct process solid silicon is consumed; (9-1) deals with gases only; for the direct process no comparable model is possible. Equilibrium Model 9-1 shows NH_3 as the single stable product, the

---

* Walther Nernst (Nobel Prize, 1920) and his students were the first to study (9-1) at high pressures.
† Mittasch [6] calls Bosch a chemist. Bosch in fact had broader training, mechanical engineering included [5, p. 123], but his prior work on nitrogen fixation at BASF had been largely chemical. The point is important because Bosch's implementation of Haber's work is a classic example of chemical engineering in its infancy.

concentration of which is calculable from thermodynamic data for the model. In the far more complex direct process, equilibrium considerations and thermodynamic calculations are not similarly helpful; here the stable end products would be undesired materials such as carbon, HCl, and $SiCl_4$.]

The unusual nature of (9-1)—a molecule of one gas reacting with three of another to give two of a third, with ideal-gas behavior often closely approached—gave it great importance in the development of chemical thermodynamics, which was the work of many illustrious men, among whom we mention only Henri Le Chatelier, Walther Nernst, and Haber. My experience suggests that students are likely to exaggerate the importance of chemical thermodynamics in the development of the Haber-Bosch process. I know that I did. I am now convinced that *catalysis, not thermodynamics*—especially not quantitative thermodynamics—was the heart of the matter. Given the experimental fact that NH_3 could not be completely decomposed into H_2 and N_2, the Haber-Bosch process could have been reached on the qualitative basis of (9-1)—although certainly less elegantly, and perhaps less speedily, than it was. But without an adequate catalyst, (9-1) could not have been verified, nor could tens of millions of tons of ammonia be made each year. Catalysis is as vital to Haber-Bosch as it is to the direct process for methylchlorosilanes.

Note also that NH_3 is removed commercially from a gas mixture *being recycled*. Obviously, a mixture with 9% NH_3 needs only a little more recycling to yield as much NH_3 as would be obtained had it contained 10% in the first place. The richer mixture is more desirable, of course; but accuracy in the thermodynamic data is of lowered importance in this situation: the making of NH_3 reproducibly at lowest cost is what counts.

Bosch and Mittasch. Discovery, Augmentation, Implementation.   With no delay BASF took hold. A small NH_3 plant went on stream at Ludwigshafen 3 years after Bosch and his associates performed "perhaps the most difficult and brilliant feat of chemical engineering ever achieved" [7, p. 1653]. Let us see what they had to do. Haber had given them *no industrially useful catalyst*. Only *experiment* could do that; there was no *theory* to help, and a *major discovery* was needed. In addition, they had to find a way of making adequately pure hydrogen at low cost. Third, they had to design and build a plant that could operate continuously and satisfactorily under conditions so demanding as to set a record for the time. We shall consider these three problems in sequence.

Experiment provided catalysts. Over a period of years, 3000 catalyst preparations were tried in 20,000 tests carried out in 24 laboratory furnaces operated night and day [6, p. 101]. And what resulted? At a grossly deceptive first glance, very little. Iron, used by Thénard in 1813 to decompose NH_3 [3, p. 570], remained the preferred metal in the "promoted catalysts" discovered by Bosch and Mittasch for the synthesis. But the *promoters,* usually two (e.g., Al_2O_3 and K_2O), added in small amount were found by Mittasch to make all the difference [3, p. 572], as they do today not only in the Haber-Bosch process, but elsewhere in modern chemical industry.

In thinking about catalysts, Bosch and Mittasch never got beyond the model stage; they never had a satisfactory theory of catalysis—all they had was "ein zutreffendes Bild" [2, pp. 127–128]. In making their catalysts, the promoters were mixed into the iron oxide, and the mixture was heated to "dissolve" the promoters, whereupon reduction by hydrogen gave iron particles so fine as usually to be pyrophoric (spontaneously igniting). The model assumed that these particles were kept from sintering (uniting) by smaller Al_2O_3 particles between them. Consequently, high catalytic activity and high surface area were created and preserved. A similar effect on the pyrophoricity of iron was observed by Magnus in 1825 [3, p. 575]. In addition to this physical effect, there are chemical effects also. Catalysts can be poisoned or improved [8, pp. 171–348 with 319 references]. Promoted catalysts are poorly understood even today.

[Mittasch's painful experience had its counterpart in the long, hard trail that began with copper (Combes, $SiHCl_3$, 1896) and ended with copper for methylchlorosilanes (Waterford, 1947). A promoter (Zn or $ZnCl_2$) was discovered en route, and much was learned about copper of various kinds. Models, free-radical mechanisms included, were only helpful guides.]

It takes energy to make hydrogen. Conditions in today's world emphasize why Bosch was concerned about the cost of hydrogen.

Haber, with typical insight, linked the success of his applied research to achieving a low "ratio of coal consumption [in the making of H_2] to [fixed] nitrogen." Today this hydrogen is often made from natural gas. When Dr. Armand Hammer bid for Libyan oil in 1966, he included "an agricultural development project and a joint ammonia-plant" [9]. The Haber-Bosch process is important in today's geopolitics, and the availability of ammonia helps determine how many people the world can support.

[The cost of energy and that of natural gas are of first importance also to Waterford. Much energy is needed to make a pound of silicon. The cost of natural gas is an important factor in the cost of CH_3Cl.]

Third, the Ludwigshafen ammonia plant set the pattern for modern chemical and petrochemical plants the world over.

[Resemblances between Ludwigshafen and Waterford are consequently not surprising. We mention only that both have explosion hazards* and that both from the beginning emphasized instrumentation and control, automated where possible and employing analytical chemistry to its maximum usefulness. Bosch's struggle with "hydrogen embrittlement" of steel, then a new, mysterious, and fatal corrosion hazard—a steel-equipment cancer [2]—deserves special mention.]

[For ammonia, as for the silicones, the research project, though it included discovery and considerable scientific augmentation, delivered a "package" unfit for large-scale implementation. Here silicones may have been the better situated because the early research and the implementation were both done within the corporation that marketed the products.]

Conclusion.   The great men associated with ammonia synthesis all deserve major credit for laying the foundation for today's chemical and petrochemical industries. Haber, Nernst, and Le Chatelier were equally at home in theory and experiment, in pure and applied research, in university and industry. Their influence on physical chemistry is comparable with that of Bosch on chemical engineering. Successful implementation of the ammonia synthesis encouraged BASF to go on to other high-pressure processes such as the synthesis of methanol and the hydrogenation of coal, achievements that made Friedrich Bergius Joint Nobel Laureate with Bosch and helped to start modern chemical engineering.† The contributions of Mittasch to catalysis are still bearing fruit. Its inherent importance, its benefit to mankind, and its ramifications give the ammonia story good claim to being the most interesting and significant in all of applied research on materials and processes.

The "true" ammonia story will probably never be told. To tell it, one would need the pertinent laboratory records, and would have to

* "At 7:30 A.M. in the morning mist of this cool day [September 21, 1921], an explosion lasting for seconds and having the force of an earthquake" demolished the mixing chamber for ammonium sulfate-nitrate, and the large silo in which this fertilizer was stored. The disaster claimed 561 lives. The quotation is from p. 94 of BASF, *In the Realm of Chemistry: Pictures from Past and Present,* Econ-Verlag, Vienna, 1965. Carl Bosch's memorable address on the disaster appears on the same page.

† This "ammonia effect" on BASF seems comparable to the "silicone effect" (increased commitment to chemical manufacturing) on the General Electric Company. The impending shortage of petroleum will force an increased use of high-pressure processes, such as the hydrogenation of coal, to make hydrocarbons and related materials. The "ammonia effect" on the world, already great, is bound to become greater.

know what the protagonists thought, not only about the scientific problems, but also of each other. The difficulties in describing the discovery
of oxygen or of the double helix are minuscule by comparison. Only a
scientific novelist could tell the ammonia story properly: doing that would
require an insight into the complex characters of the men involved and
a description of their personal conflicts—matters outside science, yet
essential to the understanding of scientific research, especially important
research that is successful: failures tend to be quietly buried.

MORE ABOUT WIRE ENAMEL

There is no discharge in that war.

Ecclesiastes 8:8

Wire enamels deserve a parting salute, appropriately given here. Although they now make up a much smaller fraction of the Company's
polymer production than in days gone by, their importance to the electrical industry continues undiminished.

The requirements for new wire enamels grow steadily stiffer: the
push for satisfactory operation at increasingly higher temperatures leads
to a more formidable wire-enamel property profile. The situation in 1954
is well described in a section of an important patent, drafted by Robert
S. Friedman and granted in 1960 to Frank M. Precopio and Daniel W.
Fox [10]. The section begins as follows:

It is well known that insulation which is to be employed for these purposes [slot
insulation and wire enamel] must be able to withstand extremes of mechanical,
chemical, electrical and thermal stresses. Thus, wires to be employed as coil
windings in electrical apparatus are generally assembled on automatic or semi-
automatic coil winding machines which, by their very nature, bend, twist, stretch
and compress the enameled wire in their operation.

The excellent treatment of the wire-enamel problem that follows this
quotation supports the earlier statement about the property profile demanded of a new wire enamel. These demands are more drastic than a
casual inspection of Table 1-1 reveals. For example, it is often not possible to upgrade one property without downgrading another. Changes in
composition often yield surprising results. Reliable models to serve as
research guides do not exist.

Formex® wire enamel (1938) has been the last great step forward in that art. After a "fluorocarbon failure"* silicones were again under study. By 1951 a blend of alkyd resins [Chapter 1] with silicones had made possible an enameled wire of improved mechanical properties but inadequate thermal stability [11]. A significant and deleterious product of thermal decomposition was the anhydride of "phthalic" acid

$$(9\text{-}2)$$

Orthophthalic acid
("Phthalic" acid)

Phthalic anhydride

one of the starting materials for making alkyd resins. Some of these resins were "oil modified"; that is, a vegetable oil, or an equivalent source of a fatty acid, had been used along with glycerine to make them.†

Meanwhile, the two isomers of "phthalic" acid, namely,

$$(9\text{-}3)$$

Metaphthalic acid
("Isophthalic" acid)

Paraphthalic acid
("Terephthalic" acid)

were becoming commercially available. Since neither isomer can form the (objectionable) anhydride, and since each has a higher melting point than "phthalic" acid, these new isomers were logical components of the complex formulations from which experimental wire enamels were then

---

* "Fluorocarbon" is a Humpty Dumptyish name; in many cases, "chlorofluorocarbon" would be better. The substance in question above was $(C_2F_3Cl)_n$, polychlorotrifluoroethylene, which of course contains chlorine as a third element. Quotation marks will alert the reader to other examples of Humpty Dumptyism.

† Glycerine, $C_3H_5(OH)_3$, having three active sites, promotes cross-linking, which leads to brittleness. As early as 1914 Callahan, Pittsfield Works Laboratory, and others had found that brittleness was reduced when some of these active sites were made to react with fatty acids, which are monobasic. Kienle [Chapter 1] used *unsaturated* fatty acids to make Glyptal® resins "air drying."

being synthesized. Frank M. Precopio was in the group doing the work; by 1953 the group was moving toward improved enamels low in "oil" content, that is, toward enamels progressively less "oil modified." As we are mainly interested in an important discovery on the road to Alkanex®, an improved polyester* wire enamel invented jointly by Precopio and Daniel W. Fox, we shall omit description of most of the work done by the group.

Let Fox tell how his memorable career in the Research Laboratory came about:

Fox was hired . . . in the Turbine Laboratory for the eventual role of insulation material consultant to assorted electrical equipment manufacturing operations. Flynn had established a relationship with Marshall which consisted of sending new PhD's out to the Knolls for 3–6 months for indoctrination. The key purpose was to [assure that] . . . the new men . . . would know whom to call when at a later time they needed technical back-up on consulting work.

When Fox joined the group [in the Chemistry Research Department, where he was closely associated with Precopio] (knowing absolutely nothing about polymers and magnet wire) Agens had already focused in on a combination of glycerine, glycol and terephthalic acid plus oil which gave *almost* everything needed [12].†

Fox, working with Precopio, was assigned a number of the formulations from which wire enamels were to be synthesized. Each synthesis took 8 hours or more.

Fox decided to save himself some time by making a master batch of glycerol-glycol-terephthalate which could then be equilibrated (catalytically) with assorted oils. The master batch (no oil) looked similar to the products with oil, so he submitted a sample for evaluation. The results were outstanding . . . [12].

Fox had done a great deal more than save his time. Oil was out! And Fox's "master batch" fortunately had contained acid, glycerine, and glycol in proportions that gave an improved wire enamel, and proved to be the basis for the Fox-Precopio Alkanex resin invention.

The first polyester wire enamels made without oil hydrolyzed objectionably when they were heated in a sealed system. During a lunch-hour

---

* "Polyester" is a logical, general classification under which both "alkyd resins" and "Glyptal®" are included.
† Note that [ethylene] glycol, $C_2H_4(OH)_2$, a well-known antifreeze, has only *two* active sites—hence tends to form polymer backbones, not skeletons.

discussion of this weakness, Donald E. Sargent asked this question:

"Wouldn't it be nice if we could start with a hydrolytically stable polymer and convert it into a wire enamel rather than trying to build hydrolytic stability into a finished product?" This was the trigger for Lexan® [12].

FROM WIRE ENAMEL TO LEXAN®

Thanks for the memory.

   Popular song

The Sargent trigger activated the Fox memory. While a postdoctoral fellow at the University of Oklahoma, Fox had tried to *hydrolyze* the carbonate ester of a phenolic compound, *and failed*. Chance helped him to this memorable failure. Fox wanted a substance not in the Oklahoma stockroom. Hydrolysis of a proper ester would give him what he needed. Fortunately (chance once more), the U.S. Army had declared surplus some drums of just such an ester and donated them to the University. Fox did his experiment and failed; the memory lingered.

   In Schenectady he wished to *make* a polycarbonate, because it might lead to a water-resistant wire enamel. Now it was the turn of a General Electric stockroom. Again (and fortunately), Fox could not find what he wanted, but "bisphenol-A"

$$\text{HO}-\!\!\overset{*}{\underset{}{\bigcirc}}\!\!-\!\!\underset{\underset{\text{CH}_3}{|}}{\overset{\overset{\text{CH}_3}{|}}{\text{C}}}\!\!-\!\!\overset{}{\underset{}{\bigcirc}}\!\!-\!\!\overset{*}{\text{OH}} \qquad (9\text{-}4)$$

2,2-Bis(4-hydroxyphenyl)propane

a compound used in making the well-known epoxy resin, was on hand.

   In (9-4) the two hydroxyl groups are "active sites" (hence asterisked) and open invitations to condensation polymerization with an appropriate substance. This process could start by linking two such molecules so that each contributes an oxygen to form the carbonate group in

$$\text{(Rest of Molecule 1)}\!-\!\text{O}\!-\!\overset{\overset{\text{O}}{\|}}{\text{C}}\!-\!\text{O}\!-\!\text{(Rest of Molecule 2)} \qquad (9\text{-}5)$$

and it could continue until the polycarbonate backbone represented by

$$\left(-O-\overset{\overset{\displaystyle O}{\|}}{C}-O-\text{⟨benzene⟩}-\overset{\overset{\displaystyle CH_3}{|}}{\underset{\underset{\displaystyle CH_3}{|}}{C}}-\text{⟨benzene⟩}-\right)_n \qquad (9\text{-}6)$$

(end groups omitted)

was complete. Today Lexan® is made by using phosgene, $COCl_2$, and the sodium salt of "bisphenol-A."

Fox proceeded in a different way. Let him tell what happened:

I . . . started making a polymer . . . and the melt became more and more viscous. Eventually, I could no longer stir it. The temperature had reached about 300°C, and I stopped at this point when the motor on the stirrer stalled. When the mass cooled down, I broke the glass off and ended up with a "mallet" made up of a semi-circular replica of the bottom of the flask with the stainless steel stirring rod sticking out of it. We kept it around the laboratory for several months as a curiosity and occasionally used it to drive nails. It was tough [13]!

Early in 1954 the Chemistry Research Department was visited by A. Pechukas, who then held a position in Pittsfield roughly comparable to that which Fox (Manager, Central Research, Plastics Business Division) holds there today. Pechukas, who knew the glass industry, tried out Fox's mallet and is said to have exclaimed, "My God, you've discovered unbreakable glass!" And so Fox had—that is to say, he had discovered Lexan®, a *transparent* polymer with *phenomenal impact strength*—and he had done it in one experiment! Was ever an early failure more effectively used? Thanks *to* the memory!

The full Lexan story, with its subsequent uncertainties, trials, tribulations, and difficulties, cannot be told here. In 1960 production began at an annual rate of 5 million lb, which has been multiplied many times since. The future is bright. Lexan windows cannot be broken. Banks love bullet-resistant Lexan sheet. Lexan 920 is a structural foam, less than 80 lb of which replaces over twice as much metal to give the new Jeep Model CJ-7 a better top [14].

Lexan has more to teach us. Its stability toward hydrolysis jibes with what Fox remembered. Its phenyl (C_6H_5) content makes for good thermal stability. Its transparency is welcome but not uncommon. So far we have an interesting polymer, but no champion. Championship status is conferred by the combination of transparency *and* high impact strength: upon being struck ever so sharply, Lexan bounces back unhurt. And

Lexan® is not even cross-linked! Why Lexan can behave thus, and other transparent polymers cannot, theory at present is powerless to explain. Polymer properties continue to be unpredictable: subtle and small changes can make large differences in the behavior of big molecules.

Several other items are worth treasuring. Chapter 1 shows that the Lexan experience had a distinguished precedent. Baekeland looked for a soluble resin and found Bakelite. Fox looked for a wire enamel and found "unbreakable glass." As with methyl silicones, as with the first oil-free master batch of wire enamel, first was best (or at least irreplaceable): though many other biphenolic compounds have been tried since Fox discovered Lexan, "bisphenol-A" continues to be the workhorse of Lexan production. Finally, there is an O. Henry twist.* On October 16, 1953, about 2 months before Fox did his first experiment, Bayer† applied for a basic patent on polycarbonate resins. When the General Electric Company, to which Bayer's activity was unknown, subsequently made application for a similar patent, an interference was declared. On April 3, 1962, U.S. Patent 3,028,365 was issued to Bayer. Did chance spice this story also?

A NEW METHOD OF POLYMERIZATION. PPO® AND NORYL® RESINS

The story begins in March, 1956, with a fire! [15]

In 1956 a fire in the Schenectady Works led the Company to sell the alkyd resin business, a setback for phthalic anhydride and the benzene ring. But the benzene ring, fresh from recent triumphs in Alkanex® and Lexan, rose like a phoenix from the ashes in a new kind of polymer, PPO® resins, made in a new way—by oxidative coupling, which is the linking of molecules through an appropriately catalyzed reaction of oxygen gas.‡ A useful new method of polymerization does not come every day!

When Allan S. Hay joined the Chemistry Research Department in 1955 after having graduated from the University of Illinois and having been an instructor in organic chemistry at the University of Alberta [13],

---

* O. Henry (William Sidney Porter, 1862–1910) was noted for the surprise endings of his short stories.
† Bayer, or Farbenfabrik Bayer Aktiengesellschaft, an important German chemical manufacturer, must not be confused with BASF or with von Baeyer.
‡ Such reactions are often called "autoxidations," a name originally assigned to oxidations taking place upon exposure to air.

he became interested in an ongoing investigation of the reactions between oxygen and hydrocarbons—pure research of top quality. Eventually the work acquired as one practical objective the achievement of a satisfactory commercial process by which xylenes (benzene rings with methyl groups replacing two hydrogens) could be oxidized to produce phthalic acids for the alkyd-resin business. When the business was sold, a satisfactory laboratory process had been found; but work thereon was stopped, and a search for new problems began.

Hay decided to see whether useful products could be made from oxygen and phenol (hydroxybenzene) by use of a catalyst (CuCl in pyridine), which had been described by others in 1955 for the oxidation of other materials. No one could know what to expect. Hay got a useless, dark brown, semisolid resin. He simplified the reaction by "blocking" two of four active sites on the benzene ring as shown (active sites are asterisked):

$$(9\text{-}7)$$

Phenol (hydroxybenzene)   2,6-Dimethylphenol (a hydroxyxylene)

Let Hay [15] describe the consequences of his simplification:

We were astonished to find that . . . the product, in high yield, was a light colored polymeric material which yielded a tough, flexible film when cast from solution. . . . The structure . . . we soon determined to be the aromatic polyether

$$(9\text{-}8)$$

Hay, as a culmination of extensive work by him and others, which enlarged and refined what was known about the oxidation of hydrocarbons, had discovered a new method of polymerization (a rare event indeed!) and an interesting new polymer, PPO®. In sharp contrast to

Kipping's distaste for the "sticky masses" of silicones, Hay quickly set out to establish whether his new polymer had properties interesting enough to warrant further work. Note the order: discovery by experiment, first; determination of properties, second. Properties could not be predicted.

The new polymer proved to be an extremely tough, durable material with an exceptionally high glass temperature* of 208°C. Surgical instruments made of PPO® could be sterilized with impunity in an operating-room autoclave. There were other promising applications. Most polymers, however, must withstand oxidation by air if they are to be used extensively enough to warrant large-scale production, and PPO could not meet this requirement. Unlike the CH_3 groups in silicones, the CH_3 groups in PPO were oxidized so readily that the high-temperature oxidative stability of the phenyl groups was wasted. In addition, there were serious processing difficulties traceable to the high glass temperature of PPO: working above 200°C on a plant scale is not easy.

Many, many, other polymers were made by oxidative coupling. To date, however, none has displaced PPO. General Electric luck held: for the third time a polymer discovered very early has proved so far to be the best in its class.

The salvation of PPO came via product development [15]. Fortunately, PPO has proved to be compatible with a number of other polymers. One of these is polystyrene† (polymerized vinyl benzene—that marvelous benzene ring once again), a hydrocarbon; blending the two gave Noryl® resin, an engineering polymer‡ in which polystyrene is advantageously upgraded at a cost the market can bear. In baseball language, Lexan® resembles a home-run king because of its high impact strength, while Noryl is a versatile utility player with acceptable skills across the board. Thus Noryl has good impact strength (but below that of Lexan) and dimensional stability, is flame retardant, and saves space and weight—all this at acceptable cost.

---

* At the glass temperature (more precisely, the "glass-transition temperature") a noncrystalline polymer changes from a brittle, glassy to a flexible, rubbery state, or vice versa. This temperature is not as sharply defined as a true melting point.

† Others [15] are our old friends polydimethylsiloxane and Lexan. As these contain oxygen, their compatibility with PPO is less surprising than that of polystyrene. Oil and water are not supposed to mix!

‡ The term "engineering polymer" (or "plastic") is gaining acceptance to describe polymers, usually manufactured in quantity, that have a desirable combination of properties (a good "property profile"; see Table 1-1), which engineers can expect them to maintain practically unchanged during adequately long life under service conditions. These polymers stand in contrast to "specialty products" such as silicones. Lexan too is an engineering polymer.

Neither the Lexan® nor the PPO® nor the Noryl® story is told here in full. Mr. John F. Welch, Jr., now a Vice President and Group Executive of the Company, made the decision to go ahead with Noryl. As with silicones, a commercial process for making the starting material had to be found for PPO [15]. The success of silicones helped the later polymers along. Subsequent new polymers became easier to champion—all the more because men with Waterford experience moved into the upper levels of management.

A COMPARATIVE SUMMARY

Table 9-1 compares apples and oranges, each of which merits a book of its own. The table will have earned its keep if, taken with what has gone before, it gives the reader a better concept of industrial research in chemistry.

The most important thread running through the table is the overriding importance of experiments. Also, there is the theme "time and chance happeneth to them all." Surely, chance is by now familiar to the reader. Time, meaning the probability of 10-year survival, needs qualification, as the footnotes to the table show. Synthetic ammonia is surest of survival; people and animals must eat, and compounds of nitrogen will always be needed. Yet Haber himself [1, p. 1049] mentioned that the Haber-Bosch process might someday be displaced by bacterial processes. With rapid recent progress in molecular biology and "genetic engineering," that day seems nearer at hand. Table 9-1 suggests also that newer projects contribute less to knowledge because the pertinent prior knowledge has become so enlarged and refined as to require less scientific augmentation. Though the table does not show it, one is safe in assuming that each project had its own significant web of personal relationships, the case of ammonia being best known because of the fame of the men— notably Haber, Bosch, Nernst, and Le Chatelier—who were involved; undeservedly, Mittasch is an almost unsung hero today [1, 2, 3, 5, 16, 17, 18, 19, 20].

In its essential simplicity, probable inevitability, and likely permanence, the ammonia synthesis resembles ductilized tungsten, Man-Made™ diamonds, and silicon for semiconductors. We have here a simple compound and three elements. The products following ammonia in Table 9-1 are different in kind. They are more complex, and their future is less certain; among them, silicones show the greatest diversity and have a broad chemistry all their own. This blend of examples gives a hint of the variety of industrial research in chemistry. There is more to come.

Table 9-1 Subjective Overview of Successful Research Projects

Project	Contribution to Knowledge	Importance of Product(s)	Initial Research Objective(s)	Objective Attained	Major Discoveries	Numbers of Products	Annual Production[a]	Probability of 10-Year Survival
Ammonia synthesis	Outstanding	Outstanding	NH$_3$ manufacture	Yes	Few	Few	Enormous	Virtual certainty
Silicones	Outstanding	High	High-temperature resin	Yes, gradually	Many[b]	Huge	Small	High
Alkanex®	Low	High	Wire enamel	Yes	Few	Few	Small (of polymer)	Moderate
Lexan®	Moderate	High	Wire enamel	No	Few	Few	Moderate	High[c]
PPO®	High	High[d]	New polymer	Yes	Few	Few	Moderate	High[e]
Noryl®	Moderate	High	Engineering polymer	Yes	Few	Few	Moderate	High[e]

[a] Most products are manufactured internationally.

[b] "Many," because the silicone "planned high road" had several forks opening on side roads that led to products of greater importance than the initial objective, which was a resin. Comparable side roads did not exist for the other projects in this table.

[c] Survival of Lexan depends on the combination of high impact strength and transparency, so far unique among polymers.

[d] "High" because of its use in Noryl.

[e] Oxidative coupling has permanent value as a new method of polymerization. PPO may find its way into materials other than Noryl. Noryl will have to compete with future engineering polymers.

OTHER DISCOVERIES VIA INDUSTRIAL RESEARCH

Most major discoveries via industrial research have unusual features that make them interesting. We shall examine an additional few.

From 1914 onward, there has been extensive work on alkyd resins. Yet it was not until 1941 that Whinfield and Dickson [21] found that terephthalic acid and ethylene glycol, each closely related to a starting material for the alkyds, react to give useful resins, from which today's Dacron®, Mylar®, and Terylene® are descended. Why should their important discovery have been made under the auspices of the Calico Printers Association, Ltd., and not by one of the big manufacturers of polymers? Why did this discovery not follow more closely upon the work of Carothers?

Let us turn now to Thomas M. Midgley, Jr. [22], a mechanical engineer employed by General Motors.* Midgley set chemists two examples in the effective use of the periodic table. Let us look at the first. In finding an additive that could control knocking in an internal combustion engine, Midgley began as Edison might have done. Intelligent exploration based on prior knowledge led Midgley to six conclusions that then made it possible to begin a "correlational procedure based on the periodic table," and to change a frustrating search, costly and prolonged, into a well-directed "fox hunt." His additive to gasoline, $Pb(C_2H_5)_4$, is now under indictment as a pollutant and is being "phased out" for this reason and others.

The periodic table also helped Midgley find a new refrigerant and open the door to the widespread manufacture and use of fluorocarbons†, which was his second great achievement. Along the road to Freon®, Midgley was prodded by a listed (and incorrect) boiling point for CF_4,

---

* Midgley's work is of added interest because of the brief comparison [Chapter 7] of General Motors with General Electric, and because he—a mechanical engineer—was awarded the Perkin Medal for progress in applied chemistry.

In his youth Midgley disagreed with his chemistry instructor's expressed belief that the periodic table was "evidence of the existence of the Deity." The ensuing argument went on "for days and weeks," and "I [Midgley] had . . . the periodic table . . . impressed upon my memory as a very useful tool in research work" [22].

† Fluorocarbons rival silicones in interest and diversity and are, like silicones, Humpty-Dumptyishly named; they often contain elements other than carbon and fluorine. Midgley's second great discovery led to the use of fluorocarbons as propellant gases in spray cans containing aerosols, an application he did not foresee. In 1976 this discovery was, like his first, under a cloud: the emitted fluorocarbons were suspected of depleting the ozone layer that exists above the earth at an altitude airplanes cannot reach. There is more on this subject in a later footnote.

This turn of events is particularly ironic in light of Midgley's concern about toxicity.

but the real excitement was still to come. To make $CHFCl_2$ (the fluoro-carbon selected for the first trial), Midgley ordered and received five 1-oz bottles of SbF_5 (believed to be the entire U.S. supply) for use as a catalyst. Everyone but Midgley expected that $CHFCl_2$ and related compounds would be too toxic for general use because they were fluorine compounds. The crucial toxicity tests resembled Russian roulette with *one chamber empty*! In Midgley's words:

One [bottle] was taken at random, and a few grams of dichloromonofluorome-thane were prepared. A guinea pig was placed under a bell jar with it and, much to the surprise of the physician in charge, didn't suddenly gasp and die. In fact it wasn't even irritated. Our predictions were fulfilled. We then took another bottle and made a few more grams and tried it again. This time the animal did what the physician expected. We repeated again but this time we smelled the material first. The answer was phosgene; a simple caustic wash was all that was needed to make it perfectly safe. Then we examined the two remaining bottles of antimony trifluoride. They were not pure. In fact, they were both badly contaminated with a double salt containing water of crystallization. This makes phosgene in ample quantities as an impurity. Of five bottles marked "antimony fluoride," one had really contained good material. We had chosen that one by accident for our first trial. Had we chosen any one of the other four, the animal would have died as expected by everyone else in the world except ourselves. I believe we would have given up what would then have seemed a "bum hunch."

And the moral of this last little story is simply this: You must be lucky as well as have good associates and assistants to succeed in this world of applied chemistry sufficiently well to receive the Perkin Medal [22].

Now a "Midgley sequel" is pertinent. In 1938 Roy J. Plunkett of Du Pont was attempting to discover improved fluorocarbon refrigerants, and he had prepared and stored in "dry ice" several cylinders of C_2F_4 to be used in his work. On April 6, 1938, his assistant noticed that one of the cylinders had lost very little weight though all the gas in it had been discharged. The reason was that almost all of the $F_2C=CF_2$ in that cylinder had polymerized to form $(—C_2F_4—)_n$, now known as Teflon®! The present surmised explanation is that polymerization must have resulted from the unintended presence of oxygen in trace amounts [23].

Plunkett's assistant had a noteworthy Du Pont predecessor, namely, Julian W. Hill, who worked with Carothers. Chance gave Hill a helping hand toward a major discovery that not only led to Nylon®, so far Du Pont's greatest research achievement, but also opened the door to other *completely* man-made (synthesis included) fibers, of which nylon* was the first.

---

* Alas, Nylon® is no more. It has been replaced by "nylon,"—now in the general vocabulary and common property. The overwhelming popularity of nylon stockings made it impossible to protect the trademark.

Carothers, as we have seen in Chapter 2, did masterly pure research on polymers. He began by studying linear polyesters, the backbones of which were formed by reacting dibasic acids and glycols as illustrated by (2-4). After Hill's discovery Carothers moved into applied research. Here is what brought this about:

Well, one day [in April 1930] one of Carothers' associates [Hill] was cleaning out a reaction vessel in which he had been making one of those polymers, and he discovered in pulling a stirring rod out of the reaction vessel that he pulled out a fiber; and he discovered its unusual flexibility, strength, and the remarkable ability of these polymers to cold draw.*[(a)]

* The sources of this too short story of nylon appear in (a), (b), and (c) below.

(a)   This quotation appears on p. 335 of Willard F. Mueller, "The Origins of Basic Inventions Underlying Du Pont's Major Product and Process Innovations," in *The Rate and Direction of Inventive Activity: Economic and Social Factors* (A Conference), Princeton University Press, Princeton, New Jersey, 1962, pp. 323–358, comments included.

The quotation is from testimony by Crawford H. Greenewalt in a complex case dealing with alleged conspiracy in restraint of trade in the chemical industry; see 100 Fed. Supp. 504 (1951) for the 90-page opinion, which is valuable as a treatise on that industry.

Greenewalt, a chemical engineer, was President of Du Pont when he testified. During a distinguished career he found time to become an authority on humming birds.

Mueller in his index (pp. 627–635) does not list either silicones or the General Electric Company. On p. 332 Mueller records of Du Pont Dulux® enamels "that the basic resin [a polyester] . . . was discovered by General Electric scientists. Du Pont acquired the rights to General Electric's discovery. . . ." Knowledge of these facts probably strengthened the determination of Marshall and Patnode [Chapter 1] to make General Electric an important manufacturer of chemicals.

(b)   E. K. Bolton, *Ind. Eng. Chem.*, **34**, 53 (1942). The address, "Development of Nylon," was delivered by Bolton, then Chemical Director of Du Pont, upon receiving the 1941 Chemistry Industry Medal.

(c)   William S. Dutton, *Du Pont: One Hundred and Forty Years*, Charles Scribner's Sons, New York, 1949, pp. 354–356, identifies Dr. Julian W. Hill.

Carothers' greatest contribution to polymer knowledge seems to have been improved understanding via scientific augmentation: he brought order out of chaos in masterly investigations that ought to have earned him the Nobel Prize had his tragic early death not foreclosed this possibility.

Dutton mentions that on April 16, 1930, Hill made a [polyester] superpolymer in a molecular still, a device presumably introduced by Carothers as part of his research program. "About two weeks later," Hill discovered cold drawing. What if he had not used a stirring rod in the way Greenewalt testified he did? How much longer would Du Pont and the world have had to wait for this chance macromolecular manipulation, of which Hill was quick to see the significance?

The unfolding of the nylon story as revealed by laboratory notebooks, especially those of Carothers and Hill, would make an invaluable addition to the history of industrial research.

The molten polymer out of which Hill so fortunately pulled his stirring rod was a polyester and commercially useless. His discovery was not primarily that the polymer could be "cold drawn" (made into a fiber without heating), but that the fiber so produced had (after reaching room temperature) properties desirable in fibers, which meant that man could hope to outdo the silkworm. The scientific basis for the change in properties was this: cold drawing produced a fiber in which most polymer crystals were lined up parallel to the fiber axis; orientation had been random before. The parallel orientation resembles that in ductilized tungsten well enough so that nylon fibers and tungsten filaments can be represented by roughly similar models.

When Hill made his discovery, nylon *in the laboratory* was still almost 5 years away. Nylon is a *polyamide*, not a *polyester*, and nylon "66" is made from these compounds

$$
\begin{array}{c}
\text{H} \quad \text{H} \quad \text{H} \quad \text{H} \quad \text{H} \quad \text{H} \\
| \quad | \quad | \quad | \quad | \quad | \\
\text{H}_2\text{N—C—C—C—C—C—C—C—NH}_2 \\
| \quad | \quad | \quad | \quad | \quad | \\
\text{H} \quad \text{H} \quad \text{H} \quad \text{H} \quad \text{H} \quad \text{H}
\end{array}
$$

Hexamethylenediamine

and

$$
\begin{array}{c}
\text{O} \quad \text{H} \quad \text{H} \quad \text{H} \quad \text{H} \quad \text{O} \\
\| \quad | \quad | \quad | \quad | \quad \| \\
\text{HO—C—C—C—C—C—C—OH} \\
| \quad | \quad | \quad | \\
\text{H} \quad \text{H} \quad \text{H} \quad \text{H}
\end{array}
$$

Adipic acid

in a reaction that resembles (2-4) with NH_2 as an active site in place of the OH on the glycol. The nylon story, one of the most important in American industrial research, has been authoritatively told by Elmer K. Bolton*[b], who said in part:

In the earlier work Carothers *had not been successful* in preparing a superpolymer [defined by Carothers as a polymer having a molecular weight above 10,000] having the necessary properties for [a commercial textile fiber]. Other research work on cellulose derivatives to obtain a distinctly new fiber had reached a stage where little hope was held of attaining the objective. A large amount of time and money had been spent on the various phases of the fiber program, but the results were *chiefly of theoretical interest*. It was therefore a *matter of considerable concern* to determine what *practical use* could be made of the scientific information that had been acquired on superpolymers. The possibility of laying the foundation for a new commercial fiber development *appeared to be remote*, with the result that research work in this field was *discontinued* for a number of months.

Carothers, however, was *encouraged to direct his work* on superpolymers *specifically* toward the development of a product which could be spun into practicable fibers. On the basis of a survey of his scientific work in this field, he *wisely decided* to resume work on the superpolyamides. . . .

Following [certain encouraging] observations, Carothers prepared polyamides from a variety of amino acids and also from dibasic acids and diamines. On February 28, 1935 (a now historic date), the superpolymer from hexamethylenediamine and adipic acid was first synthesized. The resulting polymer, polyhexamethylene adipamide, was called ''66,'' the first digit indicating the number of carbon atoms in the diamine and the second digit the number of carbon atoms in the dibasic acid'' [italics mine].

In the annals of industrial research, February 28, 1935, ranks with July 2, 1909, for synthetic ammonia, and with May 10, 1940, for silicones. However, Hill, chance, and April 1930 must not be forgotten. On that fateful day in spring, the beginning of the synthetic fiber industry hung by a thread from Hill's stirring rod.

CONCLUSIONS

With Mr. Justice Holmes [Introduction] as mentor, we shall draw conclusions and make generalizations about industrial research even though our sampling of projects has been limited.

Discoveries and Chance. Even the ''major discoveries'' (see Table 4-1) of industrial research are not all worth augmenting and implementing.* Those that are, we shall henceforth call Discoveries (capital D).

Industrial research prospers when Discoveries are frequent. How Discoveries are made is difficult to establish because one cannot look with certainty into the discoverer's mind. In addition, Discoveries are often attended by uncontrollable, unplanned, and lucky circumstances that we shall include under chance, a factor vital in some cases, unimportant in others. Let us turn to silicones for illustrative examples.

---

* Edwin Newman would approve restricting the use of ''major.'' See his *A Civil Tongue,* Bobbs-Merrill, New York, 1975–1976, especially pp. 43–47.

Why did Rochow on April 30, 1940, decide to mix CH_3Cl into the HCl gas he had been using to make CH_3Cl? To see what might happen? Because he envisioned a direct process for methylchlorosilanes? As a logical consequence of prior knowledge? In any case he had a sound reason for his experiment; chance at this point is ruled out. Nor did chance move Patnode to mix D_4 with concentrated H_2SO_4 and lead him eventually to a model for rearrangement polymerization and to the "tailoring" of silicone oils. Agens discovered silicone rubber because he had a vision of "flexible mica." Gilbert Wright arrived at $FeCl_3$ as a rearrangement catalyst in the making of silicone gums by imagining that $FeCl_3$ and the methylchlorosilanes were "fighting for water"—I suspect he saw two valiant knights jousting for the hand of an aqueous fair lady. As antecedents to the making of these Discoveries, models, visions, hunches, the "subconscious" (Patnode), and the like were important; chance was negligible.

Suppose, however, that the cheapest satisfactory source of silicon available to Combes in the 1890s had contained no copper; silicone history would not have been the same. Suppose that Corning Glass Works had not been ready in the 1930s to make glass tape an innovation; how much longer would Kipping's "sticky masses" have lain fallow? These are two examples of chance at work—away from the laboratory bench.

What we describe as "first was best (or nearly so)" also influences research. In the silicone project the expression applies to copper as catalyst,* and to the CH_3 and C_6H_5 groups as silicone components; it is applicable to the first "master batch" of Alkanex®, to "bisphenol-A" in the making of Lexan®, and to 2,6-dimethylphenol as a raw material for PPO®. Furthermore:

It is a tribute to the vision of the earlier developers of polyamides, such as Carothers, in selection of nylon-6,6, and Schlack, in selection of nylon-6, that twenty-six years later these two polyamides represent more than 98% of the total polyamide volume being converted to fibers. As a practical matter, the two polyamides are essentially interchangeable in most applications [24].

Vision, chance, prior knowledge, or how much of each? No two cases are the same, and the simplest would baffle Solomon. The nylon story, told earlier, is an illustrative example.

"First was best (or nearly so)" is desirable for many reasons, not

---

* Copper as catalyst is an interesting parallel to the chance discovery that thorium prevents "offsetting" in tungsten filaments (the "Battersea incident") and the consequent discovery ("first was best") of the thoriated-tungsten electron emitter. See Reference 30, pp. 10–11.

the least important of which is taught us by Paul Ehrlich (1908 Nobel Prize shared with Élie Metchnikoff). The Discovery of salvarsan ("606") climaxed a series that had included 605 prior experiments by Ehrlich. Premature abandonment is a risk that a project must run if it is not blessed by early success or increasing promise.

With help from *Porgy and Bess,* we summarize by saying, "A Discovery is a sometime thing."

Theories, Experiments, Models. *   Max Planck, in commemorating the great discovery that crystals give X-ray diffraction patterns, said (freely translated):

Theory and experiment, they belong together . . . Theories without experiments are empty; experiments without theories are blind. Therefore both theory and experiment demand with equal emphasis the respect to which they are entitled [25].

Although the discovery was inspired by Max von Laue's "flash of genius" and admirably supports the Planck quotation, we must remember that chance presided over the discovery of X-rays by Roentgen. Along with chance, models [26] also deserve consideration in an attempt to understand research.

Of models, we have had a great plenty: chemical bonds, the tetrahedral carbon atom, the benzene "ring," polymer backbones and skeletons, and, though less obvious, mechanisms of reactions, and equilibria such as (9-1). Models must appear first as mental constructs;† later they may be represented on paper, built in three-dimensional space, or fed into a computer.

Models, even wrong models, are more useful than is generally recognized. Their importance to industrial research is evident from the role played by the Bosch-Mittasch ammonia catalyst: here the model guided experimentation *when understanding was lacking,* one of the most important functions a model can perform. With the model to go on; Bosch and Mittasch made and tested predictions; by thus enlarging and refining

---

* Another treatment of material from this chapter shows this section in its relation to the important book *The Coming of the Post-Industrial Society,* Basic Books, New York, 1973, by Daniel Bell. See G. Wise and H. A. Liebhafsky, *J. Chem. Educ.,* late 1978.

† Kekulé was one of the world's most influential model envisioners. Chemical bonds and the benzene ring came to him while he was not fully conscious. (Was Newton fully awake when the apple struck him, if strike him it did?) Kekulé's experiences and the celebration to which they led [27] are memorable enough to warrant space in References and Notes (pp. 367–368).

knowledge, they cleared the way for theories of catalysis. A "fibrous" model of ductile tungsten did similar service for Coolidge.

The double-helix model of DNA, life's most important polymer, will always be one of the most significant in all science. Wrong models guided Crick, Wilkins, and Watson to the structure of DNA, and Miss Franklin's failure to make models in the course of her X-ray diffraction work put her under a heavy handicap.* For DNA the final model crowned the work.

The Progress of Knowledge. Research well done, whether pure or applied, always *enlarges* and usually *refines* knowledge. We shall use silicone research as a principal example to illustrate this thesis for a single field.

The *enlarging* of knowledge is far simpler than its *refining,* for refining requires that models and theories act in someone's mind upon experimental observations. Ideally in applied research, a *need* to be filled leads to a survey of *prior knowledge,* and then to the design and performance of an *experiment* that results in a *Discovery.* Experimental observations, whenever made, increase the quantity of knowledge and ought—after someone has thought about them—to lead to an improvement in its quality. We shall regard prior knowledge as a heritage formed by mental interactions of prior models, prior theories, and prior experimental observations, the last of these forming the bulk of descriptive chemistry. When an investigator, acting presumably upon the basis of prior knowledge, makes new observations, these ought to go through the same kind of mental mill to form a new heritage of prior knowledge for future use.

In pure research the state of affairs is much the same except that the original impetus to experiment is more likely to be, not a need, but curiosity, mental irritation, desire for recognition, or what-have-you.

The belief that experimental observations are the raw material for the progress of knowledge goes back, in embryonic form at least, to Robert Grosseteste (1168–1253) and his pupil Roger Bacon (1220–ca. 1292). Practice in accord with this belief is probably much older.

In Figure 9-1 we present silicone chemistry as a model for the

---

* The history of this discovery shows the desirability of documentation via original laboratory records. A balanced view requires the reading of three books: (1) James D. Watson, *The Double Helix: Being a Personal Account of the Structure of DNA,* Atheneum, New York, 1968; (2) Anne Sayre, *Rosalind Franklin & DNA,* W. W. Norton & Co., New York, 1975; (3) Robert Olby, *The Path to the Double Helix,* University of Washington Press, Seattle, 1974, especially the foreword by Crick, and pp. 438 and 439.

The 1962 Nobel Prize in Medicine and Physiology was shared by Francis H. C. Crick, James D. Watson, and Maurice H. F. Wilkins. Miss Franklin died in 1958.

progress of knowledge through research. We imagine the knowledge in this field, present and past, to be contained in three spheres that become larger and brighter as the pertinent descriptive chemistry grows and is refined. Ultimately, in fields where the human mind is capable of understanding and reconciling all experimental observations, there would be a dazzlingly white sphere, too large for the figure, with the characteristics listed under Stage 4.

Stage 1 represents what Berzelius knew when he began the work that made silicone chemistry possible. He knew that SiO_2 is not an element and that it reacts with HF. With little more than this to go on, he discovered silicon and $SiCl_4$. Furthermore, he introduced the concept of a polymer. In other words, he enlarged the sphere of Stage 1 and made it brighter.

By 1938, near the close of Kipping's activity but before the silicone project began, the descriptive chemistry applicable to silicones had been further enlarged and greatly refined: models such as chemical bonds, the tetrahedral silicon atom, and linear and cyclic polymer structures; and theories, notably the tetrahedral habit in silicone chemistry, the analogy between silicones and silicates, and the theory of optical isomerism—all these and more were now available; a too facile theory, that carbon and silicon necessarily exhibit similar chemical behavior, had been corrected. Consequently, we shall suppose that Kipping left silicone chemistry at Stage 2. The change to 1976 has been described. Chapter 8 showed the extent to which the enlarging and the refining of silicone chemistry are progressing, with theory becoming steadily more useful. One cannot predict when, if ever, we shall reach the "All is light" stage. One must have faith.

Figure 9-1, admittedly naive, is a convenient representation of a situation in which, sooner or later, industrial research in all fields will find itself. As experimental observations in a field accumulate, the chance of further Discoveries in that field decreases. If, as in the figure, predictive reliability concomitantly increases, "That's all there [are]; there [aren't] any more" may eventually be true of Discoveries. The consequences for industrial research are profound.

The silicone story is well told by Figure 9-1. Note that a decreasing frequency of Discoveries as the story unfolded did not lead to a "profitless prosperity." Nothing in the figure requires that innovations and Discoveries walk hand in hand: innovations can grow as prior knowledge increases. The spheres in Figure 9-1 will interact with similar spheres in other fields, in each of which knowledge is growing at a different rate; such interactions will cause innovations to multiply. Dow Corning's 2500 products do not connote 2500 Discoveries in silicon or silicone chemistry.

We have put today's investigator of silicone chemistry in Stage 3,

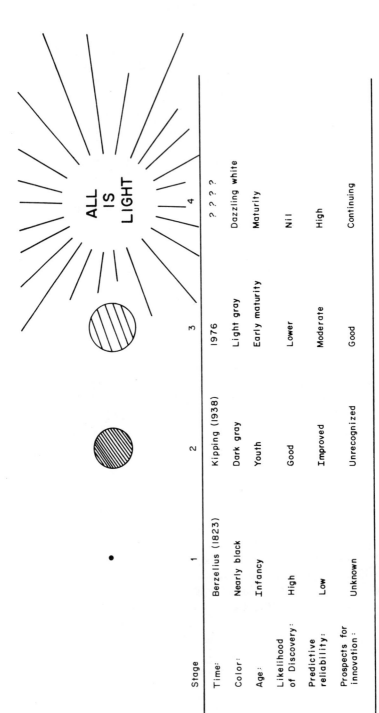

Stage	1	2	3	4
Time:	Berzelius (1823)	Kipping (1938)	1976	? ? ? ?
Color:	Nearly black	Dark gray	Light gray	Dazzling white
Age:	Infancy	Youth	Early maturity	Maturity
Likelihood of Discovery:	High	Good	Lower	Nil
Predictive reliability:	Low	Improved	Moderate	High
Prospects for innovation:	Unknown	Unrecognized	Good	Continuing

Figure 9-1 Silicone chemistry used to illustrate the enlargement and refinement of knowledge. Note that the unknown is regarded here as a scientific resource that may eventually become exhausted. As predictive reliability increases, the likelihood of Discoveries decreases, and they grow more costly; the influence of chance declines. Then research increasingly assumes the character of scientific augmentation, an activity of which no man can see the end. Stage 4, modeled upon the Pope quotation [Chapter 8], may never be reached in a given science; there may well be limits to human understanding. Industrial research will have greater need for tighter management and for planning as the knowledge it uses becomes enlarged and refined.

and he commands costly modern research tools. He is therefore in a better position than his predecessors to identify unpromising lines of research sooner and to abandon them earlier with less risk that abandonment is premature. At the same time he is likely to be tempted strongly to scientific augmentation. Self-restraint, good judgment, more planning, and tighter control by management are needed. A likely result is that, because of the need for more managers and more services (scientific augmentation, planning, and "paper work" included), the large industrial laboratory will allocate less and less of its total budget to searching for Discoveries at the laboratory bench. Berzelius, whether or not he elected industrial research, might not stand out so prominently today.

As the progress of knowledge continues according to Figure 9-1, the research manager must increasingly try to combat frustration. Enthusiasm and curiosity are priceless attributes of research workers. In Christopher Marlowe's words,

Yet should there hover in their restless heads, One thought, one grace, one wonder at the least.

For us, "restless" and "wonder" are the most important of these words. Will they apply to research men and women who must work in fields where predictive reliability is so great as to make them almost sure of what lies over the next hill? Such investigators will rarely experience the thrill of making Discoveries; they will have to be content with the quieter satisfactions of improving understanding and achieving innovations. They are as essential to our times as Berzelius was to his.

Edison and Figure 9-1.   The widespread association of Thomas A. Edison with rank empiricism is an injustice to a great man. To be sure, Edison did not found a research laboratory—rather an "inventive facility," as is indicated by a footnote in Chapter 1. But he did proceed in a way—often called "the Edisonian method"—that is of profound importance to industrial research.

What, exactly, is this Edisonian method, which he no doubt did not originate? Let him tell us in two revealing quotations:

When I want to discover something, I begin by reading up everything that has been done along that line in the past. . . . I gather the data of many thousands of experiments, and then I make [if necessary] thousands more [28, p. 113].

And, in describing his search for a better incandescent-lamp filament:

I've tried everything. I have not failed. I've just found 10,000 things that won't work [29, p. 47].

Edison thus took full advantage of prior knowledge. He did not intentionally work in the manner of Figure 9-1; though he inevitably *enlarged* knowledge, he had no interest in *refining it*. He shunned augmentation and pure research. Patents were his publications.

The Edisonian method would not have yielded tungsten ductile at room temperature. Coolidge, like Edison, took full advantage of prior knowledge, which had grown since Edison's time; and Coolidge settled upon tungsten, not carbon. Here similarity ends. Coolidge no longer needed to try "6000 different species." Prior knowledge about metals being inadequate and misleading, he was forced into applied research to ductilize tungsten; in succeeding, he enlarged and refined our knowledge, not only of that element, but of metallurgy in general [30].

But we must not dismiss the Edisonian method lightly. It becomes more powerful as prior knowledge increases—as Stage 4 of Figure 9-1 is approached. Earlier in this book, *Gmelin's Handbuch der Anorganischen Chemie* was facetiously presented as a fine-print storehouse that offers some "200,000 cc" of descriptive chemistry. *Beilstein,* the organic chemist's counterpart, is comparable. Given Edison's imagination and tenacity, how many inventions might he not have made by using modern prior knowledge as a springboard? No doubt he would have found it difficult, as do others in our time, to "begin by reading up everything . . . along that line" when "everything" in many fields has grown so huge.

INDUSTRIAL RESEARCH—WINDS OF CHANGE*

When the silicone project began in 1938, the "Whitney atmosphere" still pervaded the Research Laboratory. Two revealing letters by Whitney complete the picture of early industrial research, which was begun in Chapter 1. That picture serves as a bench mark for surveying the changes that industrial research has undergone in the last six decades.

The first letter deals with the "methods" followed by the Research Laboratory, and the second with the rewards and risks of industrial research. Mr. Little's name remains well known because of the prominent

---

* The title of the first in a series of books by Harold Macmillan, Harper and Row, New York, publishers. The winds of which he speaks are those that changed the world during the period 1914–1939. Industrial research did not escape. Note the dates of the two Whitney letters. These letters are in the Willis R. Whitney Collection, now on loan to the Schenectady Archives of Science and Technology, Union College, Schenectady, New York.

consulting firm he founded, and Dr. Herty was Advisor to the Chemical Foundation when Dr. Whitney wrote to him.

July 16, 1913

Mr. A. D. Little
President Amer. Chem. Society
Boston, Mass.

...

We see a field where it seems as tho[ugh] experimental work ought to put us ahead. We believe that we need to get into the water to learn to swim, so we go in. We start back at the academic end as far as possible, and count on knowing what to do with what we find when we find it. Suppose that we surmise that, in general, combustible insulation material could be improved upon. We try to get some work started on an artificial mica. Maybe we try to synthesize it, and soon come to a purely theoretical question; for example: Is it possible to crystallize such stuff under pressure in equilibrium with water vapor corresponding to the composition of real mica? This may lead a long way and call in a lot of pure chemistry and physical chemistry. Usually, we just keep at it, so that if you haven't seen it on the market we're probably at it yet.

...

Your inquisition on appropriations is painful, but you are safe in saying that we cost $100,000 [sic!] or more per year.

...

P.S. Of course any reference to this Research Laboratory ought to touch Dr. Coolidge's great work with ductile tungsten. As a matter of fact, tungsten lamps everywhere on earth are now made by his process.

W. R. W.

This letter was annotated "Approved by Mr. [A. G.] Davis," then the Company's ranking Patent Counsel. Dr. Whitney habitually took pains to express himself in homely, simple language, but he was never naive. The "methods" followed by the Research Laboratory, about which Little asked on January 14, are in accord with the approach to applied research during the silicone project. Of added interest is the example of "artificial mica," reminiscent of the Agens "flexible mica" that led to silicone rubber; the example shows how deeply ingrained in Research Laboratory thinking was the idea of insulating materials for high temperature.

January 18, 1928

Dr. Charles H. Herty
The Chemical Foundation
New York City

...

When it comes to selected cases, however, I don't like to weigh profit and loss, because I am accustomed to buying and selling wholesale. We may grow a peach and sell it with the gooper feathers on it, but sometimes we grow a green lemon that we "sour" fully and quietly digest. We are interested in the algebraic sum of red and black ink rather than in one color only.

...

I don't see "the nut to crack" which you mention unless I am it, but I'll try again. I can get exact figures to match my estimates, but they are entirely misleading unless they include ten lemons to each peach.

Here is the origin of the letter. Herty had heard a talk by Langmuir and sought Whitney's support on January 17 for the thesis that the benefit accruing to the public from industrial research is achieved at the expense of the company paying for the research. Herty did not get the support he sought. "Gooper feathers" (fuzz on a fresh peach?) may have been well known in Jamestown, New York, where Whitney was born when it was still a village. The Whitney "peach" is our "Discovery."

These two Whitney letters are enough to show that the founding of the Research Laboratory for "original research," with emphasis upon new materials, was a significant step forward in Lord Snow's scientific revolution enlarged to include polymers. The "four wise men" of Chapter 1 may well have realized that materials go with processes "like a horse and carriage," to borrow from a popular song. New materials usually require process innovations. The efficiencies of processes, for example, the generation of electricity by the use of turbines, are often limited because materials are unsatisfactory or uneconomical.

Earlier in the century, particularly in the United States and Germany, but to an increasing extent the world over, industrial research on materials and processes promised us a "brave new world" and a scientific "palace beautiful." For the consumer that promise was expressed in the General Electric slogan* "More goods for more people at less cost", as well as in other slogans, such as "Better things for better living through

---

* Company slogans warrant serious sociological study. They are not chosen lightly. They show what corporations think, and they are kept in tune with the times.

chemistry." *Goods* and *things*: a materialistic Lady Bountiful was expected to keep us always happy under her wing. No longer!

Changes and Their Causes. The industrial research of today differs from that of yesterday in respects other than the facing of a more critical public attitude. That changes were coming was felt in the silicone project because of the effects, already described, of World War II. Let us examine briefly some of these changes and their causes.

During World War II most research unrelated to the war was suspended for the duration. This alone made a subsequent expansion of industrial research likely, and wartime successes of industrial research increased this likelihood. One of these successes calls for special mention.

Now that the atom bomb and its successors have led to searing soul-searching and widespread fear, it is hard to believe that the success of the first of these terrible weapons could ever have contributed to the overall expansion of research. Yet it seems to have had this effect, on the basis of "reasoning" such as this: "If $2 billion gets us an atomic bomb, x billions will surely buy us a cancer cure, and y billions will certainly. . . ." Now, very little of that $2 billion went into crucial Discoveries; most of it went into the *implementation** of the Discovery of nuclear fission by Hahn and Strassman, which cost the United States nothing, and of other Discoveries that followed. Cancer, for which many cures and many Discoveries are needed, is obviously a far different story. Similarly fallacious later reasoning has taken the form, "If *we* can put a man on the moon, *we* can do research that will surely . . . ," but is gradually changing to "If *they* can put a man on the moon, why can't *they* . . . ?"

...

Another cause of change is the growing need for industrial research to play a *caretaker role* in our society. This need arises because research and its products can do harm as well as good, the harm often being unforeseeable or resulting from changes (e.g., density of automobiles) in our civilization. The poppy, among the most beautiful of flowers, carries opium in its seed. We cannot simply enjoy the flower and ignore the seed. Even in more halcyon days the public was asking science the

---

* Under government order, and with government funding, the chemical industry carried out much (more than is generally realized) of the implementation that produced the atomic bomb. The industry was more fortunate in this instance than after World War I, when "merchants-of-death" accusations became common.

question, "What have you done *for* me lately?"—a favorite query of President Johnson's. Industrial research must do what it can to keep this question from turning into "What are you doing *to* me now?"

This is not an easy assignment. Regulations aimed at eliminating harm can become nonsensical and onerous. It is easy to specify that a pollutant must be "completely" absent, but what analytical chemist can prove the absolute absence of anything? How many of those responsible for regulations realize that the cost (always borne eventually by the consumer) of removing a unit of pollutant rises, usually sharply, as the pollutant concentration decreases?

The "caretaker trade-offs" needed in these situations cannot ignore the seed of the poppy: *expected* good must be traded off against *future* harm, often in adversary proceedings. We have met trade-offs before, but these have been technical (e.g., to maximize product yield) and therefore simpler. Caretaker trade-offs are by nature political, yet industrial research cannot ignore them, for industrial research is the activity closest to materials and processes. As the best possible caretaker trade-offs will sometimes work hardships on the public, it is vital that industrial research enjoy public trust.

The most agonizing caretaker trade-offs involve nuclear energy. These being well known, we shall revert here to Midgley, honored in his time for his work on tetraethyllead and on the chlorofluorocarbons (Freons®). A caretaker trade-off has forced the gradual replacement of leaded gasoline, and Freons in spray cans are threatened by another.

The Freons make excellent propellant gases for the spray cans made by the billions each year. They also, in trace amounts, will continue to destroy ozone,* which exists some 10+ miles above the earth, absorbs ultraviolet rays from the sun, and in this way protects us from skin cancer. As yet we have no proof that Freon® in spray cans has caused the intensity of ultraviolet rays to increase on earth. Here the *convenience* provided by Freons (the good) must be traded off against a probable (but unknown) *increased risk* of skin cancer (the harm). Other propellant gases can be used. Wearing hats reduces the risk of skin cancer. What to do?†

---

* This destruction proceeds by *chain reactions*, in which a long "chain" of ozone molecules can be destroyed by "free radicals" derived from Freons because these free radicals are handed on from one "link" of the chain to the next. The laboratory evidence for these reactions is unimpeachable, but what happens 10+ miles above the earth is not established.
† "Fluorocarbon gas will be banned from most aerosol spray cans within two years under a proposed federal time table. . . . The new regulations' cost to the aerosol industry is estimated at $169 million to $267 million annually over the next four years" [*Wall Street Journal*, May 12, 1977, p. 1].

According to Figure 9-1, industrial research must change because Discoveries (Whitney peaches) are harder to come by as knowledge increases. The change favors "Discoveryless innovation," which can be both useful and profitable. The extent of change varies with the field. Barely noticeable in some life sciences, the "Figure 9-1 effect" may well prove overriding in further research on the older methods of energy conversion.

The following rough guide might facilitate the choice of research approach for a new project:

State of Prior Knowledge	Initial Research Approach
Stage 1	Whitney-Coolidge ("complete" freedom)
Stage 2	Whitney-Coolidge (restricted)[a]
Stage 3	Edisonian
Near Stage 4[b]	1. Let someone else do it.
	2. Do it, but not on own funds.
	3. Do it partially or wholly on own funds if prospects for Discoveryless innovation are good.

[a] Freedom less than "complete" because scientific augmentation is more tightly managed.
[b] Choice governed by judgment of how closely Stage 4 has been approached.

...

The instruments used in research continue to increase in kind, number, power, complexity, and cost. Often they are Discoveryless innovations based upon Discoveries in physics. Sometimes an embarrassment of riches, they usually expedite research—particularly when they make it possible to distinguish earlier a Whitney peach from a Whitney lemon.

...

By now, the reader will not be surprised to learn that the United States today has an enormous "R & D [research and development] industry." The estimated 1976 expenditure for R & D* is $38.2 billion, a number that inflation alone seems sure to increase in the future [31]. Let us equate Whitney's 1913 "$100,000 or more per year" to $600,000

* No one can say what R & D is. Until the term is better defined, the effectiveness of this multibillion dollar industry cannot be measured, nor can we be certain that our scientific manpower (more precious than money) is properly deployed.

One difficulty is that the "R," especially in industrial research, usually includes much more than just "scientific inquiry," the definition in this book. Furthermore, it is not too facetious annually in 1976 dollars. On this basis, the 1976 R & D industry would consist of almost 65,000 Research Laboratories, 1913 vintage. Are the productivities in the same ratio?

...

Industrial research has changed because government sponsors so much of it. For 1976 the government was expected to appropriate $20.2 billion for R & D, some $5.5 billion thereof for its "in-house" activities [31]. The 1976 Annual Report of the General Electric Company lists "total R & D" at $1075 million, of which $412 million came out of the Company's pocket, and the rest from contract work mainly paid for by the government.

He who pays the piper still calls the tune. Nevertheless, government funding of industrial research can at the same time promote both the public interest and that of the corporation. See the "Figure 9-1 guide" given above.

The reader will remember that silicone research in the Chemistry Section was not done on government funds. The government did buy silicone products.

...

The growth of industrial research, the changes in its character, and the increasing size of individual laboratories have all made it necessary to organize industrial research more elaborately and to manage it more tightly. Much of the "Whitney atmosphere" has had to go. His famous "Are you having fun?" is probably asked less often now.

---

to suggest that "D" is "R & D" less "R." Obviously, R & D is related to the discovery-augmentation-implementation sequence of this book, but not always in the same way.

Business Week (June 28, 1976) has wrestled bravely with the problem of defining R & D. Henry Adams would have foreseen what to expect. It is, however, *not* legitimate to conclude, as does Nicholas Valery [*New Scientist*, July 8, p. 72] that the *Business Week* article "confirms hints that the innovative power of American industry is in a serious state of decline" and that "the heyday of technological innovation in the United States is over." Money spent does not measure "innovative power."

Two concluding remarks are in order:

(1) The $38.2 billion expended for R & D is almost four times the *record sales* ($10.2 billion) reported by General Motors for the third quarter of 1976. As GM is a multinational corporation, it seems safe to say that the United States is investing in R & D at about the dollar rate at which it buys GM automobiles.

(2) In the end, definitions of R & D will have to conform to the dictum of Mr. Justice Felix Frankfurter: "This is tax language and should be read in its tax sense" [*United States* v. *Ogilvie Hardware Co.*, 330 U.S., 709,721].

Modern industrial research has an increasing need for strategic planning—an increasing need to know the answer to "Where do we go from here?" Behind this need are the closer approach to Stage 4, Figure 9-1, in many fields; increasing Adamsian multiplicity and complexity; the growing cost of research; and the greater number of industrial research laboratories now active.

Not surprisingly, points of view and language have also changed [32, 33].* Whitney's "Where might we find the next peach?" has had to become Steele's "Can we institutionalize the creation of change?" [33]. In Chapter 8 "innovation" was brought in as an elastic word to represent the first commercial introduction of a product, process, or application. Dr. Steele, who is Manager, R & D Planning, General Electric Corporate Research and Development, focuses in his Chapter 3 [33] upon the *creation* of innovations, which he regards as a "complex social process in which successful functioning requires the application of a diverse spectrum of skills . . . [that] could be collected within a group of people and coupled in the right sequence to keep the process going." Whitney's peaches and lemons would scarcely be used in describing modern industrial research.

. . .

The last chapter ends here. Industrial research is not what it was when the silicone project began. Moreover, the winds of change will continue to blow. The net good from a dollar spent on industrial research is likely to continue decreasing, but our need for that activity will not.

---

* Dr. Kusiatin's thesis [32] is valuable because it deals not only with silicones but also with other projects inside and outside General Electric, in modern language and from modern points of view.
Dr. Steele's book [33] is the best treatment I have seen of how Adamsian multiplicity and complexity are changing the relationship of R & D in a large organization, national or multinational, to the rest of the organization.

Epilogue

Watchman, what of the night?

Isaiah 21:11

It is time now to complete the Preface. Lord Snow's Rede Lecture, delivered in 1959, deserves a place in any history of modern scientific thought. He was right to be concerned about "the gulf of mutual incomprehension" between the literary intellectual culture and the scientific. How right he was is clear from the celebrated "Mental Fight"* triggered by his lecture. The literary intellectual's side of this fight was presented

---

* "I will not cease from Mental Fight,
    Nor shall my sword sleep in my hand,
    Till we have built Jerusalem,
    In England's green and pleasant Land."

WILLIAM BLAKE, *Milton,* Preface

Alas, not everyone wants the same Jerusalem, nor can a "green and pleasant Land" be blithely taken for granted today.

Blake has become a patron saint to many. About him, Priestley [3] says:

"And no doubt those who believe the society we have created during the last hundred and fifty years is essentially sound and healthy will continue to believe, if they ever think about him, that Blake was insane. But there is more profit for mind and soul in believing our

by an eminent essayist and literary critic, Professor Leavis, in his book
Nor Shall My Sword [1], published in 1972, which complements Lord
Snow's *The Two Cultures: and a Second Look* [2] to provide an illumi-
nating, albeit incomplete, record of what is now called the Snow-Leavis
controversy.

The controversy was a forerunner of more to come. We shall take
for granted that bridging the "gulf of mutual incomprehension" between
two cultures, hereafter to be called "scientific" and "nonscientific," is
necessary and desirable. We shall not pretend that this will be enough;
obviously, much more than incomprehension is at work. We shall further
take for granted that a better understanding of industrial research is
crucial to the bridging; after all, many practical consequences of the
modern scientific revolution, notably products and processes, stem directly from industrial research.

Lord Snow hoped that bridging the gulf would forestall troubles
between the rich nations and the poor.* Such troubles now seem less
threatening than the growing doubts within the rich nations about the
civilization that has grown out of the modern scientific revolution.

Concern about this civilization is felt by many, not only by literary
intellectuals. Among them are some who understand the broad aspects
of science better than many scientists, and who are free of antipathies
toward everything that "smelt of science and therefore seemed to menace
classical education" [5]. Norman Mailer edifies us with "technological
excrement over all the conduits of nature" [6]; other intellectuals [e.g.,

---

society to be increasingly insane, and Blake (as the few who knew him well always declared)
to be sane and healthy."

Blake was not insane. Even earlier, in Elizabethan times, society was attractive in many
ways that could not survive the industrial revolution and the modern scientific revolution
that followed [4]. But the old days are gone forever. Today's public is not about to turn
back the clock and will not voluntarily practice Spartan self-denial to improve what is now
called the quality of life. For the third quarter of 1976, General Motors reported a record
profit of $397 million. Inflation contributed, but so did the public.

* Relief for the poor nations may come without bridging the gulf. Because of the formation
of the Organization of Petroleum Exporting Countries (OPEC) and of similar organizations
concerned with less critical materials, the relationships between the poor nations and the
rich ones have changed, and the change continues. OPEC, the self-proclaimed leader of
the Third World, could arrange for a supply of capital to the poor nations and so narrow
the gap between them and the rich. Multinational corporations, if allowed reasonable
freedom, could help toward this end, for the multinational corporation is the best available
capitalistic device for using the world's resources for everyone's benefit.

The prime economic needs of the poor nations are the products and processes (e.g.,
synthetic fertilizers, man-made drugs, and polymers) of industrial research.

7] use more measured language. Three published scripts [8, 9, 10] of outstanding television series are somber; one has given new life to the gloomy line "Things fall apart; the centre cannot hold" [8, p. 347] that Yeats wrote some 50 years ago. We close the list, which would be easy to lengthen, with references to a ranking professor in the humanities [11], a well-known and appealing historian [12], a British columnist [13], a noted professor of history [14], the Empress of Iran [15], and the Governor of Colorado [16].

No culture, certainly not the scientific, can afford to ignore the growing disillusionment with modern civilization, or refuse to acknowledge that the modern scientific revolution must share the blame. But, to borrow from Chapter 9, we must not ignore the beauty of the poppy and think only of the opium in its seed. We must learn to cope with the Adamsian multiplicity and complexity, a dominant Leitmotiv in this book, so that we can establish and accept the trade-offs most likely to reduce our difficulties.

Henry Adams pointed out [Introduction] that the mind would need to jump if it is to cope with growing multiplicity and complexity. The elimination of Lord Snow's "gulf of mutual incomprehension" would also require such a jump, which it takes a Jacob Bronowski [10] to make. As such men are rare, it is fortunate that most of us are called upon merely to *bridge* the gulf, an easier assignment. The bridges must carry two-way traffic. Influential people from each culture must cross over and form a perceptive overview of the other side. For example, the nonscientific culture must learn enough about industrial research to judge what such research is likely and unlikely to accomplish—an undertaking that implies forming models such as Figure 9-1.

We have a representative government, not a pure democracy. Our representatives and other politicians are among the most important people who must cross the bridge from the nonscientific side. Adamsian multiplicity and complexity are especially hard on politicians [16]. Consider this apt quotation from a column by James Kilpatrick:

He [the congressman] feels the impact of technology like a stone in his stomach. Somehow he is expected to understand—he is expected to demonstrate the quality of wise leadership—in atomic fusion, solar energy, sulfuric emission, generic drugs, supersonic transport, and sequential contraceptives. He is expected to legislate on oil shale, coal degasification, the diseases of shellfish, and the disposal of atomic wastes. Does his trumpet give an uncertain sound? It does indeed [17].

It is easy to understand why Washington so often acts as if "a problem will go away if you only throw enough money at it."*

To do anything worthwhile, a politician *must survive*. Survivability, generally through re-election, is, and in our system of representative government has to be, the "overriding consideration in virtually every political judgment" [16]. Moreover, almost always such judgments must be formed, and votes must be cast, on the basis of incomplete information. The "perceptive overview" mentioned above is all that we ought to expect of a politician.

Many politicians are lawyers, and lawyers make their living with words. They are therefore well equipped to bridge the gulf under discussion. They know how to evaluate and use scientific counsel even when, as often happens, science's trumpet gives uncertain sounds. Nevertheless, it is easy to see why an exasperated senator once asked whether scientists had two hands: so much expert testimony had begun with "On the one hand . . ." and never got to the other. Nothing, of course, can absolve the politician of ultimate responsibility. He (not his scientific adviser) is elected and paid to represent the electorate.

The adequate oversight ("measurement") of a large and diversified R & D activity is among the most demanding assignments imaginable. Yet Congress has this responsibility for the enormous amount of such work that is government funded. Worse seems to be in the offing. Because of matters such as nuclear fission and fusion, environmental degradation, and fear of cancer, the public has become acutely sensitive to any harm, present or future, that it associates with research. The probability grows that Congress will be called upon to control, perhaps to ban, certain kinds of research, *no matter how financed*. To explain, let us look at what could well be the greatest Discovery (capital "D") yet made in the United States.

---

* So far, "big money" has not corrupted science, but the danger exists. Too much money chasing too few ideas, and the writing of research proposals that virtually guarantee discoveries not yet made, are both unhealthy. Perhaps research salaries ought not to be guaranteed into the next century, as those of certain able research professors working on cancer have been [18]. One investigator allegedly falsified experimental evidence related to cancer [19]; "big money" (not a guaranteed salary) seems to have been the root of this evil.

"Big money" encourages "push-pull publicity." The public's appetite for sensational research news tends to pull results prematurely out of the laboratory. This pull can be strongly reinforced by a push generated by a desire for continuing or increased funding. Headlines then follow that promise what the text cannot fulfill. Television and radio often lack the equivalent of a text.

The war against microbes began about 1850 with an attack on puerperal fever, and it continues today with the discoveries of new antibiotics, having been marked by a series of spectacular victories against disease. In the beginning the research was largely applied in character, but the availability of increasingly sophisticated methods and tools for learning about large molecules, such as X-ray diffraction, the electron microscope, and the computer, has made possible pure research that has yielded Discoveries of tremendous importance. We have mentioned the structure of DNA as one such Discovery. Even the most confirmed literary intellectual can scarcely deny that such research has done much more good than harm.

It is therefore a shock to learn that a technique for studying the living cell has aroused serious misgivings not only in the nonscientific culture, but also among life scientists of the highest rank [20]. This technique, abbreviated as RDNAT, is the "recombinant DNA technique," which leads to "genetic engineering."* This great Discovery was easier to make once the structure of DNA was known.

Perhaps unfortunately, RDNAT is beautifully simple, requires only an adequate centrifuge and easily obtainable materials and equipment, and can be practiced by laboratory assistants. *Escherichia coli (E. coli)* bacteria, which live in the human gut and can multiply at enormous rates, are burst by a detergent solution, whereupon they eject their DNA or genetic material, which is then recovered and separated by centrifuging into strands of DNA, and into tiny loops or *plasmids* thereof. A special enzyme can open the plasmids. When the enzyme operates on plasmids of two kinds, opened loops of each kind recombine to form *hybrid* plasmids, which are a *new genetic material,* man-made. Another source of "foreign" DNA can be used instead of the second plasmid. The hybrid plasmids will enter *E. coli* bacteria made porous by subjecting them to a sudden temperature rise ("heat shock") in a salt solution. The *E. coli* thus impregnated will reproduce themselves faithfully and rapidly, copying both their own DNA and the hybrid DNA in the man-made plasmid. Eventually, implementation of this Discovery will produce genetic material commercially. The University of California and Stanford University have jointly applied for a patent on the commercial use of RDNAT, and production "might be only five years away" [21]. Industrial research cannot ignore this field.

---

* On March 16, 1977, an outstanding program entitled "The Gene Engineers," produced by Station WGBH-TV, Boston, as NOVA No. 409, was transmitted for the first time over the PBS network. I know of no better or more important scientific telecast. I thank NOVA and the station for a copy of the script.

When man makes and manipulates genetic material, he comes closer than ever before to playing God. The good he might do is great; the harm, unpredictable. The fear has grown that experimentation with RDNAT might somehow bring modified *E. coli* into the human gut, spreading cancer and who knows what else. Once more, we have the beautiful poppy with the evil seed [20].

The fear of harm is not irrational, and the scientific culture has moved admirably to cope with it in various ways, among them the formulation by the National Institute of Health (NIH) of guidelines for RDNAT research. Of principal concern here is what happened in Cambridge,* the seat of Harvard University. In June 1976 the 10-man Cambridge City Council broke all precedent and jolted the scientific culture by asking the University to halt construction of a $500,000 laboratory for RDNAT research until the Council could decide whether or not to permit such work in Cambridge. Early in 1977 the Council permitted the work to proceed provided that the NIH guidelines were observed and that certain (not particularly onerous) conditions *imposed by the Council* were met.†

Cambridge, however, is not the world. One must assume that someone somewhere is likely to push RDNAT to Catton's "absolute limit." Furthermore, we may, at some time not too far in the future, face the same situation with regard to cloning (asexual reproduction) and the manufacture via organic chemistry of synthetic genes [22]. Congress will some day have "to think with wisdom" upon these matters.

So long as Congress provides the funding, it can obviously control the research by stopping the funds. Suppose, however, that Congress under pressure from the electorate decides this control is not enough, and that certain research otherwise funded must be banned? Scientists will of course protest that scientific inquiry must remain free. Constitutional guarantees, such as those of free speech and (inferentially of free thought), cover the scientist at the laboratory bench, but "The Constitution does not recognize an absolute and uncontrollable liberty" [23].

---

* At the University of Michigan, Ann Arbor, RDNAT had less trouble. After a University committee, appointed in September 1975, recommended that work proceed under adequate safeguards, Ann Arbor, according to press reports in May 1977, went along. The committee included one historian, who—as the press made clear—successfully bridged the gulf of mutual incomprehension.
† According to the WGBH television script, upon which this account is based, "The thoughtfulness of the report [by the review board of the Council] proved that lay people can participate in decisions usually left to the scientist." Another encouraging indication that the gulf of mutual incomprehension can be bridged!

This is one more reason why an improved understanding of research, especially of industrial research,* by politicians is important.

The recombinant DNA technique and related genetic matters are examples of political control that may be coming. Congress need not consider these matters now; RDNAT has not yet, and may never, hurt anyone. Energy is a different and far more urgent concern. In a recent winter, plants were closed because needed natural gas was not forthcoming; the costs of this fuel and others continue to rise. Moreover, the electorate, already hurt, must brace itself for worse to come: relief at reasonable price is not in sight. Members of Congress cannot postpone thinking about energy, and thinking with wisdom; they will have to act as though their political survivability were not at risk.

Unfortunately this risk is acute because many of the electorate, subjected to a daily force-feeding of bad news and fearful of nuclear destruction, refuse to act *now* to ameliorate a *future* energy shortage, no matter how certain. Recently, television told of a man who'd rather change his President than his automobile. Many feel no urgent need to conserve: "Science will find a way—Lady Bountiful is still with us." If the energy well will not run dry this century, why worry? It will be a brave—perhaps a reckless—legislator who advocates reduced per capita energy consumption.

Congress can easily find the information it needs to acquire a perceptive overview of the energy situation. A good starting point is "The Hard Truth about Our Energy Future" [24], a forthright summary by Arthur M. Bueche, Vice President, Research and Development, General Electric Company. The address includes a well-thought-out energy program, and it rivals in bluntness the Whitney peaches-and-lemons letter of Chapter 9; it emphasizes that peaches, even if found soon, cannot be harvested before the year 2000. In our language interim relief of energy shortages will have to come mainly from conservation and from Discovery-less innovations; implementation of Discoveries will take too long even if the enormous capital required becomes available. A reduction in our per capita energy consumption is unavoidable. There are two ameliorating circumstances: the continuing increase in energy costs, painful though it is, will of itself enforce energy savings; and the increase will bring to market supplies from hitherto uneconomic sources. Difficult trade-offs, such as those involving nuclear and breeder reactors, and

---

* Haber had a premonition [Chapter 9] that the making of fertilizers from synthetic ammonia would be displaced if man ever learned how bacteria fix nitrogen. That is one aim of RDNAT. An eventual goal might be wheat that fixes needed nitrogen as it grows. Industrial research on RDNAT for this purpose could become an important undertaking for the fertilizer industry. Will it be permitted?

large coal-fired power plants, must be made. Research, especially industrial research, is needed everywhere.

Energy and RDNAT are contrasting examples, which are alike, however, in showing that the gulf of mutual incomprehension between the scientific and the nonscientific culture must be bridged if representative government is to operate satisfactorily. The bridges must be traveled in both directions, by scientists in one direction and especially by politicians in the other. We cannot sit with folded hands and let "things fall apart"; the center must be made to hold. And we must disprove Priestley's thesis "that our society is increasingly insane."

There is no need to despair. If we can continue to avoid nuclear destruction, we shall have time to feel our way, and circumstances will force us to do so. Today's public awareness of the consequence of the scientific revolution, imperfect though it is, did not exist when Lord Snow wrote his book; and it will become a powerful force for good, albeit a force not always wisely exerted. Dread of cancer, concern for the environment, reluctance to have the passionate love affair with the automobile cool down (as it must) to a marriage of convenience, rising energy costs—all these and more will force new trade-offs to be sought, trade-offs that must be broad enough to include a balancing of public good against public harm. Properly made, these trade-offs can lead to a society necessarily more austere than ours today, but better than Blake's in most respects and more compatible with a "green and pleasant land" than what we now have. There need be no night.

References and Notes

INTRODUCTION

1. *Chem. Eng. News,* July 26, 1976, p. 17.
2. Henry Brooks Adams, *The Education of Henry Adams.* Privately published in 1907, this important book has been reprinted several times since, most recently in 1974 by the Houghton Mifflin Company, Boston, in a version edited by Ernest Samuels, to which references have been made in the text.
3. Lewis Carroll, *Through the Looking Glass and What Alice Found There,* Chapter 6.
4. *Holmes-Pollock Letters,* Vol. 2, Mark De Wolfe Howe, Ed., Harvard University Press, Cambridge, 1941, p. 13.

CHAPTER 1

1. C. P. Snow, *The Two Cultures: and a Second Look,* Cambridge University Press, London, 1964.
2. (a) L. A. Hawkins, *Adventure into the Unknown,* William Morrow & Co., New York, 1950, describes the first 50 years of the General Electric Research Laboratory. (b) A more complete history of the Laboratory is that of Kendall A. Birr, *Pioneering in Industrial Research: The Story of the General Electric Research Laboratory,* Public Affairs Press, Washington, 1957. (c) John T. Broderick, *Willis Rodney Whitney,* Fort Orange Press, Albany, New York, 1945, is particularly valuable for its account of the early days. The first part of Chapter 1 is drawn from these books and from H. A. L.'s personal knowledge.
3. H. A. Liebhafsky, *William David Coolidge: A Centenarian and His Work,* John Wiley & Sons, New York, 1974.
4. Table 1-1 is a modification of Appendix 1.1, *The Insulation of Electrical Equipment,* W. Jackson, Ed., John Wiley & Sons, New York, 1954, p. 23.

5. (a) *The New York Times*, 15:1, February 24, 1944. (b) I. Asimov, *Asimov's Biographical Encyclopedia of Science and Technology*, Doubleday and Co., Garden City, New York, 1964, p. 426. (c) W. C. Dampier, *A History of Science*, 4th ed., Cambridge University Press, Cambridge, 1966, pp. 430–431. (d) C. F. Chandler, *Ind. Eng. Chem.*, **8**, 178 (1916). Professor Chandler presented the 1916 Perkin Medal, "for the most valuable work in applied chemistry," to Dr. Baekeland.

    The Asimov account of the Baekeland-Eastman negotiations, though not further documented, seemed too good to leave out. Chandler (p. 179) speaks of "very liberal terms, which were promptly accepted."

6. R. M. Nesbit, *Your Prostate Gland*, C C Thomas, Springfield, Illinois, 1950, p. 37.

7. A. L. Marshall, *Chemical Horizons in General Electric*, CRDA, November 19, 1953.

8. A. L. Marshall, *Trans. Faraday Soc.*, **21**, 197 (1925), deals with the electrodeposition of zinc from sulfate solution, a problem suggested to A. L. M. by Professor F. G. Donnan, Sir William Ramsay Laboratory of Physical Chemistry, University College, London. The paper was first received by the journal on March 1, 1923.

9. *Trans. Faraday Soc.*, **21**, 561 (1925–1926).

10. H. A. Liebhafsky, A. L. Marshall, and F. H. Verhoek, *Ind. Eng. Chem.*, **34**, 704 (1942), and earlier publications there cited. The molecular explanation of the decrease in diffusion constant that accompanies loss of plasticizer by gaseous diffusion is in accord with modern views; see p. 142 et seq. of *Big Molecules*, cited in Reference 13. No such decrease was observed when the plasticizer escaped into stirred oil. (Unpublished work by John J. Russell, 1940 or earlier.) I think that oil diffused into the plastic and kept the polyvinyl chloride molecules apart.

11. J. D. Watson, *The Double Helix*, Atheneum Press, New York, 1968.

12. M. Goodman and F. Morehouse, *Organic Molecules in Action*, Gordon and Breach Science Publishers, New York, 1973. I thank Dr. Allan S. Hay for calling this interesting book to my attention.

13. H. Melville (Sir Harry), *Big Molecules*, The Macmillan Company, New York, 1958. The use of "big," the simplest possible adjective, is a clue to the tone of this small book.

CHAPTER 2

1. *Gmelin's Handbuch der Anorganische Chemie*, 8th ed., Verlag Chemie, Weinheim, West Germany.

2. E. G. Rochow, letter to H. A. L., 1974.

3. Of Reference 1, (a) the volume *Silicium*, Teil B, 1959, 922 pp.; (b) *Silicium*, Teil C, 1958, 487 pp.

4. J. R. Partington, *General and Inorganic Chemistry*, 4th ed., St. Martin's Press, New York, 1966. The quotation is on p. 174.

5. E. G. Rochow, (a) *The Metalloids*, D. C. Heath, Boston, 1966; (b) *The Chemistry of Silicon*, Pergamon Press, New York, reprinted with corrections, 1975, from *Comprehensive Inorganic Chemistry*, Pergamon Press, 1973.

6. L. Pauling, *The Chemical Bond*, Cornell University Press, Ithaca, New York, 1967.

7. L. Pauling and R. Hayward, *The Architecture of Molecules*, W. H. Freeman and Co., San Francisco, is a beautiful book that would make an admirable case history of pure science for literary intellectuals.

8. L. Pauling, *Science*, **123**, 255 (1956). The Nobel Prize Lecture.

9. In *The Oxford English Dictionary*, Berzelius is credited with introducing "polymer" in 1830, and the origin of the term is stated to be a Greek word meaning "having many parts."

10. *Encyclopedia of Polymer Science and Technology*, N. B. Bikales, Executive Editor, John Wiley & Sons, New York. (a) The quotation is from Vol. 11, 1969, p. 279. (b) W. A. Keutgen, Vol. 10, 1969.

11. H. K. Lichtenwalner and M. M. Sprung in the *Encyclopedia of Polymer Science and Technology*, Vol. 12, John Wiley & Sons, New York, 1970, pp. 464–569.

12. M. L. Huggins, P. Corradini, V. Desreux, O. Kratky, and H. Mark, *Polym. Lett.*, **6**, 257 (1968).

13. See *Chem. Zentralbl.* (then called *Pharmaceutisches Central-Blatt*), **4**, 39 (1833).

14. Reference 10, Vol. 1, 1964, p. 664.

15. M. Ebelmen, *Bull. Chem. Phys.*, Series 3, **16**, 129–166 (1846), which gives the compositions and properties of various compounds that form when boric and silicic acids react with "bodies of the classes of ethers," is a magnificent example of early Step 1 work in a field that might produce compounds to rival silicones.

16. What immediately follows was based on the excellent treatment of stereoisomerism in *The New Encyclopedia Britannica*, Vol. 9, 1973, p. 1035.

17. E. G. Rochow, *Allgem. Prakt. Chem.*, **22**, 157 (1971).

18. F. Wöhler, *Ann. Chem.*, **127**, 257 (1863).

19. C. Friedel and J. M. Crafts, *C. R. Acad. Sci.*, **56**, 592 (1863).

20. (a) C. Friedel and A. Ladenburg, *Ann. Chem.*, **145**, 179 (1868). (b) A. Ladenburg, *ibid.*, **164**, 300 (1872).

21. F. S. Kipping, *Proc. Roy. Soc.*, **A159**, 139 (1937). The Bakerian Lecture.

22. F. Challenger, *Dictionary of National Biography 1941–1950*, Oxford University Press, London, 1959, pp. 462–463, is the main source of the biographical information about Kipping, with whom Challenger published two papers. Two footnotes [*Trans. Chem. Soc.*, **95**, 308 (1909), and **105**, 484 (1914)] show that Kipping could be critical of coworkers.

23. Pertinent papers by Kipping are listed in H. W. Post, *Silicones and Other Organic Silicon Compounds*, Reinhold, New York, 1949. This book gives a careful and detailed account of organosilicon chemistry up to 1948, and an excellent treatment of Kipping's work.

24. Reference 21, pp. 140–141.

25. The Kipping route to silicones was discovered independently and at about the same time by Dilthey, who used the Grignard reagent to make phenylchlorosilanes; see W. Dilthey, *Chem. Ber.*, **37**, 1139 (1904). The route, perhaps a shade unjustly, has been given Kipping's name herein to avoid hyphenation and to emphasize the fact that he made the greater contribution to the silicone industry.

26. F. S. Kipping, *Proc. Chem. Soc.*, **20**, 15 (1904).

27. F. S. Kipping, *J. Chem. Soc.,* **91,** 209 (1907).

28. Leo H. Sommer, *Stereochemistry, Mechanism and Silicon,* McGraw-Hill, New York, 1965.

29. F. S. Kipping and L. L. Lloyd, *Proc. Chem. Soc.,* **15,** 174 (1899).

30. G. Martin and F. S. Kipping, *Trans. Chem. Soc.,* **95,** 302 (1909).

31. F. S. Kipping and A. G. Murray, *J. Chem. Soc.,* 1427 (1928).

32. For an admirable (and admiring) biography of Stock, see E. Wiberg, *Chem. Ber.,* **83,** XIX–LXXVI (1950). The biography gives one an interesting glimpse of chemistry in German universities at the crest of the wave.

33. Carl Wagner, *Methoden der naturwissenschaftlichen und technischen Forschung,* Bibliographisches Institut, Zurich, 1974.

34. Wilhelm Ostwald, *Grosse Männer,* Akademische Verlagsgesellschaft, Leipzig, 1909.

35. A. Stock, *Z. Elektrochem.,* **32,** 341 (1926).

36. A. Stock, *Hydrides of Boron and Silicon,* Cornell University Press, Ithaca, New York, 1933, pp. 20–37. Those who prefer English will have little need for Reference 35.

37. A. Stock, C. Somieski, and R. Wintgren, *Chem. Ber.,* **50,** 1764 (1917).

38. (a) J. F. Hyde, *J. Amer. Chem. Soc.,* **75,** 2167 (1953). (b) S. W. Kantor, *ibid.,* **75,** 2712 (1953). (c) T. Takugichi, *ibid.,* **81,** 2359 (1959). $(CH_3)_2Si(OH)_2$ melts near 100°. It does not seem to be extremely volatile.

39. (a) A. D. Petrov, B. F. Mironov, V. A. Ponomarenko, and E. A. Chernyshev, *Synthesis of Organosilicon Monomers,* Consultants Bureau, New York, 1964. Specialized, detailed, and complete; an outstanding coverage of Russian work. (b) R. J. H. Voorhoeve, *Organohalosilanes: Precursors to Silicones,* Elsevier Publishing Co., New York, 1967. The best single source of information on the precursors to silicones.

40. A. Stock and C. Somieski, *Chem. Ber.,* **52,** 695 (1919). Confirms, and gives better evidence for, the existence of H_2SiO, which was announced in Reference 37.

41. L. Pauling, *J. Amer. Chem. Soc.,* **51,** 1010 (1929).

42. L. Pauling, *College Chemistry,* 3rd ed., W. H. Freeman and Company, San Francisco, 1964.

43. W. Noll, *Chemistry and Technology of Silicones,* Academic Press, New York, 1968, is, I believe, the best and most complete relatively recent book on silicones.

44. J. E. Reynolds, *C. A.,* **4,** 426 (1910).

45. A. Bygdén, *Silicium als Vertreter des Kohlenstoffs organischer Verbindungen,* Almquist and Wiksells, Uppsala, 1916.

46. Richard Müller, *J. Chem. Educ.,* **42,** 41 (1965). Translated by Professor Rochow.

47. E. G. Rochow, *The Chemistry of the Silicones,* 2d ed., John Wiley & Sons, New York, 1951. The first edition appeared in 1946, when the silicone industry was just beginning. It has been published in five languages and has been influential the world over.

48. Please read "Some Limits to Popular Science," originally an editorial in the April 1956 issue of *Endeavour,* published by Imperial Chemical Industries, London, Trevor I. Williams, Ed.; reprinted in *Science,* **124,** 207 (1956).

CHAPTER 3

1. Sources for the first two paragraphs:
   (a) *Background of Corning meeting.* Conversation with A. L. Marshall, summer 1973. Mr. Morrow identified by E. G. Rochow. Personal recollection.
   (b) *Meeting proper.* Documents relating to following interferences in the U.S. Patent Office: No. 80,351; No. 80,406; and No. 80,812. The documents were (1) the brief for Rochow at the final hearing, which covered the three cases; (2) the three individual briefs for Hyde; and (3) decisions of the Board of Interference Examiners in the last two interferences.
   (c) *American Chemical Society meeting.* Brief for Rochow mentioned above, p. 46. Date of meeting from *Schenectady Gazette.*

2. R. R. McGregor, *Silicones and Their Uses,* McGraw-Hill, New York, 1954, p. 27.

3. Brief for Hyde, Interference No. 80,406.

4. E. G. Rochow, *Chemie in unserer Zeit,* **3(1),** 67 (1967). The address, in revised form, delivered by Professor Rochow on being awarded the honorary DSc by the Technische Hochschule Braunschweig in 1966.

5. Staff Report, *Chem. Eng. News,* **27,** 1510 (1949). The Baekeland Medal address by Professor Rochow appears with the report.

6. E. G. Rochow, *Adv. Organometall. Chem.,* **9,** 1 (1970).

7. The Research Laboratory was founded primarily because the electrical industry was going to need new materials. Navias' work was convincing evidence for the correctness of this step. The problems he faced were more sophisticated than the founders could have dreamed of.

8. Louis Navias, *Partial History of Ceramic Research and Development in the Research Laboratory of the General Electric Company,* Report 65-GP-0351M, General Electric Research and Development Center, Schenectady, New York. An outstanding account of a lifetime scientific career in an industrial laboratory.

9. E. G. Rochow, *J. Appl. Phys.,* **9,** 664 (1938).

10. The narrative here follows the brief for Rochow at the final hearing, cited in Reference 1b (1). I have verified that Rochow's notebooks agree with the brief.

11. W. Noll, *Chemistry and Technology of Silicones,* Academic Press, New York, 1968, p. 438.

12. U.S. Patents 2,258,218 to 2,258,222 inclusive.

13. G. Martin and F. S. Kipping, *J. Chem. Soc.,* **95,** 302 (1909).

14. A. Stock, *Hydrides of Boron and Silicon,* Cornell University Press, Ithaca, New York, 1933, pp. 27 and 28.

15. G. Martin, *Chem. Ber.,* **46,** 2442 (1913).

16. W. F. Gilliam, H. A. Liebhafsky, and A. F. Winslow, *J. Amer. Chem. Soc.,* **63,** 801 (1941).

17. W. F. Gilliam and M. M. Sprung, *Laboratory Report,* CRDA, January 20, 1942.

18. E. G. Rochow and W. F. Gilliam, *J. Amer. Chem. Soc.,* **63,** 798 (1941).

19. H. S. Booth and W. F. Martin, *J. Amer. Chem. Soc.,* **68,** 2655 (1946).

20. W. I. Patnode and W. J. Scheiber, *Laboratory Report,* CRDA, September 7, 1939.

21. E. G. Rochow, *Summary Laboratory Report,* CRDA, August 1, 1945.

22. H. S. Booth and W. D. Stillwell, *J. Amer. Chem. Soc.*, **56**, 1529 (1934).

23. H. A. Liebhafsky, *William David Coolidge: A Centenarian and His Work*, John Wiley & Sons, New York, 1974.

CHAPTER 4

1. H. A. Liebhafsky, *William David Coolidge: A Centenarian and His Work*, John Wiley & Sons, New York, 1974.

2. (a) K. H. Kingdon and H. C. Pollock, *Research Laboratory Bulletin*, General Electric Co., Summer 1965, p. 9. (b) K. H. Kingdon, H. C. Pollock, E. T. Booth, and J. R. Dunning, *Phys. Rev.*, **57**, 749 (1940).

3. H. S. Booth and W. D. Stillwell, *J. Amer. Chem. Soc.*, **56**, 1529 (1934).

4. H. Buff and F. Wöhler, *Ann. Chem.*, **104**, 94 (1857).

5. (a) C. Friedel and A. Ladenburg, *Ann. Chem.*, **143**, 118 (1867). See also (b) O. Ruff and K. Albert, *Chem. Ber.*, **38**, 2222 (1905).

6. C. Combes, (a) *Bull. Soc. Chim.*, 3rd Series, **7**, 242 (1892); (b) *C. R. Acad. Sci.*, **122**, 531 (1896); (c) *ibid.*, 622 (1896).

7. See W. Noll, *Chemistry and Technology of Silicones*, Academic Press, New York, 1968, p. 446.

8. A. Ladenburg, *Ann. Chem.*, **144**, 320 (1872).

9. H. K. Lichtenwalner and M. M. Sprung, *Encyclopedia of Polymer Science and Technology*, Vol. 12, John Wiley & Sons, New York, 1970, p. 531 et seq.

10. Personal communication, Loys Wright Marsden to H. A. L.

11. Personal communication, W. F. Gilliam to H. A. L.

12. Personal communication, J. Marsden to H. A. L.

13. M. Silverman, *The Saturday Evening Post*, February 26, 1955, p. 36.

14. See Reference 7, p. 339.

15. See Reference 7, pp. 342–343, for information about titanosiloxanes, modern formulation included.

16. E. G. Rochow, *J. Amer. Chem. Soc.*, **67**, 963 (1945).

17. E. G. Rochow and W. F. Gilliam, *J. Amer. Chem. Soc.*, **67**, 1772 (1945).

18. E. G. Rochow and W. F. Gilliam, *Progress Report—The Preparation of Phenyl Chlorosilanes from Chlorobenzene & Silicon*, CRDA, January 1942.

19. T. S. Kuhn, "The Structure of Scientific Revolutions," *International Encyclopedia of Unified Science*, Vol. 2, The University of Chicago Press, Chicago, 1970, p. 52 et seq.

20. H. Guerlac, *Lavoisier—The Crucial Year*, Cornell University Press, Ithaca, New York, 1961.

21. R. J. H. Voorhoeve, *Organohalosilanes: Precursors to Silicones*, Elsevier Publishing Co., New York, 1967. The index has 24 references to $SiHCl_3$. The reaction with C_6H_5 is on p. 36.

22. R. Müller, DWP 5348 (East Germany); application, June 6, 1942 (Germany). I have not seen the patent, and I do not know how far discovery preceded patent application. The date of Rochow's discovery was May 10, 1940.

Müller also arrived at his discovery via SiHCl₃. See his valuable article in *J. Chem. Educ.*, **42**, 47 (1965), which was translated into English by Rochow.

23. R. A. Laudise and A. A. Ballman in *Kirk-Othmer Encyclopedia of Chemical Technology*, Vol. 18, 2nd ed., Anthony Standen, Executive Editor, John Wiley & Sons, New York, 1969.

CHAPTER 5

1. R. J. Cordiner, *New Frontiers for Professional Managers*, McGraw-Hill, New York, 1956, p. 57 et seq.

2. (a) R. J. H. Voorhoeve, *Organohalosilanes: Precursors to Silicones*, Elsevier Publishing Co., New York, 1967, Chapter 8 and others. (b) V. Bažant, *Pure Appl. Chem.*, **19**, 473 (1969), who says on p. 488, ''I would like to state my opinion that the present state of knowledge indicates quite convincingly that for the direct synthesis of organohalogenosilanes the chemisorption [adsorption involving chemical bonds] mechanism is of decisive importance, without, however, excluding the possibility that under such high reaction temperatures the radical mechanism may also have a minor effect.'' I think that all this and more is possible in the direct process under one condition or another.

3. H. A. Liebhafsky and L. S. Wu, *J. Amer. Chem. Soc.*, **96**, 7180 (1974), deals with an oscillatory reaction, the mechanism of which has concerned me for almost 50 years. I regard this mechanism as easier to ''prove'' than that of the direct process.

4. R. O. Sauer, *Laboratory Report*, CRDA, March 14, 1941.

5. D. T. Hurd and E. G. Rochow, (a) *Laboratory Report*, CRDA, October 5, 1943. (b) *J. Amer. Chem. Soc.*, **67**, 1057 (1945).

6. H. W. Post, *Silicones and Other Organic Compounds of Silicones*, Reinhold, New York, 1949, p. 85.

7. In 1942 Patnode wrote the following three laboratory reports on work, much of which was begun in 1941: (a) *Dimethyldichlorosilane: A Preliminary Report on the Products of Its Hydrolysis*, CRDA; (b) *Linear Methylpolysiloxanes*, CRDA, September 23, 1942; and (c) *Methylpolysiloxanes*, CRDA, September 1942.

8. W. I. Patnode and D. F. Wilcock, *J. Amer. Chem. Soc.*, **68**, 358 (1946).

9. R. O. Sauer, W. J. Scheiber, and S. D. Brewer, *J. Amer. Chem. Soc.*, **68**, 962 (1946), is one more example of hydrolysis-condensation complexities.

10. C. A. Burkhard, *Pyrolysis of Methylsilicones*, CRDA, July 24, 1944.

11. J. G. E. Wright and J. Marsden, U.S. Patent 2,389,477: applied for, August 21, 1942; granted, November 20, 1945.

12. J. Marsden, Personal communication to H. A. L.

13. (a) F. J. Norton, *Gen. Elec. Rev.*, **47**, No. 8, 6 (1944). (b) E. G. Rochow, *Chemistry of the Silicones*, 2d ed., John Wiley & Sons, New York, 1951, pp. 131–140.

14. W. I. Patnode, U.S. Patent 2,306,222: applied for, November 16, 1940; granted, December 22, 1942.

15. F. J. Norton, U.S. Patent 2,386,259: applied for, July 30, 1942; granted, October 9, 1945.

16. J. R. Elliott and R. H. Krieble, U.S. Patent 2,507,200: applied for, February 10, 1945; granted, May 9, 1950.

17. Letter, dated February 12, 1975, W. I. Patnode to H. A. L. Patnode places this discovery at about the time of Wright's work on silicone rubber, or about 1943. The Canadian work he mentions may have been that of L. B. Jacques, E. Fidlar, E. T. Felsted, and A. G. MacDonald, *Can. Med. Ass. J.,* **55,** 26 (1946).

18. E. G. Rochow, *An Introduction to the Chemistry of the Silicones,* 2d ed., John Wiley & Sons, New York, 1951. Chapter 7 is an excellent summary of early work on the physical chemistry of the silicones; pp. 82–89 deal with methyl silicone oils.

19. C. B. Hurd, *J. Amer. Chem. Soc.,* **68,** 364 (1946).

20. J. Marsden, personal communication to H. A. L., is the principal source for this part of the silicone-rubber story.

21. Letter, dated February 6, 1975, E. J. Flynn to H. A. L.

22. H. A. Liebhafsky, *Anal. Chem.,* **34,** No. 7, 23A (June 1962). The Fisher Award address.

23. W. F. Gilliam, H. A. Liebhafsky, and A. F. Winslow, *J. Amer. Chem. Soc.,* **63,** 801 (1941).

24. W. F. Gilliam, E. M. Hadsell, H. A. Liebhafsky, and M. M. Sprung, *J. Amer. Chem. Soc.,* **73,** 4252 (1951).

25. W. Noll, *Chemistry and Technology of Silicones,* Academic Press, New York, 1968, p. 74.

26. Letter, dated July 15, 1974, E. M. Hadsell to H. A. L.

27. E. W. Balis, W. F. Gilliam, E. M. Hadsell, H. A. Liebhafsky, and E. H. Winslow, *J. Amer. Chem. Soc.,* **70,** 1654 (1948).

28. E. G. Rochow and W. F. Gilliam, *J. Amer. Chem. Soc.,* **63,** 798 (1941).

29. Reference 18, pp. 162–164.

30. E. W. Balis, *Eng. Exp. Stn. News,* **25,** No. 5, 38 (1953).

31. E. W. Balis, H. A. Liebhafsky, and E. H. Winslow, *Ind. Eng. Chem.,* Anal Ed., **15,** 68 (1943).

32. E. W. Balis, H. A. Liebhafsky, and L. B. Bronk, *Ind. Eng. Chem.,* Anal. Ed., **17,** 56 (1945).

33. D. F. Wilcock, *J. Amer. Chem. Soc.,* **69,** 477 (1947).

34. D. W. Scott, *J. Amer. Chem. Soc.,* **68,** 2294 (1946).

35. D. W. Scott, *J. Amer. Chem. Soc.,* **68,** 1877 (1946).

36. D. W. Scott, U.S. Patent 2,418,051: applied for, September 22, 1945; granted, March 25, 1947.

37. W. L. Roth, *J. Amer. Chem. Soc.,* **69,** 474 (1947).

38. W. L. Roth and D. Harker, *Acta Crystallogr.* **1,** 34 (1948).

39. When Roth did his X-ray work, the hydrogen atoms in the methyl groups could not be located on the electron-density map of a molecule. Nuclear magnetic resonance (NMR) can deal directly with such atoms, and Rochow appreciated early that this method would be useful for studying the motion within silicone molecules. See E. G. Rochow and H. G. Le Clair, *J. Inorg. Nucl. Chem.,* **1,** 92 (1955).

40. R. N. Lewis and A. E. Newkirk, *J. Amer. Chem. Soc.,* **69,** 71 (1947).

41. R. O. Sauer, U.S. Patent 2,381,139; August 7, 1945.

42. A. E. Newkirk, U.S. Patent 2,449,815; September 21, 1948.

43. R. O. Sauer and C. E. Reed, U.S. Patent 2,338,575; November 6, 1945.

44. R. O. Sauer, *J. Chem. Educ.*, **21**, 303 (1944).

45. A good discussion of silicone nomenclature and its problems is given by Post in Chapter 9 of his book, which is Reference 6.

46. E. W. Balis and H. A. Liebhafsky, *Ind. Eng. Chem.*, **38**, 583 (1946).

47. E. W. Balis, H. A. Liebhafsky, and D. H. Getz, *Ind. Eng. Chem.*, **41**, 1459 (1949).

48. (a) A. Stock, *Hydrides of Boron and Silicon*, Cornell University Press, Ithaca, New York, 1933, pp. 27 and 28. (b) A. Stock and C. Somieski, *Chem. Ber.*, **53**, 759 (1930). There are fine shades of difference between the two accounts. Note that the experimental evidence here is somewhat less definite than that in the case of prosiloxane because prosiloxane gas exerted the expected pressure during its fleeting existence; the methyl silicones exerted no appreciable pressure.

49. H. A. Liebhafsky, *William David Coolidge: A Centenarian and His Work*, John Wiley & Sons, New York, 1974.

50. D. M. Yost, *J. Amer. Chem. Soc.*, **72**, 3833 (1950).

CHAPTER 6

1. L. B. Bragg, *Ind. Eng. Chem.*, **33**, 279 (1941). This article and those it lists make an excellent introduction to distillation.

2. C. E. Reed and J. T. Coe, *A Fluid Dynamic Reactor for Producing Methylchlorosilanes*, July 12, 1943, Whitney Library, Corporate Research and Development Center, Schenectady, New York.

3. C. E. Reed and J. T. Coe, U.S. Patent 2,389,931: applied for, September 27, 1943; granted, November 27, 1945.

4. (a) BASF, *In the Realm of Chemistry: Pictures from Past and Present*, Econ-Verlag, Vienna, 1965 (referred to in Chapter 1). (b) *Chem. Eng. News*, March 14, 1977, p. 20. (c) W. K. Lewis, E. R. Gilliland, and W. C. Bauer, *Ind. Eng. Chem.*, **41**, 1104 (1949), lists a 1939 MIT thesis by J. M. Chambers as early reference.

5. Letter, E. G. Rochow to A. L. Marshall, dated May 27, 1942.

6. M. M. Sprung, *A Rotating Reactor for the Production of Methylchlorosilanes*, CRDA, August 27, 1942.

7. J. E. Sellers and J. L. Davis, U.S. Patent 2,449,821: applied for, December 8, 1945; granted, September 21, 1948.

8. Letter, VADM H. G. Rickover, USN, to H. A. L., dated September 13, 1973.

9. Letter, E. J. Flynn to H. A. L., dated February 6, 1975.

10. R. J. Cordiner, *New Frontiers for Professional Managers*, McGraw-Hill, New York, 1956.

11. Letter, J. Marsden to H. A. L., dated February 22, 1975.

CHAPTER 7

1. C. E. Reed, *The Industrial Chemistry, Properties, and Application of Silicones*, American Society for Testing Materials, Philadelphia, 1956.

2. *Monogram*, General Electric Company, November–December 1947, p. 2.

3. C. S. Ferguson and J. E. Sellers, U.S. Patent 2,443,902: applied for, June 27, 1947; granted, June 22, 1948.

4. D. S. Hubbell, U.S. Patent 2,420,540: applied for, June 29, 1945; granted, May 13, 1947.

5. J. E. Sellers and J. L. Davis, U.S. Patent 2,449,821: applied for, December 8, 1945; granted, September 21, 1948.

6. J. M. Dotson, U.S. Patent 3,133,109: applied for, November 28, 1960; granted, May 12, 1964.

7. H. K. Lichtenwalner and M. M. Sprung, *Encyclopedia of Polymer Science,* Vol. 12, John Wiley & Sons, New York, 1970, pp. 464–569.

8. R. J. Hengstebeck, *Distillation,* Reinhold, New York, 1961. The index is a useful guide.

9. R. J. H. Voorhoeve, *Organohalosilanes. Precursors to Silicones,* Elsevier Publishing Co., New York, 1967.

10. A. E. Schubert and C. E. Reed, U.S. Patent 2,563,557: applied for, October 20, 1948; granted, August 7, 1951.

11. W. A. Schwenker, U.S. Patent 2,758,124: applied for, April 11, 1952; granted, August 7, 1956.

12. Reference 7, p. 523.

13. Robert N. Meals in *Kirk-Othmer Encyclopedia of Chemical Technology*, Vol. 18, 2d ed., Anthony Standen, Executive Editor, John Wiley & Sons, New York, 1969, p. 235, Fig. 1(a) and (b).

14. C. E. Reed, conversation with H. A. L., August 11, 1975.

15. E. L. Warrick, U.S. Patent 2,541,137: applied for, April 7, 1949; granted, February 13, 1951.

16. E. L. Warrick, U.S. Patent 2,723,964: applied for, May 8, 1953; granted, November 15, 1955.

17. G. R. Lucas, U.S. Patent 2,938,009: applied for, April 11, 1956; granted, May 24, 1960.

18. W. Noll, *Chemistry and Technology of Silicones,* Academic Press, New York, 1968, pp. 395–399, discusses RTV rubbers and is the source of the left-hand part of Figure 7-5. See also Reference 1.

19. J. F. Hyde, U.S. Patent 2,571,039: applied for, April 12, 1950; granted, October 9, 1951.

20. (a) C. A. Berridge, U.S. Patent 2,843,555: applied for, October 1, 1956; granted, July 15, 1958. (b) J. C. Goossens, U.S. Patent 3,303,163: applied for, December 20, 1963; granted, February 7, 1967. (c) J. F. Di Paola, U.S. Patent 3,355,430: applied for, October 20, 1966; granted, November 28, 1967. (d) E. D. Brown, U.S. Patent 3,-385,822: applied for, July 3, 1967; granted, May 28, 1968.

21. R. O. Sauer and E. M. Hadsell, *Data Folder* 72253, CRDA, May 18, 1948.

22. G. Calingaert, H. Soroos, and V. Hnidza, *J. Amer. Chem. Soc.,* **62,** 1107 (1940).

23. R. O. Sauer, U.S. Patent 2,647,136: applied for, January 22, 1948; granted, July 28, 1953.

24. A. J. Barry and J. W. Gilkey, U.S. Patent 2,647,912: applied for, July 19, 1947; granted, August 4, 1953.

25. H. R. McEntee, U.S. Patent 2,786,861: applied for, April 26, 1955; granted, March 26, 1957.

26. J. R. Elliott and J. W. Eustance, *Research Laboratory Report* 128, CRDA, December 1948.

27. N. Kirk, *Ind. Eng. Chem.*, **51**, 515 (1959).

28. M. M. Sprung and F. O. Guenther, *J. Polym. Sci.*, **28**, 17 (1958).

29. W. T. Grubb, *J. Amer. Chem. Soc.*, **76**, 3408 (1954).

30. W. T. Grubb and R. C. Osthoff, *J. Amer. Chem. Soc.*, **77**, 1405 (1955).

31. D. T. Hurd, R. C. Osthoff, and M. L. Corrin, *J. Amer. Chem. Soc.*, **76**, 249 (1954).

32. S. W. Kantor, W. T. Grubb, and R. C. Osthoff, *J. Amer. Chem. Soc.*, **76**, 5190 (1954).

33. R. C. Osthoff, A. M. Bueche, and W. T. Grubb, *J. Amer. Chem. Soc.*, **76**, 4659 (1954).

34. A. R. Gilbert and S. W. Kantor, *J. Polym. Sci.*, **40**, 35 (1959).

35. J. R. Elliott and E. M. Boldebuck, *J. Amer. Chem. Soc.*, **74**, 1853 (1952).

36. A. M. Bueche, *J. Polym. Sci.*, **19**, 297 (1956).

37. A. A. Miller, *J. Amer. Chem. Soc.*, **83**, 31 (1961).

38. L. E. St. Pierre, H. A. Dewhurst, and A. M. Bueche, *J. Polym. Sci.*, **36**, 105 (1959).

39. J. F. Brown, Jr., *Am. Chem. Soc. Div. Polym. Chem. Preprints*, **2** (2), 112 (1961).

40. C. M. Huggins, L. E. St. Pierre, and A. M. Bueche, *J. Polym. Sci.*, **64**, 1304 (1960).

41. C. M. Huggins, L. E. St. Pierre, and A. M. Bueche, *J. Polym. Sci: Part A*, **1**, 2731 (1963).

CHAPTER 8

1. W. Noll, *Chemistry and Technology of Silicones*, Academic Press, New York, 1968.

2. J. M. Utterback, *Science*, **183**, 620 (1974).

3. R. Aries, *Chem. Br.*, **11**, 19 (1975).

4. H. K. Lichtenwalner and M. M. Sprung, in *Encyclopedia of Polymer Science*, Vol. 12, John Wiley & Sons, New York, 1970, pp. 464–569.

5. R. N. Meals and F. Lewis, *Silicones*, Reinhold, New York, 1959.

6. R. R. McGregor, *Silicones and Their Uses*, McGraw-Hill, New York, 1954.

7. W. L. Roth and D. Harker, *Acta Crystallogr.*, **1**, 34 (1948).

8. E. G. Rochow and H. J. Le Clair, *J. Inorg. Nucl. Chem.*, **1**, 92 (1955).

9. C. M. Huggins, L. E. St. Pierre, and A. M. Bueche, *J. Amer. Chem. Soc.*, **64**, 1304 (1960).

10. C. M. Huggins, L. E. St. Pierre, and A. M. Bueche, *J. Polym. Sci: Part A*, **1**, 2731 (1963).

11. D. W. Scott, J. F. Messerly, S. S. Todd, et al., *J. Phys. Chem.*, **65**, 1320 (1961).

12. D. W. McCall and C. M. Huggins, *Appl. Phys. Lett.*, **7**, 153 (1965). They showed also that their self-diffusion constants from **D** backbones with several hundred silicon atoms agreed with those calculated for "monomeric friction constants" drawn from an analysis by others of the viscoelastic properties of oils with such backbones.

13. E. G. Rochow, *Adv. Organomet. Chem.*, **1**, 1 (1970); see p. 15.

14. P. J. Flory, *Statistical Mechanics of Chain Molecules*, John Wiley & Sons, New York, 1969.

15. P. W. Bridgman, *Proc. Amer. Acad. Arts Sci.*, **77**, 115 (1949).

16. P. W. Bridgman, *J. Phys. Chem.*, **19**, 203 (1951).

17. C. G. Suits, *Speaking of Research*, John Wiley & Sons, New York, 1965, p. 19.

18. Dow Corning Corporation, *Prospectus, $60,000,000 9⅞% Sinking Fund Debentures due April 1, 2005*, March 25, 1975, 38 pp. *Moody's Industrial Manual*, Vol. 1, 1975, gives a virtual abstract of the prospectus on pp. 1328–1329.

19. A. Burger, *Chem. Eng. News*, September 22, 1975, p. 37. In a related example, in 1972 Dow Corning acquired Mueller Welt Contact Lenses, Inc., and its silicone-soft-contact-lens technology. Dow Corning continued the development. Early in 1975 the corporation was seeking, but had not yet obtained, approval from the U.S. Food and Drug Administration for the commercial manufacture or sale of these lenses in the United States [18, p. 10].

20. *Newsweek*, September 8, 1975, p. 77.

21. J. Willard Gibbs, *Scientific Papers*, Vol. 1, Dover Publications, New York, 1961, p. 300. This paperback is a reprint of the first (1906) edition.

22. K. A. Pigott, *Encyclopedia of Polymer Science and Technology*, Vol. 11, John Wiley & Sons, New York, 1969, pp. 506–563.

23. S. Ross and R. M. Hack, *J. Phys. Chem.*, **62**, 1260 (1964), and the two papers that follow it illustrate the complexity of the foaming problem.

24. D. L. Bailey and F. M. O'Connor, U.S. Patent 2,834,748: applied for, March 22, 1954; granted, May 13, 1958.

25. G. L. Gaines, Jr., *Monolayers, Advances in Chemistry Series*, No. 144, 1975, p. 338.

26. *Analysis of Silicones*, A. Lee Smith, Ed., John Wiley & Sons, New York, 1974. This 407-page book is Vol. 41 of the series *Chemical Analysis*.

27. *Dow Corning News*, October 24, 1974.

28. *Chem. Eng. News*, August 25, 1975, p. 26.

CHAPTER 9

1. Fritz Haber, *Naturwissenschaften*, **10**, 1041 (1922).

2. Carl Bosch, *Chem. Fabr.*, **6**, 127 (1933).

3. A. Mittasch, *Z. Elektrochem.*, **36**, 569 (1930); p. 572, the Bunsen Medal address.

4. *Chem. Eng. News*, June 7, 1976, p. 26.

5. E. Farber, *Nobel Prize Winners in Chemistry*, Henry Schuman, New York, 1953, p. 72.

6. A. Mittasch, in *Adv. Catal.*, **2**, 83 (1950).

7. J. E. Coates, *J. Chem. Soc.*, 1939, p. 1642; the Haber Memorial Lecture and an outstanding biography.

8. *Catalysis*, P. H. Emmet, Ed., Vol. 3, Reinhold, New York, 1955. The articles are by W. G. Frankenberg and by C. Bokhoven et al.

9. Anthony Sampson, *The Seven Sisters*, Viking Press, New York, 1975, p. 210.

10. F. M. Precopio and D. W. Fox, U.S. Patent 2,936,296: applied for, December 10, 1954; granted, May 10, 1960.

11. E. M. Boldebuck, M. C. Agens, F. M. Precopio, R. E. Burnett, D. W. Fox, W. F. Gilliam, and J. W. Eustance, *Heat-Stable Polyesrters*. I, Report 55-RL-1240, Whitney Library, General Electric Company, Schenectady, New York. A report by members of the Chemistry Research Department, not written until 1955, probably for lack of time.

12. D. W. Fox, letter to H. A. L., dated May 12, 1976.

13. C. G. Suits and A. M. Bueche, in *Applied Science and Technological Progress*, A Report to the U.S. House of Representatives by the National Academy of Sciences, GP-67-0399, June 1967, p. 297. At the time of this report, Dr. Bueche had succeeded Dr. Marshall (retired) and was soon to succeed Dr. Suits, who had followed Dr. Coolidge as head of the Research Laboratory.

14. *Monogram*, General Electric Company, November–December 1975, p. 14.

15. Allan S. Hay, "Polymerization by Oxidative Coupling—An Historical Review," address on receipt of 1975 International Award of the Society of Plastics Engineers; published in *Polym. Sci. Eng.*, **16**, 11 (1976).

16. Karl Jellinek, *Lehrbuch der Physikalischen Chemie*, Vol. 3, Ferdinand Enke, Stuttgart, 1930, p. 47. Jellinek did Nernst's first experiments (in 1907 or before); he was followed by Jost.

17. J. H. van't Hoff, *Chem. Ber.*, **27**, 6 (1894), gives a revealing chart that shows the origin of his important contribution to chemical thermodynamics; his equation facilitated the quantitative treatment of (9-1).

18. H. Haber, *Z. Elektrochem.*, **20**, 597 (1914).

19. W. Nernst, (a) *Z. Elektrochem.*, **16**, 96 (1910); (b) *Theoretical Chemistry*, translated by L. W. Codd and "revised in accordance with the eighteenth German edition," The Macmillan Company, London, 1923, pp. 758–760.

20. K. Mendelsohn, *The World of Walther Nernst*, University of Pittsburgh Press, Pittsburgh, 1973.

21. J. R. Whinfield and J. T. Dickson, Brit. Patent 578,079 (1946).

22. T. M. Midgley, Jr., *Ind. Eng. Chem.*, **29**, 364 (1937). The Perkin Medal address.

23. *Chem. Eng. News*, April 6, 1976, p. 182. In this, the issue memorializing the centennial year of the American Chemical Society, silicones are given somewhat less than 70 words (p. 70).

24. O. E. Snider and R. J. Richardson, in *Encyclopedia of Polymer Science and Technology*, Vol. 10, John Wiley & Sons, New York, 1969, p. 349.

25. Max Planck, *Naturwissenschaften*, **33**, 230 (1946). Planck received the 1918 Nobel Prize in physics, but he was equally at home in physical chemistry. Roentgen received this prize in 1901, and von Laue in 1914, both for work mentioned in the text.

26. Carl Wagner, *Methoden der naturwissenschaftlichen und technischen Forschung*, Bibliographisches Institut, Mannheim, West Germany, 1974.

27. The 25th Anniversary of the Benzene Ring (March 11, 1890). The benzene ring, one of the most important models in chemistry, entered science in an unusual way and had an enormous influence upon chemistry and upon the chemical industry. On the 25th anniversary of its discovery by Kekulé, he was honored by perhaps the most memorable celebration yet given any chemist (*Chem. Ber.*, **23**, 1265–1312). Highlights of the celebration were Kekulé's address and that by von Baeyer, which preceded.

Kekulé had trained von Baeyer; von Baeyer discovered the structure of indigo; indigo contains the benzene ring; synthetic indigo was instrumental in founding the modern synthetic dye industry [Chapter 1]. The two addresses are among the most noteworthy in chemistry, as the following excerpts, freely translated, show.

(1) Von Baeyer [p. 1273] said that Kekulé in 1867 had envisioned the four covalent bonds of carbon as terminating in the [four] planes of a tetrahedron. Kekulé had long been a strong advocate of tetrahedral carbon [Chapter 2].

(2) *Ibid*. [p. 1274]. [Before the discovery of optical isomerism by van't Hoff and Le Bel], Kekulé thought there could be only one carbon compound in which the carbon atom was linked to four different groups. [As late as 1890], von Baeyer had his doubts about optical isomerism in silicon compounds [Chapter 2].

(3) *Ibid*. [pp. 1276, 1286]. Kekulé initially proposed two forms of the benzene ring, and von Baeyer brilliantly deduced that these two forms were nearer the truth than any one could be. [Perhaps the origin of "chemical resonance."]

(4) *Ibid*. [p. 1287]. The Kekulé formula for benzene is the capstone of the edifice of structural chemistry. Extracts from Kekulé's remarks follow.

(5) *Ibid*. [p. 1302]. "We all stand on the shoulders of our predecessors; is it surprising that we should see further than they?" Cf. R. K. Merton, *On the Shoulders of Giants*, Harcourt Brace & World, New York, 1965, who shows learnedly and wittily that this figure of speech did not originate with Sir Isaac Newton, as is commonly believed.

(6) *Ibid*. [p. 1305]. "The structure of benzene did not spring like an armed Pallas Athene from the head of a chemical Zeus . . . the waking mind does not think in jumps."

(7) *Ibid*. [p. 1305]. Kekulé's ideas of chemical bonding and of benzene structure, though recorded immediately upon return to full consciousness, both lay unpublished for about a year after they came to him.

(8) *Ibid*. [p. 1306]. ". . . In a half-sleep, . . . atoms . . . moved before my eyes . . . everything in snakelike motion. . . . And see, what was that? One of the snakes seized its own tail and the form whirled mockingly before my eyes. I awoke [became fully conscious] as though struck by lightning; this time also [the other had been in London on his model of the chemical bonding, which came to him during reveries (*Träumereien*) on a London bus] I spent the rest of the night working out the consequences of the hypothesis.

(9) *Ibid*.[p. 1310]. "I never aimed at technology—always at pure science. . . ."

This note is an important supplement to earlier comments about discoveries: of course, relapses from full consciousness between experiments may be less rewarding for us than they were for Kekulé.

28.   F. A. Lewis, *The Incandescent Light*, new rev. ed., Sherwood Publishers, New York, 1961.

29.   *The General Electric Story*, Vol. 1, *The Edison Era 1876–1892*, The Algonquin Chapter, Elfun Society, Schenectady, New York, 1976.

30.   Herman A. Liebhafsky, *William David Coolidge: A Centenarian and His Work*, John Wiley & Sons, New York, 1974.

31.   *Probable Levels of R & D Expenditures in 1976*, Batelle-Columbus, Ohio. 1975.

32.   I. Kusiatin, *The Process and Capacity for Diversification through Internal Development*, Thesis, Graduate School of Business Administration, Harvard University, 1976.

33.   Lowell W. Steele, *Innovation in Big Business*, Elsevier North-Holland, New York, 1975.

EPILOGUE

1. F. R. Leavis, *Nor Shall My Sword,* Barnes and Noble, New York, 1972.
2. C. P. Snow, *The Two Cultures: and a Second Look,* Cambridge University Press, New York, 1964; reprinted 1965.
3. J. B. Priestley, *Literature and Western Man,* "Part Four: The Broken Web," Harper & Brothers, New York, 1960.
4. A. L. Rowse, *The Elizabethan Renaissance: The Life of the Society,* Charles Scribner's Sons, New York, 1971.
5. *The Collected Essays, Journalism and Letters of George Orwell,* Sonia Orwell and Ian Angus, Eds., Vol. 4, Harcourt, Brace and World, New York, 1968, p. 345.
6. Norman Mailer, quoted in *Time,* July 16, 1973, p. 63.
7. Fletcher Knebel, well-known author, in "The Greening of Fletcher Knebel," *The New York Times Magazine,* September 15, 1974, p. 36.
8. Kenneth Clark, *Civilisation,* Harper & Row, New York, 1969.
9. *Alistair Cooke's America,* Alfred A. Knopf, New York, 1973.
10. Jacob Bronowski, *The Ascent of Man,* Little Brown, Boston, 1974.
11. Robert Nisbet, Albert Schweitzer Professor in the Humanities, Columbia University, in "Knowledge Dethroned," *The New York Times Magazine,* September 28, 1975, says on p. 34: "In the past few years, ... disenchantment has set in, with the public concluding that the postwar promises of learning were inflated and misleading."
12. Bruce Catton, *Waiting for the Morning Train,* Doubleday, New York, 1972. The touching and beautifully written autobiographical record of a distinguished historian's early years. Part of the book appeared in *Family Weekly,* December 15, 1974, p. 3, under the title "Man's Death Sentence for Man: Our Belief in Our Own Creations." From p. 5 of the book: "The age of technology has one terrible aspect—each new technique must be exploited to its absolute limit, until man becomes the victim of his own skills."
13. Columnist *R, Encounter,* **44,** 35 (1975): "Where are they now, . . . those hordes of scientists, economists, technologists, situation analysts who . . . described the wonders which the post-industrial society had in store for us?"
14. E. E. Morison, Killian Professor of History, MIT, *From Know-How to Nowhere,* Basic Books, New York, 1974. The title masks a ray of hope.
15. Farah Pahlevi, Empress of Iran, in "Computer and Spirit: The Chasm," *The New York Times,* July 26, 1975: "How to harness the resources of science and technology without depriving mankind of his human heritage—this is the challenge we face and must surmount."
16. Richard D. Lamm, Governor of Colorado, in his address before the American Association for the Advancement of Science, February 21, 1977:

    (1) "It is my deep belief that we are faced with a crisis of crises—and that our political and scientific solutions are not keeping up with the pace of problems."
    (2) "Further, I believe that we truly live in a hinge of history—and that we will see dramatic change in the years ahead. Some of this change will be helpful—some harmful . . . .Neither the politicians nor the scientific community are prepared for the nature and extent of change."
    (3) "Science is a process which seeks truth—politics is a process which seeks *survival.*"

(4) "Pulbic policy—diamond like—takes on a new dimension *every* direction it takes. We need more than *bright* minds, we need wisdom.

"Science at best is not wisdom. It is knowledge, while wisdom is knowledge tempered with judgment. This judgment is too often missing in the accelerating pace of scientific advance."

(5) "I find that one of the great challenges of the future will be to differentiate what science and technology can do and what it cannot do."

Everyone concerned about the future of the United States ought to read this address. I am indebted to Governor Lamm for a copy.

17. James Kilpatrick, columnist, in "Statesman Bewildered," column in *The Eagle,* Bryan-College Station, Texas, January 31, 1976, p. 4A. In granting permission to quote on April 11, 1977, Mr. Kilpatrick added, "The congressional confusion I described is a situation that gets worse, not better."

18. *Schenectady Gazette,* October 17, 1975, p. 1.

19. Joseph Hixson, *The Patchwork Mouse,* Anchor Press—Doubleday and Company, New York, 1976.

20. L. F. Cavalieri, "New Strains of Life—or Death," *The New York Times Magazine,* August 22, 1976, p. 8.

21. *Schenectady Gazette,* October 28, 1976, p. 3.

22. *The New York Times,* August 29, 1976, p. 1.

23. Charles E. Hughes, *West Coast Hotel Co.* v. *Parrish,* 300 U.S., 379,391.

24. Arthur M. Bueche, "The Hard Truth about Our Energy Future," General Electric Research and Development Center, Schenectady, New York, March 1977. For copies write to the Communications Branch of the Center, P.O. Box 8, Schenectady, New York, 12301.

Index

This index has been prepared with the general reader primarily in mind. Origins, products, processes, people, history, and the nature of industrial research have been indexed to help the reader look for information. Scientifically inclined readers should find it possible to locate detailed scientific information with the index as guide.

138932